STUDENT'S RESOURCE MANUAL

Calculus

Calculus of a Single Variable

Calculus of Several Variables

STUDENT'S RESOURCE MANUAL

**THE OREGON STATE UNIVERSITY
CALCULUS CONNECTIONS PROJECT**

Calculus

Calculus of a Single Variable

Calculus of Several Variables

Thomas P. Dick
Charles M. Patton

PWS PUBLISHING COMPANY
Boston

PWS Publishing Company
20 Park Plaza, Boston, Massachusetts 02116-4324

Copyright © 1995 by PWS Publishing Company, a division of International Thomson Publishing Inc.

All rights reserved. No part of this book may be reproduced, stored in a retrieval system, or transmitted, in any form or by any means -- electronic, mechanical, photocopying, recording, or otherwise -- without the prior written permission of PWS Publishing Company.

I(T)P   International Thomson Publishing
         The trademark ITP is used under license.

For more information, contact::

PWS Publishing Company
20 Park Plaza
Boston, MA 02116-4324

International Thomson Publishing Europe
Berkshire House I68-I73
High Holborn
London SC1V 7AA
England

Thomas Nelson Australia
102 Dodds Street
South Melbourne, 3205
Australia

Nelson Canada
1120 Birchmount Road
Scarborough, Ontario
Canada M1K 5G4

International Thomson Editores
Campos Eliseos 385, Piso 7
Col. Polanco
11560 Mexico D.F., Mexico

International Thomson Publishing GmbH
Konigswinterer Strasse 418
53227 Bonn, Germany

International Thomson Publishing Asia
221 Henderson Road
#05-10 Henderson Building
Singapore 0315

International Thomson Publishing Japan
Hirakawacho Kyowa Building, 31
2-2-1 Hirakawacho
Chiyoda-ku, Tokyo 102
Japan

Printed in the United States of America by Malloy Litho

95 96 97 98 99 -- 10 9 8 7 6 5 4 3 2 1

ISBN 0-534-94454-X

This student solution supplement accompanies Dick & Patton's *Calculus*. Solutions to approximately every third problem in the textbook are included, many of them with detailed steps or explanations. At the end of the supplement you will find a study guide for differential and integral calculus and some useful programs for graphing calculators for use in Chapter 6 (for calculating Riemann sums and slope fields). The authors acknowledge Dr. Dianne Hart and Dr. Marie Franzosa for their fine work in compiling the solutions for this supplement and D'Anne Hammond for technical typing.

Contents

ANSWERS TO SELECTED EXERCISES FROM CHAPTER 0	1
ANSWERS TO SELECTED EXERCISES FROM CHAPTER 1	7
ANSWERS TO SELECTED EXERCISES FROM CHAPTER 2	15
ANSWERS TO SELECTED EXERCISES FROM CHAPTER 3	22
ANSWERS TO SELECTED EXERCISES FROM CHAPTER 4	33
ANSWERS TO SELECTED EXERCISES FROM CHAPTER 5	49
ANSWERS TO SELECTED EXERCISES FROM CHAPTER 6	67
ANSWERS TO SELECTED EXERCISES FROM CHAPTER 7	79
ANSWERS TO SELECTED EXERCISES FROM CHAPTER 8	88
ANSWERS TO SELECTED EXERCISES FROM CHAPTER 9	99
ANSWERS TO SELECTED EXERCISES FROM CHAPTER 10	114
ANSWERS TO SELECTED EXERCISES FROM CHAPTER 11	121
ANSWERS TO SELECTED EXERCISES FROM CHAPTER 12	128

ANSWERS TO SELECTED EXERCISES FROM CHAPTER 13 136

ANSWERS TO SELECTED EXERCISES FROM CHAPTER 14 140

ANSWERS TO SELECTED EXERCISES FROM CHAPTER 15 151

ANSWERS TO SELECTED EXERCISES FROM CHAPTER 16 161

ANSWERS TO SELECTED EXERCISES FROM CHAPTER 17 169

ANSWERS TO SELECTED EXERCISES FROM THE APPENDICES   **175**

STUDY GUIDE FOR DIFFERENTIAL AND INTEGRAL CALCULUS   **183**

RIEMANN SUM AND SLOPE FIELD PROGRAMS **369**

ANSWERS TO SELECTED EXERCISES FROM CHAPTER 0

Section 0.1

1. 31.

4. $\frac{7}{9}$.

7. No. Note that in the process of dividing 22 by 7 using the usual algorithm, 7 is divided into 22, and then into the numbers 10, 30, 20, 60, 40, 50 and then again into 10, 30, etc. These six numbers are the only possibilities. We can never have 70, 80 or more, because then we would have a remainder greater than or equal to 7.

10. $A = \frac{1}{2}(1)a_{99} = \frac{1}{2}(10) = 5$.

13. $A = \pi(\frac{a_n}{2})^2 = \frac{\pi}{4}(n+1)$.

16. The nth triangle overlaps the first triangle when
$\sum_{i=1}^{n} \sin^{-1}(\frac{1}{a_i}) \geq 2\pi \approx 6.2832$. This occurs when $n = 17$.

Section 0.2

1. 3. **4.** 4.

7. 1.

10. For both of the numbers $100000.xy$ and $100001.xy$, we have 10 possible digits for x and 10 possible digits for y. We cannot include 100000.00 so we have $2(10)(10) - 1$ or 199 different numbers.

13. 1×10^{-99}.

15. For the 12 decimal places of each number, 1 and 0 are fixed. We then have 10 possible digits for each of the 10 remaining decimal places. We need to exclude 10. This gives us a total of $10^{10} - 1$.

18. All three. The first student's calculator only produces real number answers. Since $\sqrt{-1}$ is non-real, the calclulator gives an error message. The third student's calculator produces real as well as non-real number answers. $(0, 1)$ is the calculator's way of displaying the number $0 + i$.

Section 0.3

1. The center of the interval is $\frac{(-8)+(-2)}{2} = \frac{-10}{2} = -5$ and the radius of the interval is $|-5-(-2)| = 3$. Hence, $|x+5| \leq 3$.

4. The center of the interval is $\frac{10+4}{2} = \frac{14}{2} = 7$, and the radius of the interval is $|7-4| = 3$. Hence, $|x-7| \geq 3$.

7. $|x + \frac{5}{2}| \leq \frac{5}{2}$.

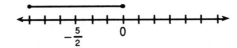

10. $|x+9| \leq 4, x \neq -13$.

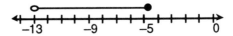

13. $x^2 - 4 \leq 5 \Rightarrow x^2 \leq 9 \Rightarrow |x| \leq 3 \Rightarrow$ solution set is $[-3, 3]$.

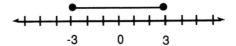

Note: The function $y = x^2 - 4$ has y values less than or equal to 5 when x is between -3 and 3, inclusive.

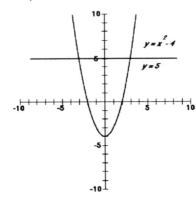

19. $x^2 + 4x + 6 \leq 1 \Rightarrow x^2 + 4x + 5 \leq 0$. This never occurs since $x^2 + 4x + 5 \geq 1$ for all real values of x.

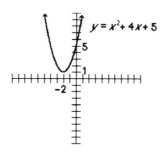

22. $[-7, -1]$.

25. $2|x - 3| > 7 \Rightarrow |x - 3| > \frac{7}{2} \Rightarrow x - 3 > \frac{7}{2}$ or $x - 3 < -\frac{7}{2} \Rightarrow x > \frac{13}{2}$ or $x < -\frac{1}{2} \Rightarrow$ the solution set is $\{x : |2x - 6| > 7\} = (-\infty, -\frac{1}{2}) \cup (\frac{13}{2}, \infty)$.

28. Closed, bounded.

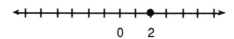

31. Both open and closed, unbounded.

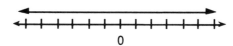

34. Open, bounded.

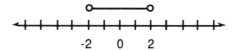

37. $[3.717, 3.723]$; $|x - 3.72| \leq 0.003$.

40. If a and b have opposite signs, then $|a| + |b| > |a + b|$.

Section 0.4

1. P represents perimeter, h hypotenuse, ℓ leg.

$h = \sqrt{\ell^2 + \ell^2} = \sqrt{2}\ell$ and $P = \ell + \ell + \sqrt{2}\ell \Rightarrow P = (2 + \sqrt{2})\ell$.

$h = \sqrt{2}\ell \geq 5 \Rightarrow \ell \geq \frac{5}{\sqrt{2}}$.

So, $P : \left[\frac{5}{\sqrt{2}}, \infty\right) \longrightarrow \mathbb{R}$

$P(\ell) = (2 + \sqrt{2})\ell$.

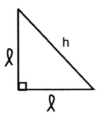

4. Function.

7. $9x^2 + 4y^2 = 36$ does not describe a function since it fails the vertical line test.

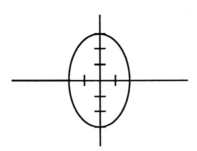

10. $8x^2 = 2y$ does describe a function.

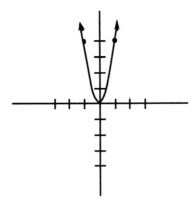

11. $xy = 4$ does describe a function.

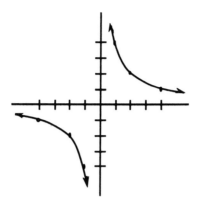

14. $(-3, 6)$ lies on $y = mx - 1 \Rightarrow 6 = m(-3) - 1 \Rightarrow m = -\frac{7}{3}$.

17. The line passing through $(-1, 2)$ and the origin has slope $\frac{2-0}{-1-0} = -2$. A perpendicular line must have slope $m = \frac{1}{2}$ (the negative reciprocal of -2).

20. Changing c moves the location of the y-intercept. If $c > 0$, the y-intercept is positive, if $c = 0$, the y-intercept is 0 and if $c < 0$, the y-intercept is negative.

23. If $d > 0$, the graph of f crosses the x-axis twice. If $d = 0$, the graph of f touches/crosses the x-axis in one place. If $d < 0$, the graph of f does not touch the x-axis at all.

26. $a = -1.5$, $b = 5$.

29. Many answers, for example: $a = -6$, $b = 18$ or $a = 6$, $b = -12$.

Section 0.5

1. The line would make a 45° angle with the x-axis.

4. The line would look like the line $y = 0$.

5. $x \approx 1.789$.

8. $x = 2.6$.

11. $x = 1$.

14. $x \approx 0.608$.

15. $x \approx 29.563 \pm 90.963i$. The solutions are complex and can be found by machine.

(For exercises 16 - 39, endpoints may have been rounded.)

16. $[-3, 3]$.

19. $(-\infty, -0.581) \cup (-0.5, 2.581)$.

23. No real solutions.

26. $(0, 1)$.

29. $[4.333, 23]$.

32. $[2.162, 4.162]$.

35. $[-1.162, 5.162]$.

38. $(-1.442, -1) \cup (0, 1)$.

41. The product of two truth functions will return an output of 1 only when both truth functions are true at the same time. Hence, multiplication is an appropriate operation to represent AND.

43. After a few zooms in horizontally (by factors of 10), the graph should appear similar to a flat line $y = c$, with $2.7 < c < 2.8$. When you continue to zoom in as many times as the number of significant digits your machine carries, then the graph should start to have jagged jumps. Continuing to zoom in will result in a graph similar to the line $y = 1$.

Here's why it happens. Suppose your calculator carries twelve significant digits. As you zoom in repeatedly, the x-values on screen become smaller and smaller in magnitude. For values x very close to 10^{-12}, the effects of the roundoff will vary quite a bit from pixel to pixel, resulting in the jagged jumps. In fact, as you zoom in, the jagged jumps should appear first to the *right* of $x = 0$, then to the left. This is because $(1 + x)$ has a leading digit in the units place for positive x, but not for negative x. For $x < 10^{-12}$, the base quantity $(1 + x)$ becomes *exactly* 1 after roundoff.

ANSWERS TO SELECTED EXERCISES FROM CHAPTER 1

Section 1.1

1. The degree of the resulting polynomial is the sum of the degrees of the original polynomials.

4. Only functions of the form $f(x) = cx$ where c is a constant, satisfy this property.

   Note: $f(a+b) = c(a+b) = ca + cb = f(a) + f(b)$.

7. The identity function is an odd function.

   Note: $f(x) = x$, $f(-x) = -x = -f(x)$.

10. $-\frac{2}{1}, -\frac{4}{3}, -\frac{1}{1}, -\frac{2}{3}$.

13. All but $-\frac{1}{1}$ and $\frac{1}{1}$.

16. $-\frac{2}{1}, -\frac{4}{3}, -\frac{3}{2}, -\frac{1}{1}, -\frac{3}{4}, -\frac{2}{3}, -\frac{1}{2}$.

19. $-\frac{2}{1}, -\frac{4}{3}, -\frac{2}{3}, \frac{2}{1}, \frac{4}{3}, \frac{2}{3}$.

22. f odd $\Rightarrow f(-x) = -f(x) \Rightarrow f(-0) = f(0) = -f(0) \Rightarrow f(0) = 0$.

25. The x-intercept for $f(x+a)$ is the x-intercept of $f(x)$ minus a. If a is negative, the original graph is translated $|a|$ units to the right. If a is positive, the original graph is translated a units to the left.

28. The y–intercept of the graph of $y = f(x) + d$ is d plus the y–intercept of the graph of $y = f(x)$. When $d < 0$, the graph of $y = f(x) + d$ is translated d units down from the graph of $y = f(x)$. When $d > 0$, the graph of $y = f(x) + d$ is translated d units up from the graph of $y = f(x)$.

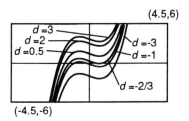

31. $-16t^2 + 1000t = 0 \Rightarrow -8t(2t - 125) = 0 \Rightarrow t = 0$ (initially) or $t = \frac{125}{2}$ sec. (One could also use a machine to find the roots.)

32.

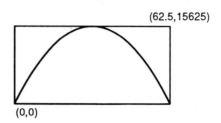

35. The maximum height is $h\left(\frac{125}{4}\right) = 15,625$ feet.

38. Different graphs can result.

Section 1.2

1. $\log_a(1) = \log_a(\frac{2}{2}) = \log_a(2) - \log_a(2) = 0$.

4. $\log_a(6) = \log_a((3)(2)) = \log_a(3) + \log_a(2) = 0.6055 + 0.9597 = 1.5652$.

7. $\log_a(9) = \log_a(3^2) = 2\log_a(3) = 2(0.9597) = 1.9194$.

10. $\log_a(15) = \log_a(\frac{(10)(3)}{2}) = \log_a(10) + \log_a(3) - \log_a(2)$
$= 2.0115 + 0.9597 - 0.6055 = 2.3657$.
(or $\log_a((5)(3)) = \log_a(5) + \log_a(3) = 1.406$ (from #3) $+0.9597 = 2.3657$.)

13. If $q(t) = A_0(2)^{-t/20}$, we want to solve $0.01 A_0 = A_0(2)^{-t/20}$ for t.
So, $0.01 = (2)^{-t/20} \Rightarrow \ln(0.01) = \ln(2)^{-t/20} \Rightarrow \ln(0.01) = \left(-\frac{t}{20}\right)\ln(2) \Rightarrow$
$t = \frac{-20\ln(0.01)}{\ln(2)} \approx 132.9$ years.

16. If $q(t) = 100(2)^{t/20}$, then $q(t) = 100(2)^{365/20} \approx 31,174,351$.

19. $1500 = 1000(1+r)^3 \Rightarrow 1.5 = (1+r)^3 \Rightarrow \sqrt[3]{1.5} = 1+r \Rightarrow \sqrt[3]{1.5} - 1 = r \Rightarrow$
$r \approx 0.14$ or 14%.

22. The number of bacteria after 12 hours is $A = A_0(2)^{12}$ which completely covers the tray. Solving $(0.1)A_0(2)^{12} = A_0(2)^t$ for t, we find that $409.6 = 2^t \Rightarrow$
$t = \frac{\ln(409.6)}{\ln(2)} \Rightarrow t \approx 8.68$ hours. At approximately 2:41 am, the tray will be 10% covered with bacteria.

24. $f(x) = -2^x$.

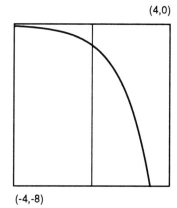

25. $f(x) = 2^{-x}$.

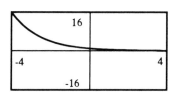

28. $f(x) = -2^{-x}$.

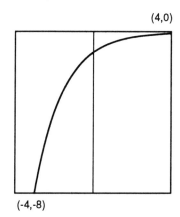

29. $f(x) = \log_2(-x), x < 0$.

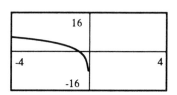

32. $f(x) = \log_2(-x), x < 0$.

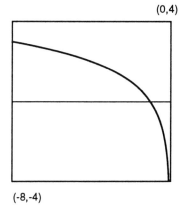

33. $a = 1, b = 1$.

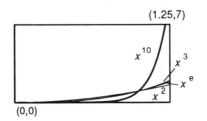

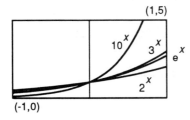
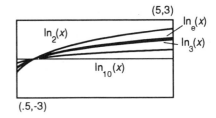

36. $(0,0)$.

39. $y = a^x$.

Section 1.3

1. $f(x) = \cos(x)$, $f(x) = \sec(x)$.

4. $f(x) = \sec(x)$, $f(x) = \csc(x)$, $f(x) = \tan(x)$, $f(x) = \cot(x)$.

7. C shifts the graph of $y = \sin(x)$ to the right (left) $|C|$ units if $C > 0 (C < 0)$.

10. $p = \frac{2\pi}{2} = \pi$.

13. The amplitude is 3.

16. $f = \frac{|B|}{2\pi}$.

19. Average output value is D.

22. The y–intercept is $A\sin(-BC) + D$.

25. a) $\{0, -0.951057, -0.587785, 0.587785, 0.951057\}$.
   b) $u(t) = -\sin(1(2\pi)t)$.

28. a) $\{0, 0, 0, 0, 0, 0, 0, 0\}$.
   b) $u(t) = \sin(4(2\pi)t)$ or $u(t) = \sin(0(2\pi)t)$.

31. It requires more than $2f$ samples per second to reconstruct a signal with frequency f.

Section 1.4

1. $D = \{x : x \neq \pm\sqrt{3}\}$.

4. $f(x) = \sec(\pi x)$ is undefined whenever $\cos(\pi x) = 0$. Therefore, $D = \{x : x \neq \frac{n}{2}, n \in \text{ Odd integers }\}$.

7. $g \circ f \circ h \circ j$.

10. $g \circ g \circ g \circ f$ or $g \circ f \circ g$.

13. $(f + g)(4) = f(4) + g(4) = -1 + 2 = 1$.

16. $(gf)(1) = g(1) \cdot f(1) = (-2) \cdot (-7) = 14$.

19. $(g \circ f)(4) = g(f(4)) = g(-1) = 3$.

22. $f(x) = g(x)$ for $x = -1$ since $f(-1) = 3 = g(-1)$.

25. $(g \circ f)(x) = 2 \Rightarrow g(f(x)) = 2 \Rightarrow f(x) = g^{-1}(2) \Rightarrow x = f^{-1}(g^{-1}(2)) \Rightarrow x = f^{-1}(4) = 0$.

27.

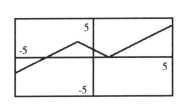

28.

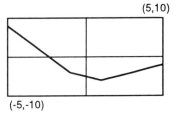

31.

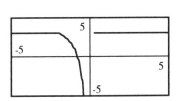

32.

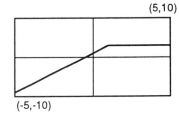

35.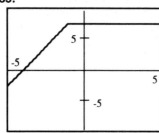

36.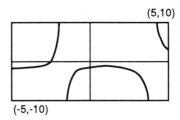

39.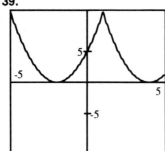

40.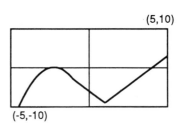

42.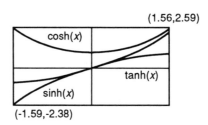

45. $f(x) = \sinh(x), f(x) = \tanh(x), f(x) = \operatorname{csch}(x), f(x) = \coth(x)$.

48. $\cosh(x) + \sinh(x) = \frac{e^x + e^{-x}}{2} + \frac{e^x - e^{-x}}{2} = \frac{2e^x}{2} = e^x$.
$\cosh(x) - \sinh(x) = \frac{e^x + e^{-x}}{2} - \frac{e^x - e^{-x}}{2} = \frac{2e^{-x}}{2} = e^{-x}$.

Section 1.5

1. f is not one-to-one since $f(0) = 0 = f(1)$; so f does not have an inverse.

4. p is a one-to-one function so it must have an inverse. However, solving $p(y) = x$, i.e., $x = y^3 + y + 1$ is not possible. To get an idea what p^{-1} looks like, graph the mirror image of p across the line $y = x$.

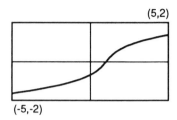

7. Solving for x, we have $y - 32 = \frac{9}{5}x$ and $x = \frac{5}{9}(y - 32)$, so $h^{-1}(y) = \frac{5}{9}(y - 32)$ where $h^{-1} : \mathbb{R} \to \mathbb{R}$.

10. Solving for x we get $y - b = mx$ and $x = \frac{y-b}{m}$. So $f^{-1}(y) = \frac{y-b}{m}$ which is a linear function.

13. $f^{-1}(2) = 3$.

16. $(f(2))^{-1} = 6^{-1} = \frac{1}{6}$.

19. $f^{-1}(g^{-1}(x)) = 0$ when $g^{-1}(x) = 4 \Rightarrow x = 2$.

22. $\arccos(-1) = \pi$.

25. $\arctan(\sqrt{3}) = \frac{\pi}{3}$.

28. $\tan(\arctan(x)) = x$ is true for all real numbers.

31. None of the above.

34. An alternate way to compute $y = \operatorname{arcsec}(x)$.

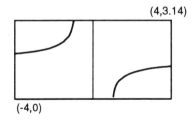

37. None of the above.

40. $f(x) = \arcsin(x), f(x) = \arctan(x)$.

44. $f(x) = \sinh^{-1}(x)$ has domain $\mathbb{R}$;
$f(x) = \cosh^{-1}(x)$ has domain $\{x : x \geq 1\}$;
$f(x) = \tanh^{-1}(x)$ has domain $\{x : |x| < 1\}$;
$f(x) = \coth^{-1}(x)$ has domain $\{x : |x| > 1\}$;
$f(x) = \text{sech}^{-1}(x)$ has domain $\{x : 0 < x \leq 1\}$;
$f(x) = \text{csch}^{-1}(x)$ has domain $\{x : |x| > 0\}$.

48. $y = \text{csch}^{-1}(x)$.

50. $y = \cosh^{-1}(x)$.

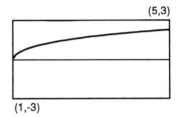

Section 1.6

1. $260 + \frac{1}{2}(293 - 260) = 276.5$.

4. $382 + \frac{1}{3}(438 - 382) = 400\frac{2}{3}$.

7. 100 miles $= 528000$ ft; $f(528000) \approx 181.659$ seconds.

10. September.

13. G.

16. D.

19. E.

22. F.

23. A or H.

ANSWERS TO SELECTED EXERCISES FROM CHAPTER 2

Section 2.1

For exercises 1-6 there are many possible solutions.

1.
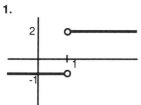

2.
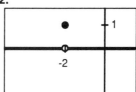

5.
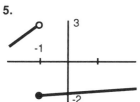

6.
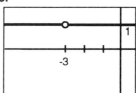

9. **10.**
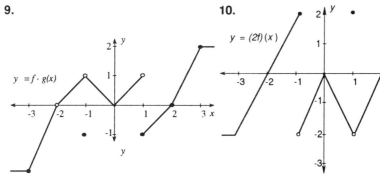

13.

$x = a$	$f(a)$	$\lim_{x \to a^-} f(x)$	$\lim_{x \to a^+} f(x)$	$\lim_{x \to a} f(x)$	Is f cont.?
-3	-1	-1	-1	-1	yes
-2	DNE	0	0	0	no
-1	1	1	-1	DNE	no
0	0	0	0	0	yes
1	1	-1	-1	-1	no
2	0	0	0	0	yes
3	1	1	1	1	yes

16.

$x=a$	$f(a)$	$\lim_{x\to a^-} f(x)$	$\lim_{x\to a^+} f(x)$	$\lim_{x\to a} f(x)$	Is f cont.?
-3	-3	-3	-3	-3	yes
-2	DNE	-2	-1	DNE	no
-1	2	0	0	0	no
0	1	1	1	1	yes
1	2	0	-2	DNE	no
2	0	-1	-2	DNE	no
3	-1	-1	-1	-1	yes

19.

$x=a$	$f(a)$	$\lim_{x\to a^-} f(x)$	$\lim_{x\to a^+} f(x)$	$\lim_{x\to a} f(x)$	Is f cont.?
-3	0	0	0	0	yes
-2	0	0	1	DNE	no
-1	1	1	1	1	yes
0	1	1	1	1	yes
1	1	1	1	1	yes
2	0	1	0	DNE	no
3	0	0	0	0	yes

22. $\lim_{x\to 0^+} f(x) = 1$, $\lim_{x\to 0^-} f(x) = 1$, $\lim_{x\to 0} f(x) = 1$.

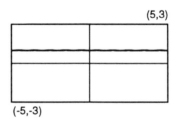

25. $\lim_{x\to 0^+} f(x)$ does not exist (∞), $\lim_{x\to 0^-} f(x)$ does not exist (∞), $\lim_{x\to 0} f(x)$ does not exist (∞).

28. $\lim_{x\to 0^+} f(x) = 0$, $\lim_{x\to 0^-} f(x) = 0$, $\lim_{x\to 0} f(x) = 0$.

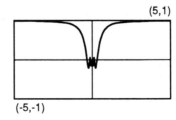

31. $\lim_{x \to 0^+} f(x) = 0$, $\lim_{x \to 0^-} f(x) = 0$, $\lim_{x \to 0} f(x) = 0$.

34. $\lim_{x \to 0^+} f(x) = \frac{\pi}{2} (\approx 1.5708)$, $\lim_{x \to 0^-} f(x) = -\frac{\pi}{2} (\approx -1.5708)$, $\lim_{x \to 0} f(x)$ does not exist.

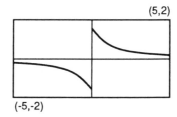

Section 2.2

1. $L = 3$.

4. $L \approx 4.27$.

7. The limit does not exist.

10. $L = 1$.

13. $L = 0$.

16. The limit does not exist.

19. $L = 0.1\overline{7}$.

22. $L = 0$.

Section 2.3

1. The function is continuous since $f(0) = 0 = \lim_{x \to 0} |x|$.

4. $\lim_{x \to 0^-} \cot(x) = -\infty$ and $\lim_{x \to 0^+} \cot(x) = \infty$. Thus, the function has a vertical asymptote at $x = 0$.

7. $f(x) = \arctan\left(\frac{1}{x}\right)$ is not defined at $x = 0$, but $\lim_{x \to 0^-} \arctan\left(\frac{1}{x}\right) = -\frac{\pi}{2}$ and $\lim_{x \to 0^+} \arctan\left(\frac{1}{x}\right) = \frac{\pi}{2}$. Thus, the graph has a jump at $x = 0$.

10. $f(x) = \cos(\frac{1}{x})$ is not defined at $x = 0$. $\lim_{x \to 0^+} \cos(\frac{1}{x})$ and $\lim_{x \to 0^-} \cos\left(\frac{1}{x}\right)$ do not exist. f has an essential discontinuity at $x = 0$ but it is not a jump discontinuity or a vertical asymptote.

13. $y = \csc(x) = \frac{1}{\sin(x)}$ has vertical asymptotes where $\sin(x) = 0$, i.e., $\{x : x = n\pi, n \text{ any integer}\}$.

16. $y = \log_{10}(x)$ has a vertical asymptote at $x = 0$.

19. Continuous at $x = -1$.

22. Discontinuous at $x = 1$ but not a jump discontinuity.

For problems 25 - 40, the roots of the denominator can be found by machine.

25. Solving $x^2 - 5x + 6 = (x-2)(x-3) = 0$, we find that f is discontinuous at $x = 2$ and $x = 3$.
$$\lim_{x \to 2^-} f(x) = -\infty, \lim_{x \to 2^+} f(x) = +\infty, \lim_{x \to 3^-} f(x) = +\infty, \lim_{x \to 3^+} f(x) = -\infty \text{ so}$$
$x = 2$ and $x = 3$ are vertical asymptotes.

28. Solving $2x^2 - 5x + 1 = 0$, we find that f is discontinuous at $x \approx 0.219$ and $x \approx 2.281$. Let a be the smaller of these two numbers and b be the larger. $\left(a = \frac{5-\sqrt{17}}{4}, b = \frac{5+\sqrt{17}}{4}\right)$.
$$\lim_{x \to a^-} f(x) = +\infty, \lim_{x \to a^+} f(x) = -\infty, \lim_{x \to b^-} f(x) = -\infty, \lim_{x \to b^+} f(x) = +\infty \text{ so}$$
$x = a$ and $x = b$ are vertical asymptotes.

31. Solving $x^2 - 6x + 9 = (x-3)^2 = 0$, we find that f is discontinuous at $x = 3$.
$$\lim_{x \to 3^-} f(x) = -\infty, \lim_{x \to 3^+} f(x) = +\infty \text{ so } x = 3 \text{ is a vertical asymptote.}$$

34. We see that f is discontinuous at $x = 2$.
$$\lim_{x \to 2} f(x) = -1 \text{ so the graph has a hole at } (2, -1).$$

37. Solving $30x - 90 = 0$, we find that f is discontinuous at $x = 3$.
$$\lim_{x \to 3^-} f(x) = \infty, \lim_{x \to 3^+} f(x) = -\infty \text{ so } x = 3 \text{ is a vertical asymptote.}$$

40. Solving $5x^2 - 7x - 6 = (5x+3)(x-2) = 0$, we find that f is discontinuous at $x = 2$ and $x = -\frac{3}{5}$.
$$\lim_{x \to 2} f(x) = \frac{3}{13}, \lim_{x \to -3/5^-} f(x) = \infty, \lim_{x \to -3/5^+} f(x) = -\infty \text{ so } x = \frac{3}{5} \text{ is a vertical}$$
asymptote and the graph has a hole at $\left(2, \frac{3}{13}\right)$.

Section 2.4

1. Since $g(1) = 2(1)^2 = 2$, we must find A such that $|1 + A| = 2$. This occurs if $A = 1$ or $A = -3$.

4. $j(0) = A^2$ and $\lim_{x \to 0^+} j(x) = (0 - 2A)^4 = 16A^4$. For j to be continuous we must have $A^2 = 16A^4$. $16A^4 - A^2 = 0 \Rightarrow A^2(16A^2 - 1) = 0 \Rightarrow$
$A^2(4A - 1)(4A + 1) = 0 \Rightarrow A = 0, \frac{1}{4} \text{ or } -\frac{1}{4}$.

7. f is not continuous on $[-1, 1]$ and therefore the Intermediate Zero Theorem does not apply.

10.

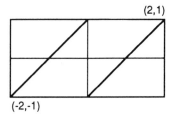

12. Think of the y-axis as representing the distance the hiker is from the bottom of the mountain and the x-axis as representing time of day. We can describe the person's location on the mountain at a certain time by a continuous curve starting at $(8\,\text{am}, 0)$. The other endpoint of the curve represents the top of the mountain 9 hours later. (If the mountain is 1500 ft high, then the endpoint is $(5\,\text{pm}, 1500)$. To describe the person's return trip, draw a continuous curve starting at the top of the mountain at 8 am (the point $(8\,\text{am}, 1500)$), and moving back to the bottom at 5 pm (the point $(5\,\text{pm}, 0)$. Notice that since these curves are continuous, they must cross, meaning that the hiker must pass one place on the mountain at exactly the same time of day going up as going down. This problem may be restated in terms of the Intermediate Value Theorem. Suppose $f(t)$ is *difference* in the hiker's location at time of day t from day 1 to day 2. This function will be continuous on the closed interval and the values of the curve at 8 am and 5 pm will have different signs. Hence, the function must have at least one zero between 8 am and 5 pm. In other words, the difference in the hiker's location at that time from day 1 to day 2 is zero. She passed this point on the mountain at exactly the same time of day going up as going down.

15. Let $f(x) = \sin(x) - 0.7$.

$x_1 = 0.75$; $\qquad f(x_1) \approx -0.018$;
$x_2 = \frac{0.75+0.8}{2} = 0.775$; $\qquad f(x_2) \approx -0.0003$;
$x_3 = \frac{0.8+0.775}{2} = 0.7875$; $\qquad f(x_3) \approx 0.009$;
$x_4 = \frac{0.775+0.7875}{2} = 0.78125$.

4 steps guarantee an approximation within 0.01 since $\frac{0.8-0.7}{2^4} = 0.00625$. However, x_2 is within 0.01 of a root.

18. Let $f(x) = x\exp(x^2) - \sqrt{x^2+1}$.

$x_1 = 0.7;$ $\qquad\qquad f(x_1) \approx -0.08;$
$x_2 = \frac{0.7+1}{2} = 0.85;$ $\qquad f(x_2) \approx 0.44;$
$x_3 = \frac{0.7+0.85}{2} = 0.775;$ $\qquad f(x_3) \approx 0.15;$
$x_4 = \frac{0.7+0.775}{2} = 0.7375.$

6 steps guarantee an approximation to within 0.01 since $\dfrac{0.6}{2^6} = 0.009375$. However, x_4 is within 0.01 of a root.

21. Let $f(x) = 3\cos(\cos(x)) - 2x$.

$x_1 = 1.65;$ $\qquad\qquad f(x_1) \approx -0.31;$
$x_2 = \frac{1.1+1.65}{2} = 1.375;$ $\qquad f(x_2) \approx 0.19;$
$x_3 = \frac{1.375+1.65}{2} = 1.5125;$ $\qquad f(x_3) \approx -0.03;$
$x_4 = \frac{1.375+1.5125}{2} = 1.44375.$

7 steps guarantee an approximation to within 0.01 since $\frac{1.1}{2^7} \approx 0.009$. However, x_6 is within 0.01 of a root.

Section 2.5

1. $\lim\limits_{x\to -\infty} \cos\left(\frac{1}{x}\right) = 1$, $\lim\limits_{x\to\infty} \cos\left(\frac{1}{x}\right) = 1$.

4. $\lim\limits_{x\to -\infty} \mathrm{arccot}(x) = \pi$, $\lim\limits_{x\to\infty} \mathrm{arccot}(x) = 0$.

7. $\lim\limits_{x\to -\infty} 2^x = 0$, $\lim\limits_{x\to\infty} 2^x = \infty$.

10. $\lim\limits_{x\to -\infty} \frac{\sin(x)}{x} = 1$, $\lim\limits_{x\to\infty} \frac{\sin(x)}{x} = 1$.

13. $\lim\limits_{x\to -\infty} 3^{-x} = \infty$, $\lim\limits_{x\to\infty} 3^{-x} = 0$.

For problems 15 - 30, viewing windows may vary.

16. Horizontal asymptote, $y = 0$; $\quad [-6500, 6400] \times [-3.1, 3.2]$.

19. Horizontal asymptote, $y = \frac{1}{2}$; $\quad [-64000, 65000] \times [-3.1, 3.2]$.

22. Horizontal asymptote, $y = -\frac{4}{9}$; $\quad [-650, 640] \times [-3.1, 3.2]$.

25. No horizontal asymptote, $y \approx \frac{1}{24}x^2$; $\quad [-100, 100] \times [-10000, 10000]$.

28. No horizontal asymptote, $y \approx x$; $\quad [-640, 640] \times [-3.1, 3.2]$.

31. This strategy works because $\lim\limits_{x\to 0^+} \frac{1}{x} = \infty$ and $\lim\limits_{x\to 0^-} \frac{1}{x} = -\infty$.

Section 2.6

For exercises 1 - 6, values for δ may vary.

1. a) $\delta = 0.001$.

b) Let $\delta = \epsilon$. Now whenever $0 < |x - 1| < \delta = \epsilon$, we have $|x - 7 + 6| = |x - 1| < \delta = \epsilon$.

4. a) $\delta = 0.001$.

b) Let $\delta = \epsilon$. Now whenever $0 < |x + 2| < \delta = \epsilon$ we have $||x + 1| - 1| = |-x - 1 - 1| = |-1||x + 2| = |x + 2| < \epsilon$. The first equality in this sequence is true since $|x + 1| = -(x + 1)$ for $x \leq -1$ and we are concerned with x values close to -2.

7. $L = 0$; $[-0.01, 0.01]$.

10. $L = 0.4$; $[-0.1, 0.1]$.

13. $\lim\limits_{x \to 0} f(x)$ does not exist.

16. $\lim\limits_{x \to 0} f(x)$ does not exist.

19. $L \approx 2.7$ $(L = e)$; $[-0.0065, 0.0065]$.

22. $\lim\limits_{x \to 0} f(x)$ does not exist.

25. Case I.   Assume $0 < \epsilon < 1$. Let $\delta = \frac{\epsilon}{7}$. (Note that $\delta < 1$.)

Now whenever $0 < |x - 3| < \delta = \frac{\epsilon}{7}$, we have

$|x^2 - 9| = |(x - 3)(x + 3)| = |x - 3||x + 3| = |x - 3||(x - 3) + 6|$
$\leq |x - 3|(|x - 3| + 6) < \delta(\delta + 6) < \frac{\epsilon}{7}(1 + 6) = \epsilon$.

Case II.   Assume $\epsilon \geq 1$. Let $\delta = \frac{1}{7}$.

Now whenever $0 < |x - 3| < \delta = \frac{1}{7}$, we have

$|x^2 - 9| = |(x - 3)(x + 3)| = |x - 3||x + 3| = |x - 3||(x - 3) + 6|$
$\leq |x - 3|(|x - 3| + 6) < \frac{1}{7}(\frac{1}{7} + 6) = \frac{43}{49} < \epsilon$.

28. $0 \leq 1 \leq 1 + \cos(x)$ since $\cos(x) \geq 0$. Multiplying the inequality by $(1 - \cos(x))$ which is non-negative, we have

$0 \leq 1 - \cos(x) \leq (1 + \cos(x))(1 - \cos(x)) = 1 - \cos^2(x) = \sin^2(x)$.

So, $0 \leq 1 - \cos(x) \leq \sin^2(x)$.

ANSWERS TO SELECTED EXERCISES FOR CHAPTER 3

Section 3.1

1. $r = \frac{40}{1.0} = 40$ mph.

4. The car is moving the fastest at approximately $t = 0.25$ hour. The car is moving the slowest at approximately $t = 0.7$ hour.

7. Instantaneous speed at $t = 0.5$ hours is approximately 40 mph and at $t = 1.0$ hours is approximately 50 mph.

11. C.

12. A.

15. A.

18. 1 min.

21. 0 min.

24. 6 min.

26. The water level rises 2 inches per minute when the plug is in. At 20 minutes, the level is 3 inches. It will take 4.5 minutes, for the water to raise 9 inches. At 24.5 minutes, the water will reach the top and then overflow.

Section 3.2

1. Find b if $y = 3x + b$. $1 = 3(-4) + b \Rightarrow b = 13 \Rightarrow y = 3x + 13$.

4. If $m = -\frac{1}{2}$ and $b = 0$, then $y = -\frac{1}{2}x$. (Recall, the slopes of the 2 parallel lines are equal.)

7. Find the intersection of f and g. Setting the equations of each line equal to each other, we have $3x + 13 = -\frac{1}{2}x - \frac{5}{2} \Rightarrow 7x = -31 \Rightarrow x = -\frac{31}{7}$ and $y = 3(-\frac{31}{7}) + 13 = -\frac{2}{7}$. The constant function is $h(x) = -\frac{2}{7}$.

10. $(f+g)(x) = (3x + 13) + (-\frac{1}{2}x - \frac{5}{2}) = \frac{5}{2}x + \frac{21}{2}$ so $y = \frac{5}{2}x + \frac{21}{2}$;
$(f-g)(x) = (3x + 13) - (-\frac{1}{2}x - \frac{5}{2}) = \frac{7}{2}x + \frac{31}{2}$ so $y = \frac{7}{2}x + \frac{31}{2}$;
$f \circ g(x) = 3(-\frac{1}{2}x - \frac{5}{2}) + 13 = -\frac{3}{2}x + \frac{11}{2}$ so $y = -\frac{3}{2}x + \frac{11}{2}$.

The slope of $f + g$ is the sum of the slopes of f and g. The slope of $f - g$ is the difference if the slopes of f and g. The slope of $f \circ g$ is the product of the slopes of f and g.

For exercises 13-18, the answers for parts d) and e) will vary depending on the numerical precision of the calculator or computer used.

13. a)

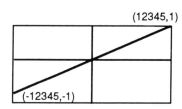

b) $y = \frac{1}{12345}x$.

c) $y = \frac{1}{12345}(x-1) + f(1) \Rightarrow y = \frac{1}{12345}(x-1) + \frac{1}{12345}$.

d) $m_1 = \frac{f(1)-f(0.9)}{1-0.9} = \frac{\frac{1}{12345} - \frac{0.9}{12345}}{0.1} = \frac{1}{12345}\left(\frac{1-0.9}{0.1}\right) = \frac{1}{12345}$;

$m_2 = \frac{f(1)-f(0.99)}{1-0.99} = \frac{\frac{1}{12345} - \frac{0.99}{12345}}{0.01} = \frac{1}{12345}\left(\frac{1-0.99}{0.01}\right) = \frac{1}{12345}$;

$m_3 = \frac{f(1)-f(0.999)}{1-0.999} = \frac{\frac{1}{12345} - \frac{0.999}{12345}}{0.001} = \frac{1}{12345}$;

$m_4 = \frac{f(1)-f(0.9999)}{1-0.9999} = \frac{\frac{1}{12345} - \frac{0.9999}{12345}}{0.0001} = \frac{1}{12345}$;

$m_5 = \frac{f(1)-f(0.99999)}{1-0.99999} = \frac{\frac{1}{12345} - \frac{0.99999}{12345}}{0.00001} = \frac{1}{12345}$.

e) If we calculate the numerator and then divide by the denominator, x_1 gives the most accurate slope value. If we simplify first as shown above, each calculation produces the exact slope.

16. a)

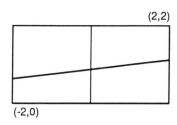

b) $y = 0.12345678(x-1) + 1 \Rightarrow y = 0.12345678x + 0.87654322$.

c) $y = 0.12345678(x-1) + 1$.

d) $m_1 = \frac{f(1)-f(0.9)}{1-0.9} = \frac{1-(0.12345678(0.9)+0.87654322)}{0.1} = 0.12345678$;

$m_2 = \frac{f(1)-f(0.99)}{1-0.99} = \frac{1-(0.12345678(0.99)+0.87654322)}{0.01} = 0.12345678$;

$m_3 = \frac{f(1)-f(0.999)}{1-0.999} = \frac{1-(0.12345678(0.999)+0.87654322)}{0.001} = 0.12345678$;

$m_4 = \frac{f(1)-f(0.9999)}{1-0.9999} = \frac{1-(0.12345678(0.9999)+0.87654322)}{0.0001} = 0.12345678$;

$m_5 = \frac{f(1)-f(0.99999)}{1-0.99999} = \frac{1-(0.12345678(0.99999)+0.87654322)}{0.00001} = 0.12345678$.

e) See note in 13e.

19. The slope of the line connecting the first two points is $m = \frac{6.11-5.32}{1} = 0.79$. The equation of the first linear piece is $y = 0.79x + 5.32$ for $0 \leq x \leq 1$. So, $f(0.5) = 0.79(0.5) + 5.32 = 5.715$.

The slope of the line connecting the second and third points is $m = \frac{2.20-6.11}{2-1} = -3.91$. The equation for the second linear piece is $y = -3.91(x-1) + 6.11$ for $1 \leq x \leq 2$. So, $f(1.6) = -3.91(1.6-1) + 6.11 = 3.764$.

The slope of the line connecting the third and fourth points is $m = \frac{4.07-2.20}{1}$. The equation for the third linear piece is $y = 1.87(x-2) + 2.20$ for $2 \leq x \leq 3$. So, $f(2.25) = 1.87(2.25-2) + 2.20 = 2.6675$.

22. Does not exist.

25. $m = 0(1) = 0$. (slope of f times slope of g)

28. $m = \frac{1}{3}$. (slope of g divided by 3)

31. $\tan\theta = \sqrt{3} \Rightarrow \theta = \frac{\pi}{3}$.

34. $\phi + 2\alpha = \frac{\pi}{2}$ (See the right triangle containing ϕ.) $\Rightarrow \phi = \frac{\pi}{2} - 2\alpha \Rightarrow \phi = \frac{\pi}{2} - 2\theta$ from #33.

37. a)

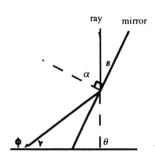

$\gamma + 2\beta = \frac{\pi}{2}$ and $\theta + \beta = \frac{\pi}{2} \Rightarrow \gamma = 2\theta - \frac{\pi}{2}$.

$\tan\gamma = \tan\left(-\frac{\pi}{2} + 2\theta\right) = \frac{\sin\left(-\frac{\pi}{2}+2\theta\right)}{\cos\left(-\frac{\pi}{2}+2\theta\right)} = \frac{\sin\left(-\frac{\pi}{2}\right)\cos(2\theta)+\cos\left(-\frac{\pi}{2}\right)\sin(2\theta)}{\cos\left(-\frac{\pi}{2}\right)\cos(2\theta)-\sin\left(-\frac{\pi}{2}\right)\sin(2\theta)} = -\frac{\cos(2\theta)}{\sin(2\theta)}$.

See #36 for the remainder of the proof.

40. There is no reflection.

43.

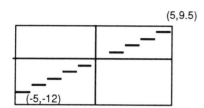

44.

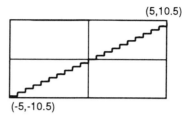

47.

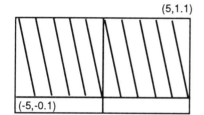

48.

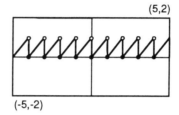

Section 3.3

2. $f'(1) = \frac{3}{2}$.

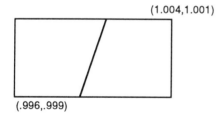

4. $f'(1) = \frac{3}{4}$.

8. $f'(1) = -\frac{1}{2}$.

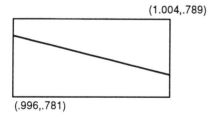

10. $f'(1) = \frac{1}{\ln(10)} = \approx 0.434$.

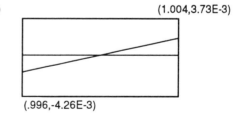

12. $f'(0) = 0$.

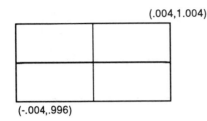

(.004, 1.004)

(−.004, .996)

14. $f'(0) = 0$.

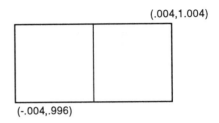

(.004, 1.004)

(−.004, .996)

18. $f'(0) = 0$.

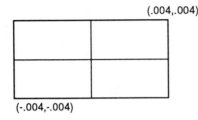

(.004, .004)

(−.004, −.004)

20. $f'(0) = 6$.

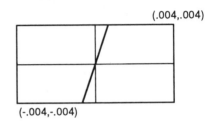

(.004, .004)

(−.004, −.004)

23. $f'(-1.5) \approx \frac{f(-1.499)-f(-1.5)}{0.001} = -.500500501$;

$f'(-1.5) \approx \frac{f(-1.501)-f(-1.5)}{-0.001} = -0.4995005$.

Averaging the results, we get $f'(-1.5) \approx \frac{-0.5005-0.4995}{2} = -0.5$.

26. $f'(1.5) = \frac{f(1.501)-f(1.5)}{0.001} = 0.4$;

$f'(1.5) = \frac{f(1.499)-f(1.5)}{-0.001} = -0.4$.

Averaging the results, we get $f'(1.5) = \frac{0.4-0.4}{2} = 0$. (The derivative does not actually exist at this point.)

The limits in exercises 29 - 36 were evaluated algebraically but could have been investigated graphically or numerically.

29. $\lim_{h \to 0} \frac{f(1.25+h)-f(1.25)}{h} = \lim_{h \to 0} \frac{(4(1.25+h)+1)-(4(1.25)+1)}{h} = \lim_{h \to 0} \frac{4h}{h} = \lim_{h \to 0} 4 = 4$.

The numerical estimate was exact.

32. $\lim_{h \to 0} \frac{f(3+h)-f(3)}{h} = \lim_{h \to 0} \frac{\sqrt{2(3+h)+3}-\sqrt{2(3)+3}}{h} = \lim_{h \to 0} \frac{\sqrt{9+2h}-\sqrt{9}}{h} \left(\frac{\sqrt{9+2h}+\sqrt{9}}{\sqrt{9+2h}+\sqrt{9}} \right)$

$= \lim_{h \to 0} \frac{2h}{h(\sqrt{9+2h}+\sqrt{9})} = \lim_{h \to 0} \frac{2}{(\sqrt{9+2h}+\sqrt{9})} = \frac{2}{2\sqrt{9}} = \frac{1}{\sqrt{9}} = \frac{1}{3}$.

The average of the numerical estimates was accurate to 8 decimal places.

35. $\lim_{h \to 0} \frac{f(6+h)-f(6)}{h} = \lim_{h \to 0} \frac{\frac{6+h-1}{4}-\frac{6-1}{4}}{h} = \lim_{h \to 0} \frac{h}{4h} = \lim_{h \to 0} \frac{1}{4} = \frac{1}{4}$.

The numerical estimates were exact.

38. $f'_-(-1) = \lim_{h \to 0^-} \frac{f(-1+h)-f(-1)}{h} = \lim_{h \to 0^-} \frac{\sqrt{(-1+h)^2-1}-\sqrt{(-1)^2-1}}{h}$

$= \lim_{h \to 0^-} \frac{\sqrt{-2h+h^2}}{h} = \lim_{h \to 0^-} \frac{\sqrt{h(-2+h)}}{h}$ which does not exist.

$f'_+(-1) = \lim_{h \to 0^-} \frac{f(-1+h)-f(-1)}{h} = \lim_{h \to 0^+} \frac{\sqrt{h(-2+h)}}{h}$.

This limit does not make sense since domain of f is $(-\infty, -1]$ and $[1, \infty)$.

41. $\frac{\frac{f(x_0+h)-f(x_0)}{h}+\frac{f(x_0)-f(x_0-h)}{h}}{2} = \frac{f(x_0+h)-f(x_0-h)}{2h}$.

Section 3.4

For problems 1-20, part c, the actual derivative is given but other answers are possible.

1. b)   **c)** $g(x) = \frac{2}{3}x^{-1/3}$.

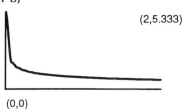
(2,5.333)
(0,0)

4. b)   **c)** $g(x) = \frac{3}{4}x^{-1/4}$.

(5,5)
(-5,-5)

7. b)   **c)** $g(x) = \frac{1}{1+x^2}$.

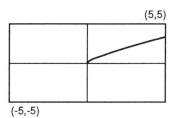
(2,1)
(0,0.079)

10. b) 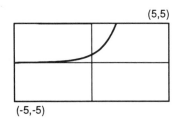　　c) $g(x) = e^x$.

13. b) 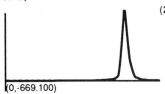　　c) $g(x) = \sec^2(x)$.

16. b) 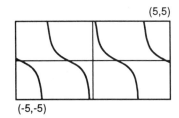　　c) $g(x) = -\csc^2(x)$.

19. b) 　　c) $g(x) = 0$.

23.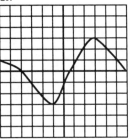

24.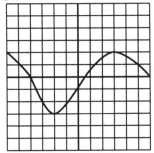

27.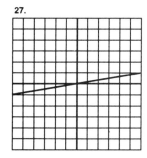

28.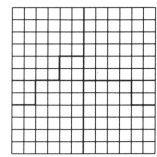

31. $f'(x) = \lim\limits_{h \to 0} \frac{f(x+h)-f(x)}{h} = \lim\limits_{h \to 0} \frac{\frac{1}{x+h} - \frac{1}{x}}{h}$
$= \lim\limits_{h \to 0} \frac{x-(x+h)}{h(x)(x+h)} = \lim\limits_{h \to 0} \frac{-1}{x(x+h)} = -\frac{1}{x^2}.$

34. $f'(x) = \lim\limits_{h \to 0} \frac{f(x+h)-f(x)}{h} = \lim\limits_{h \to 0} \frac{\frac{(x+h)^2}{4} - \frac{x^2}{4}}{h}$
$= \lim\limits_{h \to 0} \frac{2xh+h^2}{4h} = \lim\limits_{h \to 0} \frac{2x+h}{4} = \frac{x}{2}.$

37.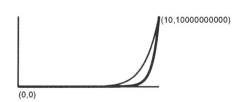

40. $f'(b) = g'(b)$ when $b = 1.13592$. (b can be found by using a root-finder on a machine.) At $x = b$, the graphs $y = f(x)$ and $y = g(x)$ have the same slope.

Section 3.5

1. $V(36") = \frac{4}{3}\pi(36)^3 = 62208\pi$ in^3;

$V(39") = \frac{4}{3}\pi(39)^3 = 79092\pi$ in^3.

The volume changes by $79092\pi - 62208\pi = 16884\pi$ in^3.

$\frac{\Delta V}{\Delta r} = \frac{16884\pi}{3} = 5628\pi$ in^3/in.

4. $\frac{\Delta V}{\Delta r} \approx \frac{4\pi r^2 \Delta r}{\Delta r} = 4\pi r^2$.

7. $P = 4x$, $\quad \frac{dP}{dx} = 4$, $\quad \left.\frac{dP}{dx}\right|_{x=5} = 4$ cm/cm.

10. $S = 6x^2$, $\dfrac{dS}{dx} = 12x$, $\dfrac{dS}{dx}\bigg|_{x=2} = 24$ mm^2/mm.

13. $s(0) = 15 - 6.729(0)^2 = 15$ m.

16. $0 = 15 - 6.729t^2 \Rightarrow t \approx 1.49$ sec.

19. $\dfrac{s(1.49)-s(0)}{1.49-0} = \dfrac{15-0}{1.49} \approx 10.07$ m/s.

22. $\dfrac{s(0.301)-s(0.3)}{0.301-0.3} \approx -4.044$ m/s.

25. $v(1.49) = 13.458(1.49) \approx -20.05$ m/s.

29. $C = 2\pi(6) - 6\theta = 12\pi - 6\theta$, $\dfrac{dC}{d\theta}\bigg|_{\theta=\pi/3} = -6$ in per radian.

Section 3.6

1. $\dfrac{dy}{dx} = (3x^2 + 6x)(2x^2 + 8x - 5) + (x^3 + 3x^2 + 1)(4x + 8)$.

4. $\dfrac{dA}{ds} = (4s - 3)(9s - 1) + (2s^2 - 3s + 1)(9) = 36s^2 - 31s + 3 + 18s^2 - 27s + 9$
$= 54s^2 - 58s + 12$.

7. $\dfrac{dy}{dx} = 2x(3x^3 - x + 1) + (x^2 - 6)(9x^2 - 1)$.

10. Let $y = (x^2 + x)(x^2 + 5x + 6)$. $\dfrac{dy}{dx} = (2x + 1)(x^2 + 5x + 6) + (x^2 + x)(2x + 5)$.

11. $\dfrac{ds}{dt} = 2t - 2t^{-3}$.   **14.** $\dfrac{dy}{dx} = 1 - x^{-2}$.

17. $\dfrac{dT}{dv} = -v^{-2} + 4v^{-3}$.   **20.** $\dfrac{dy}{dt} = -t^{-2}$.

23. $\dfrac{ds}{dt} = \dfrac{8(t^2-2t+3)-(8t+15)(2t-2)}{(t^2-2t+3)^2}$
$= \dfrac{8t^2-16t+24-16t^2-14t+30}{(t^2-2t+3)^2} = \dfrac{-8t^2-30t+54}{(t^2-2t+3)^2}$.

26. $\dfrac{ds}{dt} = \dfrac{3(6t-7)-(3t+4)6}{(6t-7)^2} = \dfrac{-45}{(6t-7)^2}$.

29. $\dfrac{dy}{dx} = \dfrac{(x+1)-x}{(x+1)^2} = \dfrac{1}{(x+1)^2}$.

32. Let $f(x) = \dfrac{1}{\sin(x)}$. So $f'(x) = \dfrac{-\cos(x)}{\sin^2(x)} = -\csc(x)\cot(x)$.

35. $f'(x) = 2\cos(x) + 7\sin(x)$.

38. $f'(x) = 2x(2^x) + x^2 \ln(2)(2^x) = 2^x(2 + x\ln(2))x$.

Section 3.7

1. $F'(3) = (7.1)f'(3) = 7.1(-0.5) = -3.55$.

4. $F'(1) = 0.6f'(1) + 5g'(1) = 0.6(-7.2) + 5(-0.7) = -7.82$.

7. $F'(1) = g'(1) - f'(1) = -0.7 - (-7.2) = 6.5$.

10. $F'(x) = -f'(g(x))g'(x) \Rightarrow$
$F'(1) = -f'(g(1))g'(1) = -f'(3)g'(1) = -(-0.5)(-0.7) = -0.35$.

13. $F'(x) = f'((g(x))^2)(2g(x))g'(x) \Rightarrow$
$F'(2) = f'((g(2))^2) \cdot 2g(2) \cdot g'(2) = f'(4)(2)(2)(-0.3) = (-4)(4)(-0.3) = 4.8$.

16. $F'(x) = \frac{g'(x)h(x) - g(x)h'(x)}{(h(x))^2} \Rightarrow$
$F'(5) = \frac{g'(5)h(5) - g(5)h'(5)}{(h(5))^2} = \frac{-0.23(18) - (1)(9)}{18^2} = -0.040\overline{5}$.

19. $F'(x) = e^{g(x)}g'(x) \Rightarrow F'(2) = e^{g(2)}g'(2) = e^2(-0.3) = -0.3e^2$.

22. $F'(x) = f'(x) - g'(x) \Rightarrow F'(-4) = f'(-4) - g'(-4) = 2 - (-1) = 3$.

25. $F'(x) = 2f'(x) - 3g'(x) \Rightarrow F'(3) = 2f'(3) - 3g'(3) = 2(0) - 3(1) = -3$.

28. $F'(x) = g'(2x - 3)(2) \Rightarrow F'(1) = g'(2(1) - 3)(2) = g'(-1)(2) = (-1)(2) = -2$.

31. $f'(x) = \frac{1}{3}x^{-2/3}, f'(27) = \frac{1}{3}(27)^{-2/3} = \frac{1}{27}$.

34. $\frac{dy}{dx} = \frac{1}{2}\left(\frac{x+3}{x+5}\right)^{-1/2}\left(\frac{x+5-(x+3)}{(x+5)^2}\right),$
$\left.\frac{dy}{dx}\right|_{x=1} = \frac{1}{2}\left(\frac{4}{6}\right)^{-1/2}\left(\frac{2}{36}\right) = \frac{\sqrt{6}}{4}\left(\frac{2}{36}\right) = \frac{\sqrt{6}}{2(36)} = \frac{\sqrt{6}}{72}$.

37. $\frac{dy}{dx} = -5x^{-4}, \quad \left.\frac{dy}{dx}\right|_{x=1} = -5$.

40. $f'(x) = \frac{1}{2}(x^2 - 2x + 1)^{-1/2}(2x - 2) \Rightarrow f'(1) = \frac{1}{2}(1 - 2 + 1)^{-1/2}(2(1) - 2) = 0$.

43. $f'(x) = (12x^2 + 2)(8 - 9x^3) + (4x^3 + 2x - 5)(-27x^2)$
$= -216x^5 - 72x^3 + 231x^2 + 16.$

46. $f'(x) = 3x^2 \sin(x^2 + 2x - 3) + (x^3 + 27) \cos(x^2 + 2x - 3)(2x + 2)$
$= 3x^2 \sin(x^2 + 2x - 3) + (2x^4 + 2x^3 + 54x + 54) \cos(x^2 + 2x - 3).$

49. $f'(x) = (2x + 6) \sec(x^2 + 6x + 8) \tan(x^2 + 6x + 8).$

52. $f'(x) = \dfrac{\frac{1}{4}(4-4x)}{(7+4x-2x^2)^{\frac{3}{4}}} = \dfrac{1-x}{(7+4x-2x^2)^{\frac{3}{4}}}.$

55. $\dfrac{d}{dx}((x^2 + 1)^4) = 4(x^2 + 1)^3(2x) = 8x(x^2 + 1)^3.$

58. $\dfrac{d}{dx}(\sin^2(x) + \cos^2(x)) = \dfrac{d}{dx}(1) = 0.$

61. $f(x) = x^2;\ x_0 = 5.$

64. $f(x) = \sqrt{x};\ x_0 = 4.$

67. $\dfrac{d}{dx}\sin(x)\{x \text{ in degrees }\} = \dfrac{d}{dx}\sin(x\tfrac{\pi}{180}) = \tfrac{\pi}{180}\cos(x\tfrac{\pi}{180}) = \tfrac{\pi}{180}\cos(x°).$

70. The car travels $2\pi(13) = 26\pi$ inches per revolution. At 1000 revolutions per minute, the car travels 26000π inches per minute. So

$$\left(\tfrac{26000\pi \text{ in}}{\text{min}}\right)\left(\tfrac{60 \text{ min}}{1 \text{ hr}}\right)\left(\tfrac{1 \text{ ft}}{12 \text{ in}}\right)\left(\tfrac{1 \text{ mi}}{5280 \text{ ft}}\right) = 24.62\overline{12}\pi \text{ mph} \approx 77.3 \text{ mph}.$$

ANSWERS TO SELECTED EXERCISES FROM CHAPTER 4

Section 4.1

1. Putting the equation of the line in slope-intercept form, we get $y = -4x + \frac{5}{2}$ and see that $m = -4$. Now, let $\frac{dy}{dx} = 3x^2 + 4x - 4 = -4$ and solve for x. $3x^2 + 4x = 0 \Rightarrow x(3x + 4) = 0 \Rightarrow x = 0$ and $x = -\frac{4}{3}$. Two points on the graph where the tangent lines are parallel to the given lines are $(0, f(0)) = (0, 5)$ and $(-\frac{4}{3}, f(-\frac{4}{3})) = (-\frac{4}{3}, \frac{311}{27})$. The equation of one line is $y = -4x + 5$. The equation of the other line is $y = -4(x + \frac{4}{3}) + \frac{311}{27}$.

4. Since $\frac{dy}{dx} = -\sin(x)$, we solve $-\sin(x) = 0 \Rightarrow x = n\pi$, n any integer. $(n\pi, \cos(n\pi)) = (n\pi, 1)$ for n even and $(n\pi, \cos(n\pi)) = (n\pi, -1)$ for n odd are points on the graph where the tangent lines are horizontal. The equations of these lines are $y = 1$ and $y = -1$.

7. $\frac{dy}{dx} = \frac{2}{3}(x - 8)^{-1/3}$ so $\frac{dy}{dx}\big|_{x=0} = -\frac{1}{3} \Rightarrow$ slope of the normal line at $x = 0$ is $m = 3$. The equation of this normal line line is $y = 3x + 5$.

10. Solving $\frac{dy}{dx} = \sec^2(x) = 1$, we find that $x = n\pi$, n any integer. $(n\pi, f(n\pi)) = (n\pi, 0)$ are the points on the graph of $y = \tan(x)$ where the tangent lines have slope 1.

12. $f'(x) = 4x - 7$, $f'(\frac{3}{2}) = -1$, and $f(\frac{3}{2}) = -3$.
    Equation of the tangent line: $y = -(x - \frac{3}{2}) - 3$.
    Equation of the normal line: $y = (x - \frac{3}{2}) - 3$.

15. $f'(x) = 5x^4$, $f'(1) = 5$, and $f(1) = 1$.
    Equation of the tangent line: $y = 5(x - 1) + 1$.
    Equation of the normal line: $y = -\frac{1}{5}(x - 1) + 1$.

18. $f'(x) = \sec^2(x)$, $f'(\frac{3\pi}{4}) = 2$, and $f(\frac{3\pi}{4}) = -1$.
    Equation of the tangent line: $y = 2(x - \frac{3\pi}{4}) - 1$.
    Equation of the normal line: $y = -\frac{1}{2}(x - \frac{3\pi}{4}) - 1$.

For problems 21 - 30, results in columns 2, 3, and 4 may have been rounded.

21. a) $f'(x) = 2x$, $f'(-3) = -6$, and $f(-3) = 9 \Rightarrow$ best linear approximation is $g(x) = -6(x+3) + 9$.

b) and c)

x	$g(x)$	$f(x)$	Error
-2	3	4	1
-4	15	16	1
-2.999	8.994	8.994001	0.000001
-3.001	9.006	9.006001	0.000001

d)

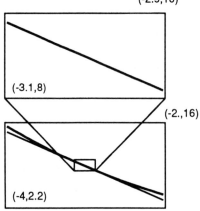

24. a) $f'(x) = 3x^2 - 4$, $f'(-1.5) = 2.75$, and $f(1.5) = 2.625 \Rightarrow$ best linear approximation is $g(x) = 2.75(x + 1.5) + 2.625$.

b) and c)

x	$g(x)$	$f(x)$	Error
-0.5	5.375	1.875	3.5
-2.5	-0.125	-5.625	5.5
-1.499	2.62775	2.62775	4.5×10^{-6}
-1.501	2.62225	2.62225	4.5×10^{-6}

d)

27. a) $f'(x) = \sec^2(x)$, $f'(\frac{\pi}{4}) = 2$, and $f(\frac{\pi}{4}) = 1 \Rightarrow$ best linear approximation is $g(x) = 2(x - \frac{\pi}{4}) + 1$.

b) and c)

x	$g(x)$	$f(x)$	Error
$\frac{\pi}{4} + 1$	3	-4.58804	7.59
$\frac{\pi}{4} - 1$	-1	-0.21796	0.78
$\frac{\pi}{4} + 0.001$	1.002	1.00200	2×10^{-6}
$\frac{\pi}{4} - 0.001$	0.998	0.99800	2×10^{-6}

d)

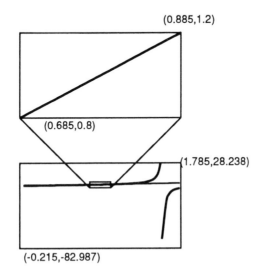

30. a) $f'(x) = 0$, $f'(132.78) = 0$, and $f(132.78) = 5.42 \Rightarrow$ best linear approximation is $g(x) = 5.42$.

b) and c)

x	$g(x)$	$f(x)$	Error
133.78	5.42	5.42	0
131.78	5.42	5.42	0
132.781	5.42	5.42	0
132.779	5.42	5.42	0

d)

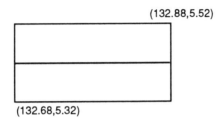

33. $\frac{dy}{dx} = \frac{x}{2}$.

When $\frac{dy}{dx}\Big|_{x=-3} = -\frac{3}{2}$, then $m = \frac{2}{3}$ and

the equation of the normal line is $y = \frac{2}{3}(x+3) + 2.25$.

When $\frac{dy}{dx}\Big|_{x=-1} = -\frac{1}{2}$, then $m = 2$ and

the equation of the normal line is $y = 2(x+1) + 0.25$.

When $\frac{dy}{dx}\Big|_{x=0} = 0$, then m is undefined and

the equation of the normal line is $x = 0$.

When $\frac{dy}{dx}\Big|_{x=2} = 1$, then $m = -1$ and

the equation of the normal line is $y = -(x-2) + 1$.

When $\frac{dy}{dx}\Big|_{x=4} = 2$, then $m = -\frac{1}{2}$ and

the equation of the normal line is $y = -\frac{1}{2}(x-4) + 4$.

36. $(0, 1)$.

39. $y = (\frac{a^2-4}{4a})(x-a) + \frac{a^2}{4}$. Let $x = 0$ to find the y-intercept:
$y = (\frac{a^2-4}{4a})(-a) + \frac{a^2}{4} \Rightarrow y = -\frac{a^2}{4} + 1 + \frac{a^2}{4} \Rightarrow y = 1$.

Section 4.2

All approximations are accurate to 12 digits.

1. $f(x) = x^2 - 3.2$; $f'(x) = 2x$.

$g(x) = x - \frac{x^2-3.2}{2x}$ is the iterating function.
$x_0 = 1.5$, $\quad x_1 = 1.81\overline{6}$, $\quad x_2 = 1.78906727829$,
$x_3 = 1.78885439467$, $\quad x_4 = 1.788854382$, $\quad x_5 = 1.788854382$.

4. $f(x) = x^5 + 3.6x^4 - 2.51x^3 - 22.986x^2 - 28.24x - 10.4$;

$f'(x) = 5x^4 + 14.4x^3 - 7.53x^2 - 45.972x - 28.24$.

$g(x) = x - \frac{f(x)}{f'(x)}$ is the iterating function.
$x_0 = 2.5$, $\quad x_1 = 2.61126564673$, $\quad x_2 = 2.60012352999$,
$x_3 = 2.60000001506$, $\quad x_4 = 2.6$, $\quad x_5 = 2.6$.

7. $f(x) = \ln(x^2 + \ln(x))$; $f'(x) = \frac{2x + \frac{1}{x}}{x^2 + \ln(x)}$.

$g(x) = x - \frac{\ln(x^2+\ln(x))}{\frac{2x+\frac{1}{x}}{x^2+\ln(x)}}$ is the iterating function.

$x_0 = 0.9875$, $\quad x_1 = 0.99978835581$, $\quad x_2 = 0.99999994026$,
$x_3 = 1$, $\quad x_4 = 1$, $\quad x_5 = 1$.

10. $f(x) = \sqrt[4]{x} + x^3 - x - 0.5$; $f'(x) = \frac{1}{4}x^{-3/4} + 3x^2 - 1$.

$g(x) = x - \frac{\sqrt[4]{x}+x^3-x-0.5}{\frac{1}{4}x^{-3/4}+3x^2-1}$ is the iterating function.

$x_0 = 0.7$, $\quad x_1 = 0.627585040703$, $\quad x_2 = 0.609576604098$,
$x_3 = 0.608455610599$, $\quad x_4 = 0.608451352913$, $\quad x_5 = 0.608451352856$.

13. If $x_n = 7$ for some n then $x_{n+1} = 7 + 7b(7-7) = 7$ and $x_{n+2} = 7$, etc. Therefore, if we can find values of b that produce $x_n = 7$, then we certainly have the desired convergence.

We're given $x_0 = 1$, so $x_1 = 1 + 6b$. If $b = 1$, then the sequence converges to 7. In any case, if $x_1 = 1 + 6b$, then $x_2 = (1+6b) + b(1+6b)(7-(1+6b)) = -36b^3 + 30b^2 + 12b + 1$.

If $b = \frac{1}{3}$, then $x_2 = 7$ and the sequence converges to 7. Any values in the interval $(0, \frac{2}{7})$ will also work.

16. If $x_0 = 1$, we get $x_1 = g(1) = 1 + b(7-1) = 1 + 6b$. If $x_1 = 1 + 6b$, then $x_2 = g(x_1) = g(1+6b) = 1 + 6b + b(1+6b)(7-(1+6b)) = 1 + 6b + (b+6b^2)(6-6b)$
$= 1 + 12b + 30b^2 - 36b^3$. Solve $x_2 = 1 + 12b + 30b^2 - 36b^3 = 1$ or
$-36b^3 + 30b^2 + 12b = 0 \Rightarrow -6b(6b^2 - 5b - 2) = 0 \Rightarrow b = 0$ or
$b = \frac{5 \pm \sqrt{25+48}}{12} = \frac{5 \pm \sqrt{73}}{12}$.

For $b = \frac{5 \pm \sqrt{73}}{12}$, Newton's method approximations do not converge.

19. By graphing $y = x + \tan(x)$ we see that the three smallest positive roots occur near $x = 2$, $x = 4.9$ and $x = 8$.

The iterative function is $g(x) = x - \frac{x+\tan x}{1+\sec^2 x}$.

Using $x_0 = 2$ yields $x \approx 2.02875783811$.

Using $x_0 = 4.9$ yields $x \approx 4.91318043943$.

Using $x_0 = 8$ yields $x \approx 7.97866571241$.

22. Let $f(x) = x^2 - x - 1$. $x = \frac{1 \pm \sqrt{5}}{2}$ using the quadratic formula.

25. $x_0 = 1 = \frac{1}{1}$, $\quad x_1 = 2 = \frac{2}{1}$, $\quad x_2 = 1.66666666667 = \frac{3}{2}$,
$x_3 = 1.61904761905 = \frac{34}{21}$, $\quad x_4 = 1.61803444782 = \frac{1597}{987}$,
$x_5 = 1.61803398875 = \frac{514229}{317811}$.

Section 4.3

1. a)   c) $f'(x) = 2x$.

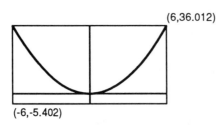
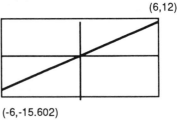

b) f is increasing over $(0, 6)$ and decreasing over $(-6, 0)$.

d) f' is positive over $(0, 6)$ and f' is negative over $(-6, 0)$.

e) $x = 0$ is the critical value of f.

f) $x = 0$ represents a local minimum.

4. a) and c)

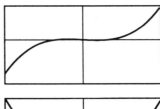

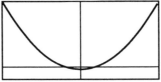

b) f is increasing over $(-6, -\frac{2}{\sqrt{3}})$ and $(\frac{2}{\sqrt{3}}, 6)$ and f is decreasing over $(-\frac{2}{\sqrt{3}}, \frac{2}{\sqrt{3}})$. (Students will probably use decimal approximations for the endpoints of the intervals since they are gathering this information graphically.)

c) $f'(x) = 3x^2 - 4$.

d) f' is positive over $(-6, -\frac{2}{\sqrt{3}})$ and $(\frac{2}{\sqrt{3}}, 6)$ and f' is negative over $(-\frac{2}{\sqrt{3}}, \frac{2}{\sqrt{3}})$.

e) $x = \frac{2}{\sqrt{3}}$ and $x = -\frac{2}{\sqrt{3}}$ are the critical values of f.

f) $x = -\frac{2}{\sqrt{3}}$ represents a local maximum and $x = \frac{2}{\sqrt{3}}$ represents a local minimum.

6. a) and c)

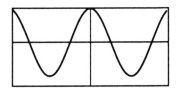

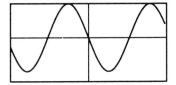

b) f is increasing over $(-\pi, 0)$ and $(\pi, 6)$ and f is decreasing over $(-6, -\pi)$ and $(0, \pi)$.

c) $f'(x) = -\sin(x)$.

d) f' is positive over $(-\pi, 0)$ and $(\pi, 6)$ and f' is negative over $(-6, -\pi)$ and $(0, \pi)$.

e) $x = -\pi$, $x = 0$, $x = \pi$ are the critical values of f.

f) $x = -\pi$ and $x = \pi$ represent local minima and $x = 0$ represents a local maximum.

9. a)  c) $f'(x) = \frac{1}{5}x^{-4/5}$.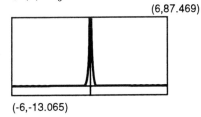

b) f is increasing over $(-6, 0)$ and $(0, 6)$.

d) f' is positive over $(-6, 0)$ and $(0, 6)$

e) $x = 0$ is the critical value of f.

f) $x = 0$ represents neither a local maximum nor a local minimum.

12. a) and c)

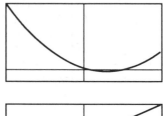

b) f is decreasing over $(-6, 1.75)$ and f is increasing over $(1.75, 6)$.

c) $f'(x) = 4x - 7$.

d) f' is negative over $(-6, 1.75)$ and f' is positive over $(1.75, 6)$.

e) $x = 1.75$ is the critical value of f.

f) $x = 1.75$ represents a local minimum.

15. a)   c) $f'(x) = 5x^4$.

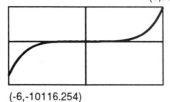
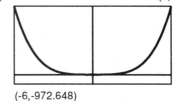

b) f is increasing over $(-6, 0)$ and $(0, 6)$.

d) f' is positive over $(-6, 0)$ and $(0, 6)$.

e) $f'(x) = 5x^4 = 0 \Rightarrow x = 0$ is the critical value of f.

f) $x = 0$ represents neither a local minimum nor a local maximum.

18. a) and c)

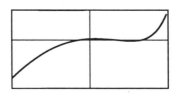

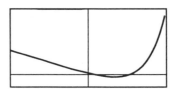

b) f is increasing over $(-6, \approx 0.485)$ and $(\approx 3.212, 6)$ and decreasing over $(\approx 0.485, \approx 3.212)$.

c) $f'(x) = \ln 2(2^x) - 2x$.

d) f' is positive over $(-6, \approx 0.485)$ and $(\approx 3.212, 6)$ and f' is negative over $(\approx 0.485, \approx 3.212)$.

e) $x \approx 0.485$ and $x \approx 3.212$ are critical values of f.

f) $x \approx 0.485$ represents a local maximum and $x \approx 3.212$ represents a local minimum.

21. The critical values at which $f'(x)$ changes from positive to negative represent local maxima.

24. Suppose x_0 is a critical value of f. If $f'(x)$ changes sign from positive to negative at x_0, then a local maximum occurs at x_0. If $f'(x)$ changes sign from negative to positive at x_0, then a local minimum occurs at x_0. If $f'(x)$ does not change sign at x_0, then the critical value is the location of neither a local minimum nor a local maximum.

27.

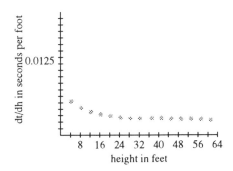

30. The derivative is positive during the winter and the spring and negative during the summer and the fall.

33. The lowest level occurs at $t = 0$ and the highest level occurs at $t = 6.4$.

36. $\frac{dy}{dt}$ is positive over $(0,2)$ and $(4.8, 6.4)$ and $\frac{dy}{dx}$ is negative over $(2, 4.8)$.

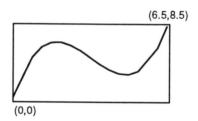

39.

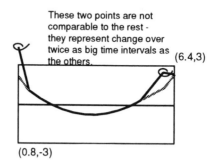

These two points are not comparable to the rest - they represent change over twice as big time intervals as the others.

Section 4.4

1. $f'(x) = 3x^2 - 1 = 0 \Rightarrow x = \pm\sqrt{1/3}$ are the critical values. Since $f'(x) > 0$ over $(-\infty, -\frac{1}{\sqrt{3}})$ and $f'(x) < 0$ over $(-\frac{1}{\sqrt{3}}, \frac{1}{\sqrt{3}})$, $f(-\frac{1}{\sqrt{3}}) \approx 0.385$ is a local maximum. Since $f'(x) < 0$ over $(-\frac{1}{\sqrt{3}}, \frac{1}{\sqrt{3}})$ and $f'(x) > 0$ over $(\frac{1}{\sqrt{3}}, \infty)$, $f(\frac{1}{\sqrt{3}}) \approx -0.385$ is a local minimum.

4. $f'(x) = 0$. The critical values of f are all real numbers. Each critical value represents both a local minimum and a local maximum.

7. $f'(x) = \sec(x)\tan(x) \Rightarrow x = n\pi, n$ an integer, are critical values of f. Checking the signs of f' on each side of each critical point, we find that $f(n\pi) = 1$ if n is even are local minima and $f(n\pi) = -1$ if n is odd are local maxima.

10. $f'(x) = 0$. The critical values of f are all real numbers. Each critical value represents both a local minimum and a local maximum.

13. $f'(x) = -2x - 3 = 0 \Rightarrow x = -1.5$ is the critical value of f. However, it is not in the interval. Evaluating $f(1) = 2$, and $f(2) = -4$, we see that $f(1) = 2$ is the absolute maximum and $f(2) = -4$ is the absolute minimum over $[1,2]$.

16. $f'(x) = \frac{1}{3}\cos(\frac{x}{3}) \neq 0$ over $(\frac{\pi}{4}, \frac{\pi}{3})$. So there are no critical values for f over $[\frac{\pi}{4}, \frac{\pi}{3}]$. Evaluating $f(\frac{\pi}{4}) \approx 0.259$ and $f(\frac{\pi}{3}) \approx 0.342$ we find that $f(\frac{\pi}{4}) \approx 0.259$ is the absolute minimum of f and $f(\frac{\pi}{3}) \approx 0.342$ is the absolute maximum of f over $[\frac{\pi}{4}, \frac{\pi}{3}]$.

19. $s'(t) = \frac{(90-t^2/30)}{|90-t^2/30|}(-\frac{t}{15}) = 0 \Rightarrow t = \sqrt{2700}$ is the critical value in $[0, 60]$. Evaluating $s(0) = 90$, $s(\sqrt{2700}) = 0$, $s(60) = 30$, we find that $s(0) = 90$ is the maximum height and $s(\sqrt{2700}) = 0$ is the minimum height over $[0, 60]$.

22. $x = -4$ represents a local maximum since $f'(x) > 0$ over $(-6, -4)$ and $f'(x) < 0$ over $(-4, 6)$.

25. $x \approx -\frac{2}{3}$ represents a local maximum since $f'(x) > 0$ over $(-6, -\frac{2}{3})$ and $f'(x) < 0$ over $(-\frac{2}{3}, \frac{11}{4})$.

$x \approx \frac{11}{4}$ represents a local minimum since $f'(x) < 0$ over $(-\frac{2}{3}, \frac{11}{4})$ and $f'(x) > 0$ over $(\frac{11}{4}, 6)$.

27. $f'(x) \geq 0$ for x over $(-6, 6)$ so there are no local maxima or minima.

Section 4.5

1. $S = 2\pi rh + 2\pi r^2$ where r and h are the radius and the height of the cylinder, respectively.
$(2r)^2 + h^2 = 10^2 \Rightarrow r^2 = \frac{100-h^2}{4} \Rightarrow r = \frac{1}{2}\sqrt{100-h^2}$.
Therefore, $S = 2\pi(\frac{1}{2}\sqrt{100-h^2})h + 2\pi(\frac{1}{2}\sqrt{100-h^2})^2$
$= \pi h\sqrt{100-h^2} + (\frac{\pi}{2})(100-h^2)$ for $h \in (0, 10)$.

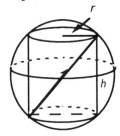

4. $S = 2x^2 + 4xh$ where x is the length and the width of the base and h is the height of the box. See picture from #3. $x = \sqrt{\frac{144-h^2}{2}}$ from #3.
Therefore, $S = 2(\sqrt{\frac{144-h^2}{2}})^2 + 4\sqrt{\frac{144-h^2}{2}}(h)$
$= 144 - h^2 + h\sqrt{1152-8h^2}$ for $h \in (0, 12)$.

7. Let $\alpha = \frac{\theta}{2}$.

$$\sin(\alpha) = \frac{4}{y-4}, \qquad y = \frac{4}{\sin(\alpha)} + 4, \qquad \tan(\alpha) = \frac{x}{y}.$$

$$A = xy = \tan(\alpha)y^2 = \tan(\alpha)\left(\frac{16}{\sin^2(\alpha)} + \frac{32}{\sin(\alpha)} + 16\right).$$

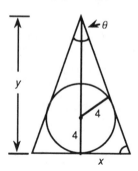

10. $V = \frac{\pi}{3}r^2 h$ where r and h are the radius and height of the cone. The circumference of the cone will be the circumference of the circle minus the arc length of the cut out sector or $C = 12\pi - 6\theta \Rightarrow 2\pi r = 12\pi - 6\theta \Rightarrow r = 6 - \frac{3\theta}{\pi}$.

$(6 - \frac{3\theta}{\pi})^2 + h^2 = 6^2 \Rightarrow h = \sqrt{36 - (6 - \frac{3\theta}{\pi})^2}$.

Therefore, $V = \frac{\pi}{3}(6 - \frac{3\theta}{\pi})^2 \sqrt{36 - (6 - \frac{3\theta}{\pi})^2}$

$$= \frac{\pi}{3}(6 - \frac{3\theta}{\pi})^2 \sqrt{36 - 36 + \frac{36\theta}{\pi} - \frac{9\theta^2}{\pi^2}}$$

$$= \frac{\pi}{3}(6 - \frac{3\theta}{\pi})^2 \sqrt{\frac{9}{\pi^2}(4\pi\theta - \theta^2)}$$

$$= (6 - \frac{3\theta}{\pi})^2 \sqrt{4\pi\theta - \theta^2} \text{ for } \theta \in (0, 2\pi).$$

13. See picture in problem 3.

Maximize $V(h) = 72h - \frac{h^3}{2}$ for $h \in (0, 12)$.

$V'(h) = 72 - \frac{3}{2}h^2$

$V'(h) = 0$ when $h^2 = 48$ or $h = \pm 4\sqrt{3} \Rightarrow$ the critical value is $h = 4\sqrt{3}$.

$V'(h) > 0$ for $0 < h < 4\sqrt{3}$ and $V'(h) < 0$ for $4\sqrt{3} < h < 12 \Rightarrow$

$V(4\sqrt{3}) \approx 332.55$ ft^3 is the maximum volume.

16. See picture in problem 6.

Maximize $A(\theta) = 64\cos^3(\frac{\theta}{2})\sin(\frac{\theta}{2})$ for $\theta \in (0, \pi)$.
$A'(\theta) = 64(3\cos^2(\frac{\theta}{2})(-\sin(\frac{\theta}{2}))(\frac{1}{2}))\sin(\frac{\theta}{2}) + 64\cos^3(\frac{\theta}{2})\cos(\frac{\theta}{2})(\frac{1}{2})$
$= 32\cos^2(\frac{\theta}{2})(-3\sin^2(\frac{\theta}{2}) + \cos^2(\frac{\theta}{2}))$.
$A'(\theta) = 0$ when $\cos^2(\frac{\theta}{2}) = 0$ or $-3\sin^2(\frac{\theta}{2}) + \cos^2(\frac{\theta}{2}) = 0 \Rightarrow$
$\theta = \pi$ or $\tan(\frac{\theta}{2}) = \frac{1}{\sqrt{3}} \Rightarrow \theta = \frac{\pi}{3}$ is the critical value.
$A'(\theta) > 0$ for $0 < \theta < \frac{\pi}{3}$ and $A'(\theta) < 0$ for $\frac{\pi}{3} < \theta < \pi \Rightarrow$
$A(\frac{\pi}{3}) = 12\sqrt{3}$ cm^2 is the maximum area.

19. See picture in problem 9.

$\ell = 24$ is the maximum of the diagonal..

Section 4.6

In some cases, roots have been found by machine.

1. a) and b)

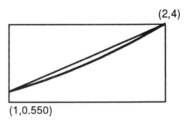

c) $f'(x) = 2x$ and $\frac{f(2)-f(1)}{2-1} = 4 - 1 = 3$, so $2x = 3 \Rightarrow x = \frac{3}{2}$.

4. a) and b)

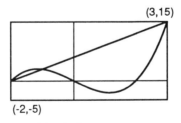

c) $f'(x) = 3x^2 - 4$ and $\frac{f(3)-f(-2)}{3-(-2)} = \frac{15-0}{5} = 5$, so $3x^2 - 4 = 5 \Rightarrow 3x^2 = 9 \Rightarrow x = \pm\sqrt{3}$.

7. a) and b)

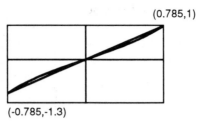

c) $f'(x) = \sec^2(x)$ and $\frac{f(\pi/4) - f(-\pi/4)}{\pi/4 - (-\pi/4)} = \frac{1-(-1)}{\pi/2} = \frac{4}{\pi}$, so $\sec^2(x) = \frac{4}{\pi} \Rightarrow \cos^2(x) = \frac{\pi}{4} \Rightarrow \cos(x) \approx \pm 0.87 \Rightarrow x \approx 0.48, x \approx -0.48$. ($x \approx \pm 2.65$ is outside the interval.)

10. a) and b)

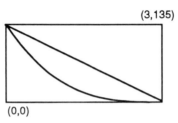

c) $f'(x) = -15(3-x)^2$ and $\frac{f(3)-f(0)}{3} = \frac{0-135}{3} = -45$, so $-15(3-x)^2 = -45 \Rightarrow (3-x)^2 = 3 \Rightarrow 3 - x = \pm\sqrt{3} \Rightarrow x = 3 - \sqrt{3}$ or $x = 3 + \sqrt{3}$. Only $x = 3 - \sqrt{3}$ is in the interval.

13. a) and b)

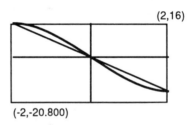

c) $f'(x) = 3x^2 - 12$ and $\frac{f(2)-f(-2)}{2-(-2)} = \frac{-16-16}{4} = -8$, so $3x^2 - 12 = -8 \Rightarrow 3x^2 = 4 \Rightarrow x = \pm\frac{2}{\sqrt{3}}$.

16. a) and b)

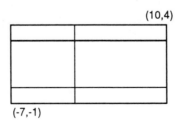

c) $f'(x) = 0$ and $\frac{f(10)-f(-7)}{10-(-7)} = 0$, so $0 = 0$ which is true for all x in the interval.

19. a) and b)

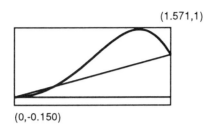

(1.571,1)

(0,-0.150)

c) $f'(x) = 2x\cos(x^2)$ and $\frac{f(\pi/2)-f(0)}{\pi/2-0} \approx \frac{0.624}{\pi/2} \approx 0.397$, so $2x\cos(x^2) = 0.397 \Rightarrow x \approx 0.20$ and $x \approx 1.18$.

22. $m_{\text{sec}} = -\frac{12}{11}$. At $x \approx -3$ the tangent line has this same slope.

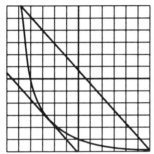

25. $m_{\text{sec}} = \frac{2}{11}$. At no points in the interval $(-6, 6)$ does the tangent line have this same slope. Hypothesis 2 of the Mean Value Theorem is violated.

28. None.

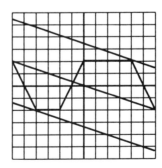

31.

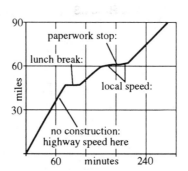

34. It is possible to have an average speed of 45 and yet violate a 45 mph speed limit. For example, one could drive 40 mph for a while and 50 mph for the same length of time producing an average speed of 45 mph.

37. Those positions where the graph appears to be horizontal are likely to correspond to red traffic lights.

40. Sue's average speed $= \frac{3-\frac{1}{2}}{3\frac{1}{2}} = \frac{5}{7}$ mi per min or ≈ 43 mph. Jamal's average speed $= \frac{2\frac{1}{2}}{5-1\frac{1}{3}} = \frac{15}{22}$ mi per min or ≈ 41 mph.

ANSWERS TO SELECTED EXERCISES FROM CHAPTER 5

Section 5.1

1.

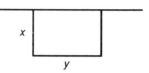

Maximize $A = xy$ where $2x + y = 100$ or $y = 100 - 2x$.

$A(x) = x(100 - 2x) = 100x - 2x^2$ for $x \in (0, 50)$.

$A'(x) = 100 - 4x$.

$A'(x) = 0$ when $x = 25$ which is the critical value from the domain.

$A'(x) > 0$ for $0 < x < 25$ and $A'(x) < 0$ for $25 < x < 50 \Rightarrow A(25)$ is the maximum area.

The dimensions should be 25 m × 50 m.

4.

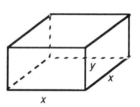

Minimize $C(x) = 0.06(x^2) + 0.03(4xy)$ where $x^2 y = 270$ or $y = \frac{270}{x^2}$.

$C(x) = 0.06x^2 + 0.12x(\frac{270}{x^2}) = 0.06x^2 + \frac{32.4}{x}$ for $x \in (0, \infty)$.

$C'(x) = 0.12x - \frac{32.4}{x^2}$.

$C'(x) = 0$ when $x^3 = \frac{32.4}{0.12}$ or $x = 3\sqrt[3]{10} \approx 6.46$ so $x \approx 6.46$ is the critical value from the domain.

$C'(x) < 0$ for $0 < x < 6.46$ and $C'(x) > 0$ for $x > 6.46 \Rightarrow x = 3\sqrt[3]{10}$ represents a minimum.

The dimensions should be approximately 6.46 in × 6.46 in × 6.46 in.

7. Minimize $A = (x + 2S)(y + 2T) = xy + 2Tx + 2Sy + 4ST$
where $xy = 24$ or $y = \frac{24}{x}$.
$A(x) = 24 + 2Tx + 2S(\frac{24}{x}) + 4ST$ for $x \in (0, \infty)$.
$A'(x) = 2T - \frac{48S}{x^2}$.
$A'(x) = 0$ when $x = \pm\sqrt{\frac{24S}{T}}$ so $x = \sqrt{\frac{24S}{T}}$ is the critical value from the domain.
$A'(x) < 0$ for $0 < x < \sqrt{\frac{24S}{T}}$ and $A'(x) > 0$ for $x > \sqrt{\frac{24S}{T}} \Rightarrow x = \sqrt{\frac{24S}{T}}$ represents a minimum.

The overall dimensions of the page should be
$(\sqrt{\frac{24S}{T}} + 2S)$ in × $(\frac{24\sqrt{T}}{\sqrt{24S}} + 2T)$ in or $(\sqrt{\frac{24S}{T}} + 2S)$ in × $(\sqrt{\frac{24T}{S}} + 2T)$ in.

10.

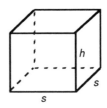

Maximize $V = s^2 h$ where
$3.00 = 0.10(s^2) + 0.06(4sh) = 0.10s^2 + 0.24sh$ or $h = \frac{300-10s^2}{24s} = \frac{150-5s^2}{12s}$.
$V(s) = s^2(\frac{150-5s^2}{12s}) = \frac{150}{12}s - \frac{5}{12}s^3$ for $s \in (0, \sqrt{30})$.
$V'(s) = \frac{150}{12} - \frac{15}{12}s^2$.
$V'(s) = 0$ when $s = \pm\sqrt{10}$ so $s = \sqrt{10}$ is the critical value from the domain.
$V'(s) > 0$ for $0 < s < \sqrt{10}$ and $V'(s) < 0$ for $\sqrt{10} < s < \sqrt{30} \Rightarrow$
$s = \sqrt{10}$ in represents a maximum.

13.

Maximize $V(x) = (20 - 2x)(30 - 2x)x = 600x - 100x^2 + 4x^3$ for $x \in (0, 10)$.
$V'(x) = 600 - 200x + 12x^2 = 4(3x^2 - 50x + 150)$.
$V'(x) = 0$ when $x \approx 3.924$ or $x \approx 12.74$ so $x \approx 3.924$ is the critical value from the domain.
$V'(x) > 0$ for $0 < x < 3.924$ and $V'(x) < 0$ for $3.924 < x < 10 \Rightarrow x \approx 3.924$ represents a maximum.

The squares should be approximately 3.924 cm × 3.924 cm.

16.

Maximize $V = \pi r^2 h$ where $7.35 = 0.05(\pi r^2) + 0.03(2\pi rh)$ or $h = \frac{7.35 - 0.05\pi r^2}{0.06\pi r}$.

$V(r) = \pi r^2 \left(\frac{7.35 - 0.05\pi r^2}{0.06\pi r}\right) = 122.5r - \frac{5}{6}\pi r^3$ for $r \in (0, \sqrt{\frac{147}{\pi}})$.

$V'(r) = 122.5 - \frac{5}{2}\pi r^2$.

$V'(r) = 0$ when $r = \frac{7}{\sqrt{\pi}}$ which is the critical value from the domain.

$V'(r) > 0$ for $0 < r < \frac{7}{\sqrt{\pi}}$ and $V'(r) < 0$ for $\frac{7}{\sqrt{\pi}} < r < \sqrt{\frac{147}{\pi}} \Rightarrow r = \frac{7}{\sqrt{\pi}}$ represents a maximum.

The dimensions should be $r = \frac{7}{\sqrt{\pi}}$ and $h = \frac{35}{3\sqrt{\pi}}$ in
or $r \approx 3.95$ in, $h \approx 6.58$ in.

19.

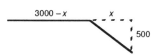

Minimize $C(x) = 20(3000 - x) + 80\sqrt{500^2 + x^2}$ for $x \in [0, 3000]$.

$C'(x) = -20 + \frac{80x}{\sqrt{500^2 + x^2}}$.

$C'(x) = 0$ when $400 = \frac{6400x^2}{500^2 + x^2}$ or $x \approx 129$ which is the critical value from the domain.

$C(0) = 100,000.$ $C(129) \approx 98,730.$ $C(3,000) \approx 243,311.$

Lay the line ≈ 2871 m under ground and then ≈ 516 m under water.

22.

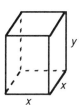

Maximize $V = x^2 y$ where $8x + 4y = 6$ or $y = \frac{3 - 4x}{2}$.

$V(x) = x^2 \left(\frac{3 - 4x}{2}\right) = \frac{1}{2}(3x^2 - 4x^3)$ for $x \in (0, \frac{3}{4})$.

$V'(x) = 3x - 6x^2 = 3x(1 - 2x)$.

$V'(x) = 0$ when $x = 0$ and $x = \frac{1}{2}$ so $x = \frac{1}{2}$ is the critical value from the domain.

$V'(x) > 0$ for $0 < x < \frac{1}{2}$ and $\frac{1}{2} < x < \frac{3}{4} \Rightarrow x = \frac{1}{2}$ represents a maximum.

The wire should be cut every $\frac{1}{2}$ m.

22. continued

Maximize $S = 2x^2 + 4xy$ where $8x + 4y = 6$ or $y = \frac{3-4x}{2}$

$S(x) = 2x^2 + 6x - 8x^2$ for $x \in (0, \frac{3}{4})$.

$S'(x) = 6 - 12x$.

$S'(x) = 0$ when $x = \frac{1}{2}$ which is the critical value from the domain.

$S'(x) > 0$ for $0 < x < \frac{1}{2}$ and $S'(x) < 0$ for $\frac{1}{2} < x < \frac{3}{4} \Rightarrow x = \frac{1}{2}$ represents a maximum.

The wire should be cut every $\frac{1}{2}$ meter.

25. Minimize $I(x) = \frac{200}{x} + \frac{300}{50-x}$ for $x \in (0, 50)$.

$I'(x) = -\frac{200}{x^2} + \frac{300}{(50-x)^2}$.

$I'(x) = 0$ when

$300x^2 = 200(50-x)^2$

$100x^2 + 20000x - 500000 = 0$

$x^2 + 200x - 5000 = 0$

$x = \frac{-200 \pm \sqrt{40000+20000}}{2}$ or $x = -100 \pm 50\sqrt{6}$ so

$x = -100 + 50\sqrt{6}$ is the critical value from the domain.

$I'(x) < 0$ for $0 < x < -100 + 50\sqrt{6}$ and $I'(x) > 0$ for $-100 + 50\sqrt{6} < x < 50 \Rightarrow x = -100 + 50\sqrt{6}$ represents a minimum.

The total illumination is the least approximately 22.47 feet from the light rated 200.

28.

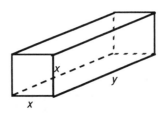

Maximize $V = x^2y$ where $y + 4x = 108$ or $y = 108 - 4x$.

$V = x^2(108 - 4x) = 108x^2 - 4x^3$ for $x \in (0, 27)$.

$V'(x) = 216x - 12x^2 = 12x(18 - x)$.

$V'(x) = 0$ when $x = 0, 18$ so $x = 18$ is the critical value from the domain.

$V'(x) > 0$ for $0 < x < 18$ and $V'(x) < 0$ for $18 < x < 27 \Rightarrow x = 18$ represents a maximum.

The dimensions should be 18 in $\times$ 18 in $\times$ 36 in.

30. Minimize $C(g) = \frac{1200}{g}(100) + \frac{g}{2} + 200$ for $g \in \{200, 240, 300, 400, 600, 1200\}$.

(Note: $\frac{1200}{g}$ must be an integer.)

Evaluate C at each of the values in the domain.

Order either 400 or 600 gallons each time.

33. Maximize $B(x) = (8+x)(200-10x)$ where x is the number of additional wells and $x \in [0, 20]$.
$B'(x) = (200 - 10x) + (8+x)(-10) = 120 - 20x$.
$B'(x) = 0$ when $x = 6$ which is the critical value from the domain.
$B(0) = 1600. \quad B(6) = 1960. \quad B(20) = 0$.

Platypus should drill 6 additional wells.

36. Maximize $V_1(x) = (9-2x)(12-x)x$ where $x \in \left(0, \frac{9}{2}\right)$ and
$V_2(x) = (12-2x)(9-x)x$ where $x \in (0, 6)$. Then determine which expression yields the largest volume for the scoop.
$V_1'(x) = (-2)(12x - x^2) + (9-2x)(12-2x) = -24x + 2x^2 + 108 - 42x + 4x^2$
$\quad = 6x^2 - 66x + 108 = 6(x^2 - 11x + 18) = 6(x-9)(x-2)$.
$V_1'(x) = 0$ when $x = 9$ or $x = 2$ so $x = 2$ is the critical value from the domain.

$V_1'(x) > 0$ for $0 < x < 2$ and $V_1'(x) < 0$ for $2 < x < \frac{9}{2} \Rightarrow x = 2$ represents a maximum.
$V_2'(x) = (-2)(9x - x^2) + (12-2x)(9-2x)$
$\quad = -18x + 2x^2 + 108 - 42x + 4x^2$
$\quad = 6x^2 - 60x + 108 = 6(x^2 - 10x + 18)$.
$V_2'(x) = 0$ when $x = \frac{10 \pm \sqrt{28}}{2}$ so $x = \frac{10 - \sqrt{28}}{2} \approx 2.35$ is the critical value from the domain.

$V_2'(x) > 0$ for $0 < x < 2.35$ and $V_2'(x) < 0$ for $2.35 < x < 6 \Rightarrow x \approx 2.35$ represents a maximum.

$V_1(2) = 5(10)(2) = 100$ in^3.

$V_2(2.35) \approx 114.08$ in^3.

The best strategy is to cut the two squares from one of the 12 inch sides. The length and width of each square should be ≈ 2.35 inches.

39. $D = (0, \infty); 5 \times 20 \times 2, 2 \times 8 \times 12.5, 1 \times 4 \times 50$.

42. The total cost $C(x) = 2.88x^2 + 1.35x + \frac{145}{x} + \frac{4}{x^2}$.

46. See the triangle shown in Figure D. in problem 45 and assume the length of the material is unknown. Side c of the triangle shown is $c = 2r$ for a given radius r. Side $a = L - 2r$ where L is the length of the material. Side $b = \sqrt{(2r)^2 - (L-2r)^2}$. The width of the rectangular material is $b + 2r = \sqrt{(2r)^2 - (L-2r)^2} + 2r$.

Minimize the area of the material, i.e.

Minimize $A(L) = L(\sqrt{(2r)^2 - (L-2r)^2}) + 2r$ for $L \in (0, 4r)$.
$A'(L) = ((2r)^2 - (L-2r)^2)^{1/2} + L(\frac{1}{2})((2r)^2 - (L-2r)^2)^{-1/2}(-2)(L-2r)$
$= \frac{(4Lr - L^2) - L(L - 2r)}{((2r)^2 - (L-2r)^2)^{1/2}}$.

$A'(L) = 0$ when $6rL - 2L^2 = 0 \Rightarrow L = 3r$ is the critical value from the domain.

$A'(L) < 0$ for $0 < L < 3r$ and $A'(L) > 0$ for $3r < L < 4r \Rightarrow L = 3r$ represents a minimum.

$A(3r) = 3(\sqrt{3} + 2)r^2 \approx 11.2r^2$ is the minimum area when cutting as in plan B.

The area of the material used with plan A would be $(4r)(2r) = 8r$ so the most economical plan is plan A.

47. Maximize $A = \frac{1}{2}(4.3 + (4.3 + 2x))h$ where x is the length of the base of the right triangle shown in the picture and h is the height of the trapezoid. Writing x and h in terms of θ, we have $\sin(\theta) = \frac{x}{3}$ or $x = 3\sin(\theta)$ and $\cos(\theta) = \frac{h}{3}$ or $h = 3\cos(\theta)$.
$A(\theta) = \frac{1}{2}(8.6 + 6\sin(\theta))3\cos(\theta)$ for $\theta \in (0, \frac{\pi}{2})$.
$A'(\theta) = \frac{3}{2}(6\cos(\theta))\cos(\theta) + \frac{3}{2}(8.6 + 6\sin(\theta))(-\sin(\theta))$
$= 9\cos^2(\theta) - 12.9\sin(\theta) - 9\sin^2(\theta)$
$= 9(1 - \sin^2(\theta)) - 12.9\sin(\theta) - 9\sin^2(\theta)$
$= 9 - 12.9\sin(\theta) - 18\sin^2(\theta)$.

$A'(\theta) = 0$ when $\sin(\theta) \approx -1.15$ or $\sin(\theta) \approx 0.43$ so $\theta \approx 0.45$ is the critical value from the domain.

Graphically, we see $\theta \approx 0.45$ represents a maximum.

The maximum area is approximately 15.1 in^2.

50. Observe the following picture.

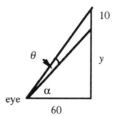

$\tan(\theta + \alpha) = \frac{10+y}{60}$ and $\tan(\alpha) = \frac{y}{60}$.

$\theta + \alpha = \arctan(\frac{10+y}{60}) \Rightarrow \theta = \arctan(\frac{10+y}{60}) - \arctan(\frac{y}{60})$ for $y \in [0, \infty)$.

$\theta'(y) = \frac{1}{1+(\frac{10+y}{60})^2}(\frac{1}{60}) - \frac{1}{1+(\frac{y}{60})^2}(\frac{1}{60})$

$= \frac{1}{60}(\frac{3600}{3600+(10+y)^2}) - \frac{1}{60}(\frac{3600}{3600+y^2})$.

$\theta(y)$ is a decreasing function so its maximum value occurs when $y = 0$.

The sign should be at eye level for the best view.

Section 5.2

Use the following graphs for problems 1 -10.

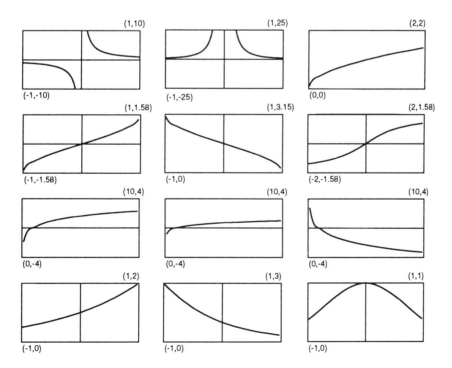

1. $y = \frac{1}{x^2}$, $y = 2^x$, $y = 3^{-x}$, $y = e^{-x^2}$.

4. $y = \frac{1}{x}$, $y = \arccos(x)$, $y = \log_{1/2}(x)$, $y = 3^{-x}$.

7. $y = \arcsin(x)$, $y = \arccos(x)$, $y = \arctan(x)$.

8. $y = \sqrt{x}$, $y = e^{-x^2}$.

Use the following graphs for problems 11 - 20.

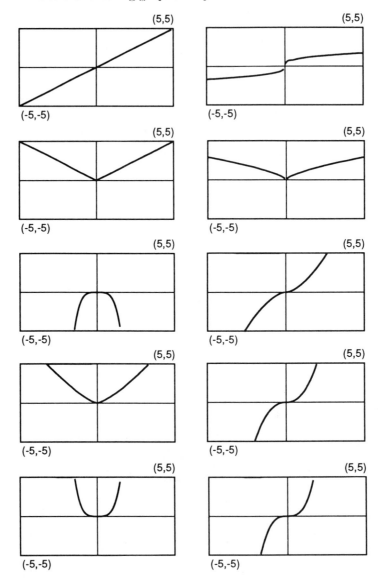

11. All but $y = x$.

14. $y = x^{1/3}$ since $y' = \frac{1}{3x^{2/3}}$. $y = x^{2/3}$ since $y' = \frac{2}{3x^{1/3}}$.

17. $y = x$ since $y'' = 0$ for all x.

$y = -x^4$ since $y'' = -12x^2 < 0$ for all $x \neq 0$.

$y = x^4$ since $y'' = 12x^2 > 0$ for all $x \neq 0$.

$y = |x|$ since $y'' = 0$ for all $x \neq 0$.

20. $y = x^{4/3}$ since $y'' = \frac{4}{9x^{2/3}} > 0$ for all $x \neq 0$.

$y = x^{2/3}$ since $y'' = -\frac{2}{9x^{4/3}} < 0$ for all $x \neq 0$.

23. Inflection points at $x = -4, \frac{1}{2}$.

Concave up on $(-6, -4)$ and $(\frac{1}{2}, 6)$. Concave down on $(-4, \frac{1}{2})$.

26. No inflection points. Concave up on $(0, 6)$.

29. For $y = \sin(x)$, $f'(x) = \cos(x)$, $f''(x) = -\sin(x)$.

$f''(x) = 0$ for $x = 0, \pm\pi, \pm 2\pi \ldots$.

Points of inflection occur at $x = n\pi$, n an integer.

For $y = \cos(x)$, $f'(x) = -\sin(x)$, $f''(x) = -\cos(x)$.

$f''(x) = 0$ for $x = \pm\frac{\pi}{2}, \pm\frac{3\pi}{2}, \ldots$.

Points of inflection occur at $x = \frac{n\pi}{2}$, n an odd integer.

For $y = \tan(x)$, $f'(x) = \sec^2(x)$, $f''(x) = 2\sec(x)\sec(x)\tan(x) = \frac{2\sin(x)}{\cos^3(x)}$.

$f''(x) = 0$ for $x = 0, \pm\pi, \pm 2\pi, \ldots$.

Points of inflection occur at $x = n\pi$, n an integer. (The graph of $y = \tan(x)$ also changes concavity over the asymptotes $x = \frac{n\pi}{2}$, n an integer.)

For $y = \sec(x)$, $f'(x) = \sec(x)\tan(x)$,

$f''(x) = \sec(x)\tan(x)\tan(x) + \sec(x)\sec^2(x)$

$= \sec(x)\tan^2(x) + \sec^3(x) = \frac{\sin^2(x)}{\cos^3(x)} + \frac{1}{\cos^3(x)}$.

$f''(x) = 0$ for no values of x. By observing the graph, we see changes in concavity only over the asymptotes.

For $y = \csc(x)$, $f'(x) = -\csc(x)\cot(x)$,

$f''(x) = \csc(x)\cot(x)\cot(x) - \csc(x)(-\csc^2(x)) = \frac{\cos^2(x)}{\sin^3(x)} + \frac{1}{\sin^3(x)}$.

$f''(x) = 0$ for no values of x. By observing the graph, we see changes in concavity only over the asymptotes.

For $y = \cot(x)$, $f'(x) = -\csc^2(x)$,

$f''(x) = -2\csc(x)(-\csc(x)\cot(x)) = \frac{2\cos(x)}{\sin^3(x)}$.

$f''(x) = 0$ for $x = \frac{n\pi}{2}$, n an odd integer.

Points of inflection occur at $x = \frac{n\pi}{2}$, n an odd integer. (The graph of $y = \cot(x)$ also changes concavity over the asymptotes $x = n\pi$, n an integer.)

32. $f'(x) = \cos(x)$, $f''(x) = -\sin(x)$, $f'''(x) = -\cos(x)$,
$f^{(4)}(x) = \sin(x)$, $f^{(1000)}(x) = \sin(x)$.
$g'(x) = -\sin(x)$, $g''(x) = -\cos(x)$, $g'''(x) = \sin(x)$,
$g^{(4)}(x) = \cos(x)$, $g^{(1000)}(x) = \cos(x)$.
$k'(x) = e^x$, $k''(x) = e^x$, $k^{(1000)}(x) = e^x$.
$q'(x) = \ln(2)2^x$, $q''(x) = (\ln 2)^2(2^x)$, $q^{(1000)}(x) = (\ln 2)^{(1000)}(2^x)$.

35. $v(t) = -32t$, $\quad a(t) = -32$.

a) $a(0) = -32$ ft/sec^2.

b) $a(t)$ is never 0.

c) The velocity is non-positive and decreasing over $(0, 60)$ so its magnitude is at its greatest when $t = 60$. We have $s(60) = 6400$.

d) $|a(t)| = |-32| = 32$ ft/sec^2 for all t in $[0, 60]$.

e)

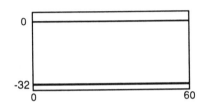

38. $v(t) = \frac{\pi}{10} \cos \frac{\pi t}{20}$, $\quad a(t) = -\frac{\pi^2}{200} \sin \frac{\pi t}{20}$.

a) $a(0) = 0$ ft/sec^2.

b) $a(t) = 0$ when $t = 0, 20, 40, 60$ s.

c) $|v(t)|$ is at its greatest when $t = 0, 20, 40, 60$ sec. At each of these times, $s = 0$.

d) $|a(t)|$ is at its greatest when $t = 10, 30, 50$ sec.

e)

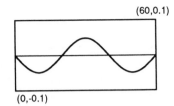

41. x_0 may represent an inflection point or it may represent a local minimum or local maximum (as in $f(x) = x^4$).

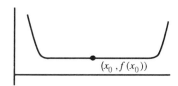

44. a) $f'(x) = 3x^2+6x-9 = 0 \Rightarrow 3(x^2+2x-3) = 0 \Rightarrow 3(x+3)(x-1) = 0 \Rightarrow x = -3$, $x = 1$ are critical values of f.

b) $f''(x) = 6x + 6$. $f''(-3) = -12 \Rightarrow x = -3$ is the location of a local maximum. $f''(1) = 12 \Rightarrow x = 1$ is the location of a local minimum.

c) $f''(x) = 6x + 6 = 0 \Rightarrow x = -1$. A point of inflection occurs at $(-1, 9)$ since $f''(x) < 0$ for $x < -1$ and $f'' > 0$ for $x > -1$.

d) The graph of f is concave up over $(-1, \infty)$ since $f''(x) > 0$ over $(-1, \infty)$.

e) The graph of f is concave down over $(-\infty, -1)$ since $f''(x) < 0$ over $(-\infty, -1)$.

47. a) $f'(x) = 6x^2 + 4x - 2 = 0 \Rightarrow 2(3x^2 + 2x - 1) = 0 \Rightarrow 2(3x - 1)(x + 1) = 0 \Rightarrow x = -1$ and $x = \frac{1}{3}$ are the critical values of f.

b) $f''(x) = 12x + 4$. $f''(-1) = -8 \Rightarrow$ a local maximum occurs at $x = -1$. $f''(\frac{1}{3}) = 8 \Rightarrow$ a local minimum occurs at $x = \frac{1}{3}$.

c) $f''(x) = 12x + 4 = 0 \Rightarrow x = -\frac{1}{3}$. $\left(-\frac{1}{3}, -\frac{5}{27}\right)$ is an inflection point since $f''(x) < 0$ for $x < -\frac{1}{3}$ and $f''(x) > 0$ for $x > -\frac{1}{3}$.

d) The graph of f is concave up over $\left(-\frac{1}{3}, \infty\right)$ since $f''(x) > 0$ over $\left(-\frac{1}{3}, \infty\right)$.

e) The graph of f is concave down over $\left(-\infty, -\frac{1}{3}\right)$ since $f''(x) < 0$ over $\left(-\infty, -\frac{1}{3}\right)$.

48. a) $f'(x) = -12x^3 - 24x^2 - 12x = 0 \Rightarrow -12x(x^2 + 2x + 1) = 0 \Rightarrow -12x(x+1)(x+1) = 0 \Rightarrow x = 0, x = -1$ are the critical values of f.

b) $f''(x) = -36x^2 - 48x - 12$. $f''(0) = -12 \Rightarrow x = 0$ is the location of a local maximum. $f''(-1) = 0$ so the second derivative test fails. $f'(x) < 0$ for $x < -1$ and $f'(x) > 0$ for $x \in (-1, 0) \Rightarrow x = -1$ is the location of a local minimum.

c) $f''(x) = -36x^2 - 48x - 12 = 0 \Rightarrow -12(3x^2 + 4x + 1) = 0 \Rightarrow -12(3x+1)(x+1) = 0 \Rightarrow x = -\frac{1}{3}, x = -1$. A point of inflection occurs at $(-1, 2)$ since $f''(x) > 0$ for $x < -1$ and $f''(x) < 0$ for $x \in (-1, -\frac{1}{3})$.

A point of inflection occurs at $(-\frac{1}{3}, \frac{70}{27})$ since $f''(x) < 0$ for $x \in (-1, -\frac{1}{3})$ and $f''(x) > 0$ for $x > -\frac{1}{3}$.

d) The graph of f is concave up over $(-\infty, -1)$ and $(-\frac{1}{3}, \infty)$ since $f''(x) > 0$ over these intervals.

e) The graph of f is concave down over $(-1, -\frac{1}{3})$ since $f''(x) < 0$ over this interval.

Section 5.3

1. $8x + 18y\frac{dy}{dx} = 0 \Rightarrow \frac{dy}{dx} = -\frac{4x}{9y}$.

$\left.\frac{dy}{dx}\right|_{(-\frac{3}{2}, \sqrt{3})} = \frac{6}{9\sqrt{3}} = \frac{2}{3\sqrt{3}}$.

Equation of tangent line: $y = \frac{2}{3\sqrt{3}}(x + \frac{3}{2}) + \sqrt{3}$.

Equation of normal line: $y = \frac{-3\sqrt{3}}{2}(x + \frac{3}{2}) + \sqrt{3}$. .

4. $\frac{2}{3}x^{-1/3} + \frac{2}{3}y^{-1/3}\frac{dy}{dx} = 0 \Rightarrow \frac{dy}{dx} = -\frac{y^{1/3}}{x^{1/3}}$.

$\left.\frac{dy}{dx}\right|_{(-27, -8)} = -\frac{2}{3}$.

Equation of tangent line: $y = -\frac{2}{3}(x + 27) - 8$.

Equation of normal line: $y = \frac{3}{2}(x + 27) - 8$.

7. $e^x + e^y\frac{dy}{dx} = -y - x\frac{dy}{dx} \Rightarrow (e^y + x)\frac{dy}{dx} = -y - e^x \Rightarrow \frac{dy}{dx} = \frac{-y - e^x}{e^y + x}$.

$\left.\frac{dy}{dx}\right|_{(0,0)} = \frac{-0 - e^0}{e^0 + 0} = -1$.

Equation of tangent line: $y = -x$.

Equation of normal line: $y = x$.

10. $-\dfrac{1}{x^2} - \dfrac{3}{y^2}\dfrac{dy}{dx} = 0 \Rightarrow \dfrac{dy}{dx} = -\dfrac{y^2}{3x^2}$.

$\left.\dfrac{dy}{dx}\right|_{(2,6)} = -\dfrac{36}{3(4)} = -3$.

Equation of tangent line: $y = -3(x-2) + 6$.

Equation of normal line: $y = \tfrac{1}{3}(x-2) + 6$.

13. $2xy^2 + x^2(2y)\dfrac{dy}{dx} = 2x + 2y\dfrac{dy}{dx} \Rightarrow (2x^2y - 2y)\dfrac{dy}{dx} = 2x - 2xy^2 \Rightarrow$

$\dfrac{dy}{dx} = \dfrac{2x - 2xy^2}{2x^2y - 2y} = \dfrac{x(1-y^2)}{y(x^2-1)}$. $\left.\dfrac{dy}{dx}\right|_{(0,0)}$ is undefined.

16. By implicit differentiation, we obtain

$2x - 2y - 2x\dfrac{dy}{dx} + 2y\dfrac{dy}{dx} = 0$.

Substituting $x = 2$ and $y = 0$, we have

$4 - 4\dfrac{dy}{dx} = 0 \Rightarrow \left.\dfrac{dy}{dx}\right|_{(2,0)} = 1$.

By implicit differentiation again, we obtain

$2 - 2\dfrac{dy}{dx} - 2\dfrac{dy}{dx} - 2x\dfrac{d^2y}{dx^2} + 2\left(\dfrac{dy}{dx}\right)^2 + 2y\dfrac{d^2y}{dx^2} = 0$.

Substituting $x = 2$ and $y = 0$ and $\dfrac{dy}{dx} = 1$, we have

$2 - 2 - 2 - 4\dfrac{d^2y}{dx^2} + 2 = 0 \Rightarrow \left.\dfrac{d^2y}{dx^2}\right|_{(2,0)} = 0$.

19. Let $\sec(y) = x$, then $\sec(y)\tan(y)\dfrac{dy}{dx} = 1 \Rightarrow \dfrac{dy}{dx} = \dfrac{1}{\sec(y)\tan(y)}$. The range of the $\text{arcsec}(x)$ is $0 < y < \dfrac{\pi}{2}$ and $\dfrac{\pi}{2} < y < \pi$.

Since $\sec^2(y) - 1 = \tan^2(y)$, we have $\tan(y) = \pm\sqrt{\sec^2(y) - 1}$. Note that $\tan(y) > 0$ for $x > 1$, so we choose the plus sign and $\tan(y) < 0$ if $x < -1$ so we choose the minus sign. Therefore, $\dfrac{dy}{dx} = \dfrac{1}{\pm x\sqrt{x^2-1}}$ or $\dfrac{dy}{dx} = \dfrac{1}{|x|\sqrt{x^2-1}}$.

22. $\dfrac{dy}{dx} = x\left(\dfrac{3}{5}\right)(x+2)^{-2/5} + (x+2)^{3/5} = \dfrac{3x + 5(x+2)}{5(x+2)^{2/5}} = \dfrac{8x + 10}{5(x+2)^{2/5}} \Rightarrow$

$\dfrac{dx}{dy} = \dfrac{5(x+2)^{2/5}}{8x + 10}$.

$\dfrac{dx}{dy} = 0$ when $x = -2 \Rightarrow$ the tangent line is vertical at the point $(-2, 0)$.

25. $\dfrac{dy}{dx} = \dfrac{1}{3}(x-5)^{-2/3} \Rightarrow \dfrac{dx}{dy} = 3(x-5)^{2/3}$.

$\dfrac{dx}{dy} = 0$ when $x = 5 \Rightarrow$ the tangent line is vertical at the point $(5, 0)$.

28. $\dfrac{dy}{dx} = \begin{cases} 1 - 3x^2 & x < -1 \text{ and } 0 < x < -1 \\ 3x^2 - 1 & -1 < x < 0 \text{ and } x > 1. \end{cases}$

So $\dfrac{dx}{dy} = \begin{cases} \frac{1}{1-3x^2} & x < -1 \text{ and } 0 < x < -1 \\ \frac{1}{3x^2-1} & -1 < x < 0 \text{ and } x > 1. \end{cases}$

There are no points where the graph has a vertical tangent.

31. $\dfrac{2x}{a^2} - \dfrac{2y}{b^2}\dfrac{dy}{dx} = 0 \Rightarrow \dfrac{dy}{dx} = \dfrac{xb^2}{ya^2}$. The tangent line is vertical when $\dfrac{dx}{dy} = \dfrac{ya^2}{xb^2} = 0$ or when $y = 0 (a \neq 0, b \neq 0)$, i.e., at the points $(\pm a, 0)$.

34. $f'(x) = \dfrac{1}{\tan(\sqrt{x})}\sec^2(\sqrt{x})\left(\dfrac{1}{2\sqrt{x}}\right) = \dfrac{\sec^2(\sqrt{x})}{2\sqrt{x}\tan(\sqrt{x})}$.

37. $f'(x) = \dfrac{6x - 3e^{-3x}}{3x^2 + e^{-3x}}$.

40. $f'(x) = \dfrac{1}{\sqrt{1-(2x)^2}}(2)(\sin(2x)) + \arcsin(2x)\cos(2x)(2)$

$= \dfrac{2\sin(2x)}{\sqrt{1-4x^2}} + 2\arcsin(2x)\cos(2x)$.

43. $f'(x) = \dfrac{1}{\left(\frac{2-x}{2+x}\right)^{1/2}\ln(2)} \cdot \left(\dfrac{1}{2}\right)\left(\dfrac{2-x}{2+x}\right)^{-1/2}\left(\dfrac{-1(2+x)-(2-x)}{(2+x)^2}\right)$

$= \dfrac{-4(2+x)}{2\ln(2)(2+x)^2(2-x)} = \dfrac{-2}{\ln 2(2+x)(2-x)}$.

46. $f'(x) = \dfrac{1}{1+(\sqrt{x})^2}\left(\dfrac{1}{2\sqrt{x}}\right) = \dfrac{1}{2(1+x)\sqrt{x}}$.

Section 5.4

1. The circumference $C = 2\pi r$ where r is the radius. $\dfrac{dC}{dt} = 2\pi\dfrac{dr}{dt}$. If $\dfrac{dr}{dt} = 3$ cm/sec, then $\dfrac{dC}{dt} = 2\pi(3) = 6\pi$ cm/sec.

4. $\dfrac{dE}{dt} = -\dfrac{1}{2\sqrt{V}}\dfrac{dV}{dt}$.

$\dfrac{dE}{dt} \approx -\dfrac{1}{2\sqrt{10000}}(6734) = -33.67$ at the moment of explosion.

7. $\dfrac{df}{dt} = 15x^2\dfrac{dx}{dt} - \dfrac{4}{\sqrt{1-y^2}}\dfrac{dy}{dt}$.

10. $\dfrac{df}{dt} = 2\pi\sin(\theta)\cos(\theta)\dfrac{d\theta}{dt} - 2\pi\cos(\theta)\sin(\theta)\dfrac{d\theta}{dt} = 0$.

13. $v = \frac{1}{3}\pi r^2 h$ for a cone. Since $h = 2r$, $v = \frac{1}{3}\pi(2r^3)$ and $\frac{dv}{dt} = 2\pi r^2 \frac{dr}{dt}$.

$10 = 2\pi\left(\frac{5}{2}\right)^2 \frac{dr}{dt} \Rightarrow \frac{20}{25\pi} = \frac{dr}{dt} \Rightarrow \frac{dr}{dt} = \frac{4}{5\pi}$ cm/sec when $h = 5$ cm.

15.

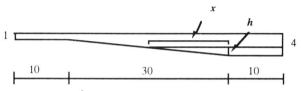

For $0 < h \leq 3$, $V = \frac{1}{2}hx(25) + 10(25)h$ where x and h are as shown in the picture above. We know $\frac{h}{x} = \frac{3}{30} \Rightarrow x = 10h$.

Recall, 1 gallon ≈ 0.13368 ft^3 and 1 meter ≈ 3.2808 ft. Convert gallons per minute to m^3/min.

$250 \frac{\text{gal}}{\text{min}} \times \frac{0.13368 \text{ ft}^3}{1 \text{ gal}} \times \frac{1 \text{ m}^3}{(3.2808)^3 \text{ ft}^3} \approx 0.946$ m^3/min.

$V = \frac{1}{2}h(10h)(25) + 10(25)h = 125h^2 + 250h$.

$\frac{dV}{dt} = 250h\frac{dh}{dt} + 250\frac{dh}{dt}$.

$-0.946 \approx 250(1)\frac{dh}{dt} + 250\frac{dh}{dt} \Rightarrow \frac{dh}{dt} \approx -0.00189$ m/min when $h = 1$ m.

$-0.946 = 250(2)\frac{dh}{dt} + 250\frac{dh}{dt} \Rightarrow \frac{dh}{dt} \approx -0.00126$ m/min when $h = 2$ m.

$\frac{dh}{dt}$ is not well-defined at $h = 3$.

18. We have $y^2 = 90^2 + x^2 - 2(90)(x)\cos(45°)$ using the law of cosines where y is the distance between first base and the ball and x is the distance between the ball and home plate.

We also have $2y\frac{dy}{dt} = 2x\frac{dx}{dt} - 180(\cos(45°))\frac{dx}{dt}$ where $\frac{dx}{dt} = -100(\frac{5280 \text{ ft}}{3600 \text{ sec}}) = -146.\overline{6}$ ft/sec.

We want $\frac{dy}{dt}$ when $x = 0$ and $y = 90$. If $\frac{dy}{dt} = \frac{x}{y}\frac{dx}{dt} - \frac{90(\cos(45°))}{y}\frac{dx}{dt}$, then $\frac{dy}{dt} = \frac{146.\overline{6}(90)\cos(45°)}{90} \approx 103.71$ ft/sec.

21. $V = 6y\left(5 + \frac{11}{4}y\right)$ and $\frac{dV}{dt} = 30\frac{dy}{dt} + 33y\frac{dy}{dt}$.

$15 \approx 30\frac{dy}{dt} + 33(1.41)\frac{dy}{dt} \Rightarrow \frac{dy}{dt} \approx 0.20$ ft/min. (See solution to exercise 20 for calculations.)

24. $S = 4\pi r^2$ and $V = \dfrac{4}{3}\pi r^3$. After 7 minutes $V = 70 = \dfrac{4}{3}\pi r^3$ or $r = \sqrt[3]{\dfrac{210}{4\pi}}$.

Find $\dfrac{dr}{dt}$ if $\dfrac{dV}{dt} = 4\pi r^2 \dfrac{dr}{dt}$: $10 = 4\pi \left(\sqrt[3]{\dfrac{210}{4\pi}}\right)^2 \dfrac{dr}{dt} \Rightarrow \dfrac{dr}{dt} = \dfrac{10}{4\pi}\left(\sqrt[3]{\dfrac{4\pi}{210}}\right)^2$

Now find $\dfrac{dS}{dt}$ after 7 minutes if $\dfrac{dS}{dt} = 8\pi r \dfrac{dr}{dt}$:

$\dfrac{dS}{dt} = 8\pi \left(\dfrac{210}{4\pi}\right)^{1/3} \left(\dfrac{10}{4\pi}\right) \left(\dfrac{4\pi}{210}\right)^{2/3} = \dfrac{8\pi(10)}{(4\pi)^{2/3}(210)^{1/3}} \approx 7.82$ feet/min.

27. $V = \pi r^2(4)$ and $\dfrac{dV}{dt} = 8\pi r \dfrac{dr}{dt}$.

$5000(1000) = 8\pi(30000)\dfrac{dr}{dt} \Rightarrow \dfrac{dr}{dt} = \dfrac{125}{6\pi}$ cm/sec.

30. $V = \dfrac{1}{3}\pi r^2 h$ and $\dfrac{r}{h} = \dfrac{4}{11}$ or $r = \dfrac{4}{11}h$. So $V = \dfrac{1}{3}\pi\left(\dfrac{4}{11}h\right)^2 h$ and

$\dfrac{dV}{dt} = \dfrac{16\pi}{121}h^2\dfrac{dh}{dt}$.

After 10 min, $V = 150 = \pi\left(\dfrac{1}{3}\right)\left(\dfrac{4}{11}\right)^2 h^3$ so $h = \sqrt[3]{\dfrac{150}{\pi}\left(\dfrac{363}{16}\right)}$ ft.

$15 = \dfrac{16\pi}{121}\left(\dfrac{150(363)}{16\pi}\right)^{2/3}\dfrac{dh}{dt}$ after 10 min $\Rightarrow \dfrac{dh}{dt} = \dfrac{15(121)}{16\pi}\left(\dfrac{16\pi}{150(363)}\right)^{2/3}$

≈ 0.34 ft per min.

Section 5.5

1. b) $\dfrac{dy}{dx} = \dfrac{3\cos(t)}{-2\sin(t)}$.

  c) $\dfrac{dy}{dx}\bigg|_{t_0=\pi/3} = \dfrac{3\cos(\frac{\pi}{3})}{-2\sin(\frac{\pi}{3})} = \dfrac{3(\frac{1}{2})}{-2(\frac{\sqrt{3}}{2})} = \dfrac{3}{-2\sqrt{3}} = -\dfrac{\sqrt{3}}{2}$,

  $x(\frac{\pi}{3}) = 2\cos(\frac{\pi}{3}) = 1$, $y(\frac{\pi}{3}) = 3\sin(\frac{\pi}{3}) = \dfrac{3\sqrt{3}}{2}$.

  Equation of tangent line: $y = -\dfrac{\sqrt{3}}{2}(x-1) + \dfrac{3\sqrt{3}}{2}$.

  Equation of normal line: $y = \dfrac{2}{\sqrt{3}}(x-1) + \dfrac{3\sqrt{3}}{2}$.

4. b) $\dfrac{dy}{dx} = \dfrac{3}{2}$.

  c) $\dfrac{dy}{dx}\bigg|_{t_0=5} = \dfrac{3}{2}$, $x(5) = 13$, $y(5) = 17$.

  Equation of tangent line: $y = \dfrac{3}{2}(x-13) + 17$.

  Equation of normal line: $y = -\dfrac{2}{3}(x-13) + 17$.

6. b) $\frac{dy}{dx} = \frac{1}{\frac{1}{1+t^2}} = 1+t^2$.

c) $\frac{dy}{dx}\bigg|_{t_0=1} = 2$, $x(1) = \frac{\pi}{4}$, $y(1) = 1$.

Equation of tangent line: $y = 2(x - \frac{\pi}{4}) + 1$.

Equation of normal line: $y = -\frac{1}{2}(x - \frac{\pi}{4}) + 1$.

9. b) $\frac{dy}{dx} = \frac{\cosh(t)}{\sinh(t)}$.

c) $\frac{dy}{dx}\bigg|_{t_0=2} = \frac{\frac{e^2+e^{-2}}{2}}{\frac{e^2-e^{-2}}{2}} = \frac{e^2+e^{-2}}{e^2-e^{-2}} \approx 1.04$, $x(2) = \frac{e^2+e^{-2}}{2} \approx 3.76$, $y(2) \approx 3.63$.

Equation of the tangent line: $\approx y = 1.04(x - 3.76) + 3.63$.

Equation of the normal line: $\approx y = -0.96(x - 3.76) + 3.63$.

12. Solve $x_1(t) = x_2(s)$ and $y_1(t) = y_2(s)$, ie. solve $2\cos(t) = 1 + \sin(s)$ and $3\sin(t) = \cos(s) - 3$.

$\cos(t) = \frac{(1+\sin(s))}{2}$ and $\sin(t) = \frac{\cos(s)-3}{3}$.

Now, $\cos^2(t) + \sin^2(t) = 1 \Rightarrow (\frac{1+\sin(s)}{2})^2 + (\frac{\cos(s)-3}{3})^2 = 1 \Rightarrow$

$s \approx 0.44$ or $s \approx 4.712$ (using a root-finder). Evaluating x_1, y_1, x_2, and y_2 at these s values, we find that at $s \approx 4.712$ we have a collision point and at $s \approx 0.44$ we do not have a collision point. This non-collision point is $(x_2(0.44), y_2(0.44)) \approx (1.43, -2.10)$.

15. $x = \sin(y) \Rightarrow x = \sin(t)$, $y = t$ for $-\frac{\pi}{2} \le t \le \frac{\pi}{2}$.

18. $x = \cot(y) \Rightarrow x = \frac{\cos(t)}{\sin(t)}$, $y = t$ for $0 < t < \pi$.

21. $x = \sin(t)$, $y = \cos(t)$ for $0 \le t < 2\pi$.

24. $x = t$, $y = 2\sin(t)\cos(t)$ for $-\pi \le t \le \pi$.

27. First, $\sin(72°) \approx 0.951$ and $\cos(72°) \approx 0.309$.

$x(t) \approx 1100(0.309)t \approx 339.9t$, $y(t) \approx 1100(0.951)t - 16t^2 = 1046.1t - 16t^2$.

a) $\frac{dy}{dt} = 1046.1 - 32t = 0 \Rightarrow t \approx 32.69$ seconds, the time at which the maximum height occurs. The maximum height is approximately $y(32.69) \approx 17099$ feet.

b) $y(t) = 1046.1t - 16t^2 = 0 \Rightarrow t(1046.1 - 16t) = 0 \Rightarrow t = 0$ sec (launch time) and $t \approx 65.38$ sec (return time).

c) $x(65.38) = 339.9(65.38) \approx 22,223$ feet.

27. continued

d) $\dfrac{dx}{dt}\bigg|_{t\approx 65.38} \approx 339.9$ ft/sec,

$\dfrac{dy}{dt}\bigg|_{t\approx 65.38} \approx 1046.1 - 32(65.38) \approx -1046.1$ ft/sec.

29. $x(t) = 1000(\cos(\theta))t$ will be at a maximum when $\theta = 0$.

$y(t) = 1100t\sin(\theta) - 16t^2$ will be at a maximum when $\theta = \frac{\pi}{2}$.

32. See 31 for added information. The ball hits the ground when $y(t) \approx 37.97t - 16t^2 + 5 = 0$, ie, when $t \approx 2.5$ sec (use a root-finder).

$\dfrac{dy}{dt} \approx 37.97 - 32t$, $\quad \dfrac{dy}{dt}\bigg|_{t\approx 2.5} \approx -42.03$ ft/sec. So the downward velocity is ≈ 42.03 ft/sec.

35. $y(t) = \sin(0°)(146.7)t - 16t^2 + 5 = 0 \Rightarrow -16t^2 + 5 = 0 \Rightarrow t^2 = \frac{5}{16}$

$\Rightarrow t \approx 0.56$ sec. $\quad x(0.31) \approx \cos(0)(146.7)(0.56) \approx 82$ feet.

36. 100 mph $= 100(\frac{5280 \text{ ft}}{3600 \text{ sec}}) = 146.\overline{6}$ ft per sec, so $x(t) = 146.\overline{6}\cos(45°)t$, $y(t) = 146.\overline{6}\sin(45°)t - 16t^2 + 5$. When is the horizontal position 500 feet? $x(t) = 146.\overline{6}\cos(45°)t = 500 \Rightarrow t \approx 4.82$ sec.

We then have $y(4.82) = 146.\overline{6}\sin(45°)(4.82) - 16(4.82)^2 + 5 \approx 133.16$ ft.

The ball will leave the stadium.

ANSWERS TO SELECTED EXERCISES FROM CHAPTER 6

Section 6.1

1. $\frac{20+0}{2}(0.25) = 2.5$ miles.

4. $\frac{60+45}{2}(0.25) = 13.125$ miles.

7. $\frac{55+30}{2}(0.25) = 10.625$ miles.

9.

Minute of trip	Approx. dist.	Minute of trip	Approx. dist.
16	$28(1) \div 60 \approx 0.4667$	24	0
17	$24(1) \div 60 = 0.4$	25	.0167
18	0.3333	26	.0167
19	0.2667	27	0.0333
20	0.2	28	0.05
21	0.15	29	0.0667
22	0.1	30	0.1
23	0.05	Total	≈ 2.25 miles

10. Between 18 and 19 minutes.

13. ≈ 0.85 hours.

16. $t \approx 0.85$ hrs, $t \approx 1.4$ hrs, $t \approx 1.65$ hrs, and $t \approx 1.85$ hrs.

19. The car traveling 40 mph passes the other one 7 times between $t = 0.25$ hr and $t = 0.75$ hr, 3 times around $t = 1.5$ hr, and once near the end of the 2 hour period.

22. The figure is a triangle so $A = \frac{1}{2}(1.5)(3) = 2.25$.

25. The figure consists of 2 triangles so
$A = \frac{1}{2}\left(\frac{13}{3}\right)(13) + \frac{1}{2}\left(\frac{5}{3}\right)(5) = \frac{169}{6} + \frac{25}{6} = \frac{194}{6}$.

28. The figure is $\frac{1}{4}$ of a circle with radius 3 so $A = \frac{1}{4}\pi(3)^2 = \frac{9\pi}{4}$.

31. $A = \frac{1}{2}(4+5)(1) + \frac{1}{2}(2.5)(5) + \frac{1}{2}(1.5)(3) = 4.5 + 6.25 + 2.25 = 13$.

34. $A = \frac{1}{2}(3)(3) + \frac{1}{2}(6)(3) + \frac{1}{2}(1)(1) = 14$. (3 triangles.)

37. $\sum_{n=3}^{7}(2n-5) = (6-5) + (8-5) + (10-5) + (12-5) + (14-5)$
$= 1 + 3 + 5 + 7 + 9 = 25$.

40. $\sum_{n=0}^{5}(2n^2 - 3n + 1)$
$= (1) + (2 - 3 + 1) + (2(2)^2 - 3(2) + 1) + (2(3)^2 - 3(3) + 1)$
$= +(2(4)^2 - 3(4) + 1) + (2(5)^2 - 3(5) + 1) = 1 + 0 + 3 + 10 + 21 + 36 = 71.$

43. $\sum_{n=0}^{100}(-1)^n = (-1)^0 + (-1)^1 + (-1)^2 + (-1)^3 + \cdots + (-1)^{100}$
$= 1 - 1 + 1 - 1 + \cdots + 1 = 1.$

46. $\sum_{n=1}^{1000} n = 500500.$

49. $L_{10} = \sum_{n=0}^{9}\left(1 + \frac{n}{10}\right)^2\left(\frac{1}{10}\right) = 2.185;$

$U_{10} = \sum_{n=1}^{10}\left(1 + \frac{n}{10}\right)^2\left(\frac{1}{10}\right) = 2.485.$

Section 6.2

1. $\Delta x = \frac{10 - (-2)}{6} = \frac{12}{6} = 2.$
Partition points: $-2, 0, 2, 4, 6, 8, 10.$
a) $-2, 0, 2, 4, 6, 8.$
b) $0, 2, 4, 6, 8, 10.$
c) $-1, 1, 3, 5, 7, 9.$

4. $\Delta x = \frac{3 - (-1)}{10} = 0.4.$
Partition points: $-1, -0.6, -0.2, 0.2, 0.6, 1.0, 1.4, 1.8, 2.2, 2.6, 3.0.$
a) $-1, -0.6, -0.2, 0.2, 0.6, 1.0, 1.4, 1.8, 2.2, 2.6.$
b) $-0.6, -0.2, 0.2, 0.6, 1.0, 1.4, 1.8, 2.2, 2.6, 3.0.$
c) $-0.8, -0.4, 0, 0.4, 0.8, 1.2, 1.6, 2.0, 2.4, 2.8.$

7. $L_5 = \left(\arctan(-1) + \arctan\left(-\frac{3}{5}\right) + \arctan\left(-\frac{1}{5}\right) + \arctan\left(\frac{1}{5}\right) + \arctan\left(\frac{3}{5}\right)\right)\left(\frac{2}{5}\right)$
$\approx -0.3142.$
$U_5 = \left(\arctan\left(-\frac{3}{5}\right) + \arctan\left(-\frac{1}{5}\right) + \arctan\left(\frac{1}{5}\right) + \arctan\left(\frac{3}{5}\right) + \arctan(1)\right)\left(\frac{2}{5}\right)$
$\approx 0.3142.$
Average $= \frac{L_5 + U_5}{2} = 0.$
Largest error $\approx \frac{U_5 - L_5}{2} = 0.3142.$

10. $\int_a^c f(x)dx = \int_a^b f(x)dx + \int_b^c f(x)dx = 2.5 + (-5.0) = -2.5.$

13. $\int_a^c 2f(x)dx = 2\int_a^b f(x)dx + 2\int_b^c f(x)dx = 2(2.5) + 2(-5.0) = -5.$

16. $\int_c^c f(x)dx = 0.$

19. $R_1 = 0,\ R_2 = 0,\ R_4 = 0,\ R_8 = 0.$

22. $\int_1^5 f(x)dx \approx -23,\ \int_{-3}^3 f(x)dx \approx -28.5,\ \int_6^{-2} f(x)dx \approx 43.$

25. $\int_1^5 f(x)dx \approx 6\tfrac{1}{2},\ \int_{-3}^3 f(x)dx \approx -1\tfrac{1}{4},$ function not defined over $(5,6]$.

28. $\int_1^5 f(x)dx \approx 7,\ \int_{-3}^3 f(x)dx = 4,\ \int_6^{-2} f(x)dx = -8.$

31. $\int_{-3}^5 f(x)dx = \tfrac{1}{2}(2+5)(3) + \tfrac{1}{2}(2.5)(5) - \tfrac{1}{2}(2.5)(5) = 10.5.$

34. $\int_{-3}^4 2g(x) + f(x)dx = 2\int_{-3}^4 g(x) + \int_{-3}^4 f(x)dx$
$= 2[\tfrac{1}{2}(1)(1) - \tfrac{1}{2}(6)(3)] + (-\tfrac{1}{2})(\tfrac{1}{2})(1) + \tfrac{1}{2}(1\tfrac{1}{2})(3) + 5(3) = -17 + 17 = 0.$

37. $R = 0.$

40. The definite integral does not exist. The upper Riemann sum will always equal 1 and the lower Riemann sum will always equal 0 for any size partition of $[0,1]$.

Section 6.3

1. See # 9 in section 6.1, ≈ 2.25 miles.

2.

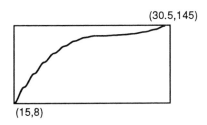

5.

(5,1)

(1,0)

6.

(-1,2)

(-5,-2)

9.

(1.57,1)

(-1.57,0)

10.

(3.142,2)

(0,-2)

13.

(1,1)

(-1,0)

14.

(1,2)

(-1,-2)

17.

(2,1)

(1,-0.1)

18.

(1,1)

(-2,-1)

21.

22.

25.
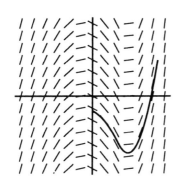

26.
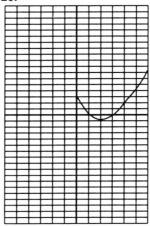

Section 6.4

1. $\frac{x^3}{3} + \frac{x^2}{2} + x + C.$

4. $\sqrt{17}x + C.$

7. $6\sqrt{x} + C.$

10. $\frac{5}{6}x^{6/5} + C.$

13. $x + C.$

16. $\frac{1}{4}\ln|x| + C.$

19. $\frac{x^2}{2} - \sec(x) + C.$

22.
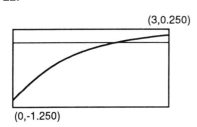
(3, 0.250)
(0, -1.250)

23.
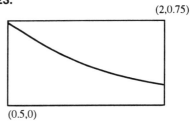
(2, 0.75)
(0.5, 0)

24. The values in the table below were calculated by machine and rounded to 4 decimal places using the formula given in problem 24 in the text. (They were not calculated from the table from problem 21 as was requested.)

b	G(b)	b	G(b)	b	G(b)
0	0.9967	1.0	0.4758	2.0	0.1923
0.1	0.9973	1.1	0.4308	2.1	0.1779
0.2	0.9406	1.2	0.3904	2.2	0.1650
0.3	0.8905	1.3	0.3545	2.3	0.1534
0.4	0.8314	1.4	0.3225	2.4	0.1428
0.5	0.7677	1.5	0.2940	2.5	0.1333
0.6	0.7031	1.6	0.2688	2.6	0.1247
0.7	0.6401	1.7	0.2463	2.7	0.1168
0.8	0.5807	1.8	0.2262	2.8	0.1096
0.9	0.5258	1.9	0.2083	2.9	0.1031

The graph looks like the graph $y = \frac{1}{1+x^2}$ over $[0, 3]$.

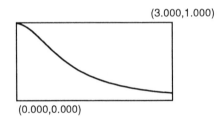

28. $A(3) \approx 0.1419 = \arctan(3) - \arctan(2)$.

32. $b = 1$.

35. $f(x) = \frac{1}{x}$.

36. The values in the table below were calculated by machine and rounded to 4 decimal places using the formula given in problem 36 in the text. (They were not calculated from the table from problem 31 as was requested.)

b	G(b)	b	G(b)	b	G(b)	b	G(b)
		1.0	0.9531	2.0	0.4879	3.0	0.3279
		1.1	0.8701	2.1	0.4652	3.1	0.3175
0.2	4.0547	1.2	0.8004	2.2	0.4445	3.2	0.3077
0.3	2.8768	1.3	0.7411	2.3	0.4256	3.3	0.2985
0.4	2.2314	1.4	0.6899	2.4	0.4082	3.4	0.2899
0.5	1.8232	1.5	0.6454	2.5	0.3922	3.5	0.2817
0.6	1.5415	1.6	0.6062	2.6	0.3774	3.6	0.2740
0.7	1.3353	1.7	0.5716	2.7	0.3637	3.7	0.2667
0.8	1.1778	1.8	0.5407	2.8	0.3509	3.8	0.2598
0.9	1.0536	1.9	0.5129	2.9	0.3390	3.9	0.2532

36. continued

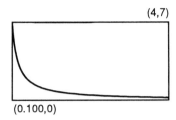

Section 6.5

1. $F(x) = \int_1^x 2^{-t^2} \, dt$.

4. $F(x) = \int_{-1}^x \frac{1}{t^3+8} \, dt$.

7. $\frac{dy}{dx} = \frac{\sin(2x)}{1+x^2}$.

10. $\frac{dy}{dx} = \frac{G'(x)x^2 - G(x)(2x)}{x^4} = \frac{1}{x^4}\left(\frac{x^2 \sin(2x)}{1+x^2} - 2x \int_2^x \frac{\sin(2t)}{1+t^2} \, dt\right)$

$= \frac{\sin(2x)}{x^2(1+x^2)} - \frac{2}{x^3} \int_2^x \frac{\sin(2t)}{1+t^2} \, dt$.

13. $\int_{-1}^{3}(x^2 + x + 1)dx = \frac{x^3}{3} + \frac{x^2}{2} + x \Big]_{-1}^{3} = \left[\frac{27}{3} + \frac{9}{2} + 3\right] - \left[-\frac{1}{3} + \frac{1}{2} - 1\right] = 17\frac{1}{3}$.

16. $\int_{\pi/2}^{\pi}(5\sin(x) - 3\cos(x))dx = -5\cos(x) - 3\sin(x)\Big]_{\pi/2}^{\pi}$

$= (-5\cos(\pi) - 3\sin(\pi)) - \left(-5\cos\left(\frac{\pi}{2}\right) - 3\sin\left(\frac{\pi}{2}\right)\right) = 5 + 3 = 8$.

19. $\int_0^1 \frac{1}{4+4x^2} dx = \frac{1}{4}\arctan(x)\Big]_0^1 = \frac{1}{4}(\arctan(1) - \arctan(0)) = \frac{1}{4}\left(\frac{\pi}{4}\right) = \frac{\pi}{16}$.

21.-28. The graphs of these area functions look similar to the antiderivatives sketched in problems 21 - 28 in Section 6.3. Remember they may differ by a constant.

29. The answer is unreasonable since $y = \frac{1}{x^2} > 0$ for all x so the value of the definite integral if it exists should be positive. $f(x) = \frac{1}{x^2}$ is not continuous over $[-1, 1]$, so the Fundamental Theorem of Calculus does not apply.

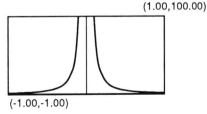

Section 6.6

1.

n	left	right	midpoint
2	1.05902	1.26612	1.14039
4	1.09970	1.20326	1.14595
8	1.12283	1.17460	1.14733
16	1.13508	1.16097	1.14768
32	1.14138	1.15432	1.14776

4.

n	left	right	midpoint
2	1.98289	4.98289	3.76169
4	2.87229	4.37229	3.69633
8	3.28431	4.03431	3.67819
16	3.48125	3.85625	3.67350
32	3.57737	3.76487	3.67231

7.

n	TRAP	SIMP
2	1.16257	1.14785
4	1.15148	1.14778
8	1.14871	1.14779
16	1.14802	1.14779
32	1.14785	1.14779

$$\int_0^1 \sqrt{1+x^2}\, dx \approx 1.1478;\ \text{SIMP gave the best estimate.}$$

10.

n	TRAP	SIMP
2	3.48289	3.66876
4	3.62229	3.67165
8	3.65931	3.67190
16	3.66875	3.67191
32	3.67112	3.67191

$$\int_1^4 \log_2(x)\, dx \approx 3.6719;\ \text{SIMP gave the best estimate.}$$

13. For $\Delta t = \frac{1}{60}$, the trapezoidal rule gives the distance
$\frac{1}{120}[f(15) + 2f(16) + \ldots 2f(29) + f(30)] = 2.08\overline{3}$ miles.

16. $S_2 \approx 3.14156862745$, $\quad S_4 \approx 3.14159250246$, $\quad S_6 \approx 3.1415926403$,

$S_8 \approx 3.14159265123$, $\quad S_{10} \approx 3.14159265297$.

Sequence converges to π. S_{10} is accurate to 8 decimal places.

$T_{50} \approx 3.1452598694$.

19. $x^2(x-1)(x-2)(x-3)(x-4) = x^6 - 10x^5 + 35x^4 - 50x^3 + 24x^2$ so

$$\int_0^4 x^6 - 10x^5 + 35x^4 - 50x^3 + 24x^2 \, dx$$

$$= \frac{1}{7}x^7 - \frac{10}{6}x^6 + \frac{35}{5}x^5 - \frac{50}{4}x^4 + \frac{24}{3}x^3 \Big]_0^4$$

$$= \frac{1}{7}x^7 - \frac{5}{3}x^6 + 7x^5 - \frac{25}{2}x^4 + 8x^3 \Big]_0^4$$

$$= \frac{1}{7}(4)^7 - \frac{5}{3}(4)^6 + 7(4)^5 - \frac{25}{2}(4)^4 + 8(4)^3$$

$$\approx -6.095238097.$$

22. $\int_1^5 f(x)dx \approx \frac{1}{2}(-5\frac{1}{3} + 2(-5\frac{1}{2}) + 2(-5\frac{2}{3}) + 2(-5\frac{3}{4}) + (-5\frac{7}{8})) \approx -22.5$.

25. $\int_1^5 f(x)dx \approx \frac{1}{2}(1)(-\frac{5}{4} + 2(-2) + 2(\frac{2}{3}) + 2(\frac{10}{3}) + 6) = 4\frac{3}{8}$.

28. $\int_1^5 f(x)dx \approx \frac{1}{2}(2 + 2(2) + 2(2) + 2(2) + 0) = 5$.

31. In the first picture, construct the tangent line. Two congruent triangles are formed. The area of the lower triangle plus the shaded area (in picture 1.) under the tangent line make up the area of the rectangle when using the midpoint estimate. In this same picture, the area of the upper triangle plus the shaded area under the tangent line make up the area of the trapezoid formed. Since the triangles are all congruent, both estimates are the same. A similar argument holds if we construct the tangent line in picture 3.

34. a) Left rectangle rule, midpoint rule, trapezoidal rule   b) Right rectangle rule   c) none   d) Simpson's rule.

37. a) Right rectangle rule, trapezoidal rule   b) Left rectangle rule
c) none   d) Midpoint rule, Simpson's rule.

40. a) Left rectangle rule   b) Right rectangle rule   c) Midpoint rule, trapezoidal rule, Simpson's rule   d) none.

43. a) Right rectangle rule   b) Left rectangle rule   c) Midpoint rule, trapezoidal rule, Simpson's rule   d) none.

Section 6.7

1. Let $u = x^2$, $du = 2x\,dx$.
$$\int x\cos(x^2)\,dx = \tfrac{1}{2}\int \cos(x^2)(2x\,dx)$$
$$= \int \tfrac{1}{2}\cos(u)\,du = \tfrac{1}{2}\sin(u) + C = \tfrac{1}{2}\sin(x^2) + C.$$

4. Let $u = 0.05x - \pi$, $du = 0.05\,dx$.
$$\int (0.05x - \pi)^{132}\,dx = \tfrac{1}{0.05}\int (0.05x - \pi)^{132}(0.05\,dx) = \tfrac{1}{0.05}\int u^{132}\,du$$
$$= \tfrac{20}{133}u^{133} + C = \tfrac{20}{133}(0.05x - \pi)^{133} + C.$$

7. Let $u = -x^2$, $du = -2x\,dx$.
$$\int xe^{-x^2}\,dx = -\tfrac{1}{2}\int e^{-x^2}(2x\,dx) = -\tfrac{1}{2}\int e^u\,du = -\tfrac{1}{2}e^u + C = -\tfrac{1}{2}e^{-x^2} + C.$$

10. Let $u = \tan(x)$, $du = \sec^2(x)\,dx$.
$$\int \tan^3(x)\sec^2(x)\,dx = \int u^3\,du = \tfrac{1}{4}u^4 + C = \tfrac{1}{4}\tan^4(x) + C.$$

13. Let $u = x^3 + 1$, $du = 3x^2\,dx$.
$$\int 3x^2(x^3 + 1)^4\,dx = \int u^4\,du = \tfrac{1}{5}u^5 + C = \tfrac{(x^3+1)^5}{5} + C.$$

16. Let $u = x^2$, $du = 2x\,dx$.
$$\int x\sin(x^2)\,dx = \tfrac{1}{2}\int 2x\sin(x^2)\,dx = \tfrac{1}{2}\int \sin(u)\,du$$
$$= -\tfrac{1}{2}\cos(u) + C = -\tfrac{1}{2}\cos(x^2) + C.$$

19. Let $u = 4x^3$, $du = 12x^2\,dx$.
$$\int x^2 e^{4x^3}\,dx = \tfrac{1}{12}\int e^{4x^3}(12x^2\,dx) = \tfrac{1}{12}\int e^u\,du = \tfrac{1}{12}e^u + C = \tfrac{1}{12}e^{4x^3} + C.$$

22. Let $u = \sqrt{x}$, $du = \tfrac{1}{2\sqrt{x}}\,dx$.
$$\int \tfrac{\cos(\sqrt{x})}{\sqrt{x}}\,dx = 2\int \tfrac{\cos(\sqrt{x})}{2\sqrt{x}}\,dx = 2\int \cos(u)\,du = 2\sin(u) + C = 2\sin(\sqrt{x}) + C.$$

25. Let $u = e^x$, $du = e^x\,dx$.
$$\int \tfrac{e^x}{1+e^{2x}}\,dx = \int \tfrac{1}{1+u^2}\,du = \arctan(u) + C = \arctan(e^x) + C.$$

28. Let $u = \tan(x)$, $du = \sec^2(x)\,dx$.
$$\int \tfrac{e^{\tan(x)}}{\cos^2(x)}\,dx = \int e^{\tan(x)}\sec^2(x)\,dx = \int e^u\,du = e^u + C = e^{\tan(x)} + C.$$

31. Let $u = \cos^2(x)$, $du = -2\cos(x)\sin(x)\,dx$.

$$\int \sin(x)\cos(x)e^{\cos^2(x)}\,dx = -\frac{1}{2}\int e^{\cos^2(x)}(-2\cos(x)\sin(x))\,dx$$
$$= -\frac{1}{2}\int e^u\,du = -\frac{1}{2}e^u + C = -\frac{1}{2}e^{\cos^2(x)} + C.$$

34. Let $u = \ln(x)$, $du = \frac{1}{x}\,dx$.

$$\int \frac{\cos(\ln(x))}{x}\,dx = \int \cos(u)\,du = \sin(u) + C = \sin(\ln(x)) + C.$$

37. Let $u = \ln(x)$, $du = \frac{1}{x}\,dx$.

$$\int \frac{\sec(\ln(x))\tan(\ln(x))}{x}\,dx = \int \sec(u)\tan(u)\,du = \sec(u) + C = \sec(\ln(x)) + C.$$

40. Let $u = 3x^3 - x$, $du = 9x^2 - 1\,dx$.

$$\int (3x^3 - x)^5(18x^2 - 2)\,dx = 2\int u^5\,du = 2\left(\frac{u^6}{6}\right) + C = \frac{1}{3}(3x^3 - x)^6 + C.$$

43. See #3 for substitution. $x = -1 \Rightarrow u = (-1)^5 - 17 = -18$ and $x = 1 \Rightarrow u = (1)^5 - 17 = -16$.

$$\int_{-1}^{1}\frac{x^4}{x^5-17}\,dx = \frac{1}{5}\int_{-18}^{-16}\frac{1}{u}\,du = \frac{1}{5}\ln|u|\Big]_{-18}^{-16} = \frac{1}{5}(\ln|-16| - \ln|-18|) \approx -0.024.$$

46. See # 6 for substitution. $x = 0 \Rightarrow u = \cos(0) + 3 = 5$ and $x = \frac{\pi}{2} \Rightarrow u = 2\cos\left(\frac{\pi}{2}\right) + 3 = 3$.

$$\int_0^{\pi/2}\sin(x)\sqrt{2\cos(x)+3}\,dx = -\frac{1}{2}\int_5^3 u^{1/2}\,du \approx 1.995.$$

49. See # 9 for substitution. $x = e \Rightarrow u = \ln(e) = 1$ and $x = 1 \Rightarrow u = \ln(1) = 0$.

$$\int_1^e \frac{\ln(x)}{x}\,dx = \int_0^1 u\,du = \frac{u^2}{2}\Big]_0^1 = \frac{1}{2} - 0 = \frac{1}{2}.$$

52. $\int \cot(x)\,dx = \int \frac{\cos(x)}{\sin(x)}\,dx = \int \frac{1}{u}\,du = \ln|u| + C = \ln|\sin(x)| + C.$

55. Let $u = -\arctan(x)$, $du = -\frac{1}{1+x^2}\,dx$.

$$\int \frac{e^{-\arctan(x)}\cos(4\arctan(x))}{1+x^2}\,dx = -\int \frac{e^{-\arctan(x)}\cos(4\arctan(x))}{-(1+x^2)}\,dx$$
$$= -\int e^u \cos(-4u)\,du = -\frac{e^u}{1+16}(\cos(-4u) - 4\sin(-4u)) + C$$
$$= -\frac{e^{-\arctan(x)}}{17}(\cos(4\arctan(x)) - 4\sin(4\arctan(x))) + C.$$

58. $\frac{d}{dx}\cosh(x) = \frac{d}{dx}\left(\frac{e^x+e^{-x}}{2}\right) = \frac{e^x-e^{-x}}{2} = \sinh(x).$

$\int \cosh(x)\,dx = \int \frac{e^x+e^{-x}}{2}\,dx = \frac{e^x-e^{-x}}{2} = \sinh(x) + C$

61. $\frac{d}{dx}\tanh(x) = \frac{d}{dx}\left(\frac{\sinh(x)}{\cosh(x)}\right) = \frac{\cosh(x)\cosh(x)-\sinh(x)\sinh(x))}{\cosh^2(x)}$

$= \frac{\left(\frac{e^x+e^{-x}}{2}\right)^2 - \left(\frac{e^x-e^{-x}}{2}\right)^2}{\cosh^2(x)} = \frac{\frac{e^{2x}+2+e^{-2x}-e^{2x}+2-e^{-2x}}{4}}{\cosh^2(x)} = \frac{1}{\cosh^2(x)} = \operatorname{sech}^2(x).$

$\int \tanh(x)\,dx = \int \frac{\sinh(x)}{\cosh(x)}\,dx = \int \frac{1}{u}\,du$ (letting $u = \cosh(x)$)

$= \ln|u| + C = \ln|\cosh(x)| + C.$

64. Let $u = e^x,\ du = e^x\,dx.$

$\int e^x \operatorname{sech}(e^x)\tanh(e^x)\,dx = \int \operatorname{sech}(u)\tanh(u)\,du$

$= -\operatorname{sech}(u) + C = -\operatorname{sech}(e^x) + C.$

ANSWERS TO SELECTED EXERCISES FROM CHAPTER 7

Throughout this chapter, limits of integration and definite integrals may have been found by machine.

Section 7.1

1. $\int_0^5 x\,dx = \frac{x^2}{2}\Big]_0^5 = \frac{25}{2}.$

4. $\int_0^3 e^{x-1}\,dx = e^{x-1}\Big]_0^3 = e^2 - e^{-1} \approx 7.02.$

7. $2\int_1^2 (1-(2-x))\,dx = 2\int_1^2 (x-1)\,dx = 2(\frac{x^2}{2} - x)\Big|_1^2 = 1.$

10. $\int_{-3}^1 (x+3)^{1/3}\,dx = \frac{3}{4}(x+3)^{4/3}\Big]_{-3}^1 = \frac{3}{4}(4)^{4/3} \approx 4.7622.$

13. $\int_0^1 \text{arccot}(x) - \arctan(x)\,dx \approx 0.693.$

16. Find a such that $4\int_0^a (a^2 - x^2)\,dx = 72.$

$4(a^2 x - \frac{x^3}{3})\Big]_0^a = 4(a^3 - \frac{a^3}{3}) = \frac{8}{3}a^3 = 72 \Rightarrow a^3 = 27$ so $a = 3.$

19. $y = \frac{7}{5}\sqrt{25 - x^2}, \quad A = 4(\frac{7}{5})\int_0^5 \sqrt{25 - x^2}\,dx = \frac{28}{5}(\frac{1}{4}\pi(5)^2) = 35\pi.$

For problems 21 - 28, answers may vary.

22. 51.5.   **25.** 21.   **28.** 18.

Section 7.2

1. $\int_0^5 x^2\,dx = \frac{x^3}{3}\Big]_0^5 = \frac{125}{3}.$

4. $\pi \int_0^3 e^{2x-2}\,dx = \frac{\pi}{2} e^{2x-2}\Big]_0^3 = \frac{\pi}{2}(e^4 - e^{-2}) \approx 27.23.$

7. x-axis:
$$\int_1^3 \pi(1-(2-x)^2)\,dx = \int_1^3 \pi(4x-x^2-3)\,dx = \pi(2x^2 - \tfrac{x^3}{3} - 3x)\Big]_1^3 = \tfrac{4}{3}\pi.$$
y-axis:
$$\int_0^1 \pi((2+y)^2 - (2-y)^2)\,dy = \int_0^1 \pi(8y)\,dy = 4\pi y^2\Big]_0^1 = 4\pi.$$

10. x-axis:
$$\int_{-3.63796}^{-1.86236} \pi(\cos^2(x) - (\tfrac{x+1}{3})^2)\,dx + \int_{-1.86236}^{-1.5708} \pi((\tfrac{x+1}{3})^2 - \cos^2(x))\,dx$$
$$+ \int_{-1.5708}^{-1.428} \pi(\tfrac{x+1}{3})^2\,dx + \int_{-1.428}^{-1} \pi \cos^2(x)\,dx$$
$$+ \int_{-1}^{0.88947} \pi(\cos^2(x) - (\tfrac{x+1}{3})^2)\,dx \approx 5.703.$$
y-axis:
$$\int_{-3.63796}^{0} 2\pi|x|\left|\cos(x) - \tfrac{(x+1)}{3}\right|\,dx \approx 10.932.$$

13. x-axis:
$$\int_0^1 \pi((x+3)^{2/3})\,dx \approx 7.237.$$
y-axis:
$$\int_0^1 2\pi x(x+3)^{1/3}\,dx \approx 4.842.$$

16. Area of hexagon with diameter d is $A = \tfrac{3}{16}\sqrt{3}d^2$ and diameter of handle at x is $\tfrac{1}{8}x$.
$$\text{Volume} = \int_4^8 \tfrac{3}{16}\sqrt{3}(\tfrac{x}{8})^2\,dx = \tfrac{\sqrt{3}}{1024}x^3\Big]_4^8 = \tfrac{7}{16}\sqrt{3} \approx 0.758 \text{ in}^3.$$

19. $\pi \int_{-r}^{r} (R+\sqrt{r^2-x^2})^2 - (R-\sqrt{r^2-x^2})^2\,dx$
$$= 4\pi R \int_{-r}^{r} \sqrt{r^2-x^2}\,dx = 4\pi R(\tfrac{1}{2}\pi r^2) = 2\pi^2 Rr^2.$$
(Definite integral represents area of $\tfrac{1}{2}$ circle of radius r.)

22. $V = \tfrac{1}{3}\pi(3)^2(3) = 9\pi.$ (Solid is a cone.)

Section 7.3

1. $\tfrac{1}{4-2}\int_2^4 x\,dx = \tfrac{1}{2}(\tfrac{x^2}{2})\Big]_2^4 = 3.$

4. $\frac{1}{3}\int_1^4 e^x\,dx = \frac{1}{3}(e^4 - e^1) \approx 17.293.$

7. $\frac{1}{\frac{4\pi}{6}}\int_{\pi/6}^{5\pi/6} \sin(x) - \frac{1}{2}dx = \frac{6}{4\pi}\left[-\cos(x) - 0.5x\right]_{\pi/6}^{5\pi/6}$

$= \frac{6}{4\pi}[-\cos(\frac{5\pi}{6}) - 0.5(\frac{5\pi}{6}) + \cos(\frac{\pi}{6}) + 0.5(\frac{\pi}{6})] \approx 0.327.$

10. $\frac{1}{10-\frac{1}{2}}\int_{1/2}^{10} e^x - \ln(x)dx = \frac{2}{19}(e^x - \ln(x) + x)\Big]_{1/2}^{10}$

$= \frac{2}{19}(e^{10} - e^{1/2} - \ln((\sqrt{2})(10^{10})) + 9\frac{1}{2}) \approx 2316.9415.$

13. $\left(\frac{\pi(81) - \pi(49)}{\pi(81)}\right) = \frac{32}{81} \approx 0.395.$

16. $\left(\frac{\pi(9) - \pi}{\pi(81)}\right) = \frac{8}{81} \approx 0.099.$

19. The actual integral value is $\frac{1}{4}$.

22. See 1-10 for evaluation of the definite integrals shown below. In some cases, a machine was used to compute.

(1) $\frac{1}{2}\int_2^4 x\,dx = 3 = f(x) = x \Rightarrow x = 3.$

(4) $\frac{1}{3}\int_1^4 e^x\,dx = \frac{1}{3}(e^4 - e^1) = f(x) = e^x \Rightarrow x \approx 2.850.$

(7) $\frac{6}{4\pi}\int_{\pi/6}^{5\pi/6} \sin(x) - \frac{1}{2}dx = \frac{3\sqrt{3}}{2\pi} - \frac{1}{2} = f(x) = \sin(x) - \frac{1}{2} \Rightarrow x \approx 0.974$ or
$x \approx \pi - 0.974 \approx 2.168.$

(9) Integral is undefined.

(10) $\frac{2}{19}\int_{1/2}^{10} e^x - \ln(x)dx \approx 2316.9415 = f(x) = e^x - \ln(x) \Rightarrow x \approx 7.749.$

25. $\dfrac{\int_{x-3}^{x} u\,du}{3} = \frac{1}{6}u^2\Big]_{x-3}^{x} = \frac{1}{6}(x^2 - (x-3)^2) = \frac{1}{6}(6x - 9) = x - \frac{3}{2}.$

29. The center of the coaster must land within 2 inches of the side of the tile. $p = \frac{4(4)}{8(8)} = \frac{1}{4}.$

32. $p = (\frac{1}{2\pi})(2)\int_0^\pi \frac{1}{2}\sin(y)\,dy = \dfrac{-\cos(y)\Big]_0^\pi}{2\pi} = \frac{1}{\pi}$.

Section 7.4

1. $v(t) = \frac{t^2}{2} - \frac{t^3}{3} + C$, $\quad v(0) = 0 + C = -2 \Rightarrow C = -2$ so
$v(t) = \frac{t^2}{2} - \frac{t^3}{3} - 2$.
$s(t) = \frac{t^3}{6} - \frac{t^4}{12} - 2t + C$, $\quad s(0) = 0 + C = 3 \Rightarrow C = 3$ so
$s(t) = \frac{t^3}{6} - \frac{t^4}{12} - 2t + 3$.

4. $v(t) = 3\arctan(t) + C$, $\quad v(0) = 0 + C = 3.5 \Rightarrow C = 3.5$ so
$v(t) = 3\arctan(t) + 3.5$.
$s(t) = 3t\arctan(t) - \frac{3}{2}\ln(1+t^2) + 3.5t + C$,
$s(0) = 0 + C = 1.25 \Rightarrow C = 1.25$ so
$s(t) = 3t\arctan(t) - \frac{3}{2}\ln(1+t^2) + 3.5t + 1.25$.

7. Net and Total distance: $\int_{-1}^{1} e^{-t}\,dt = -e^{-t}\Big]_{-1}^{1} = -e^{-1} + e^1 \approx 2.350$.

10. Net and Total: $\int_0^2 \sqrt{4-t^2}\,dt = \frac{1}{4}(\pi(2^2)) = \pi$.
(Integral represents the area of a quarter-circle of radius 2.)

12. $a(t) = -k$.
$v(t) = -kt + C$, $\quad v(0) = 0 + C = 88 \Rightarrow C = 88$.
$v(t) = -kt + 88$.
$s(t) = -\frac{kt^2}{2} + 88t$.
Find k when $v(t) = 0$ and $s(t) = 150$.
Solve: $-kt + 88 = 0$ and $-\frac{kt^2}{2} + 88t = 150$.
$-kt + 88 = 0 \Rightarrow t = \frac{88}{k}$.
$-\frac{k}{2}(\frac{88}{k})^2 + 88(\frac{88}{k}) = 150$
$-\frac{3872}{k} + \frac{7744}{k} = 150$
$3872 = 150k$
$25.81 \approx k$.
Find t when $v(t) = 30$ mph $= 44$ ft/sec: $\quad -25.81t + 88 = 44 \Rightarrow t \approx 1.70$ s.
Find $s(1.70)$: $\quad s(1.70) \approx -\frac{25.81}{2}(1.70)^2 + 88(1.70) \approx 112.3$ feet.

Section 7.5

Most of the definite integrals in this section have been computed by machine and the answers have been rounded to 3 decimal places.

1. $\int_{\pi/4}^{\pi/3} \sqrt{1+\cos^2(x)}\, dx \approx 0.307$.

4. $\int_{\pi/4}^{\pi/3} \sqrt{1+\sec^2(x)\tan^2(x)}\, dx \approx 0.645$.

7. $\int_{1/4}^{1/3} \sqrt{1+\frac{1}{1-x^2}}\, dx \approx 0.121$.

10. $\int_{1/4}^{1/3} \sqrt{1+(\frac{1}{1+x^2})^2}\, dx \approx 0.113$.

13. $\int_{1}^{4} \sqrt{1+\operatorname{sech}^4(x)}\, dx \approx 3.026$.

16. $\int_{1}^{4} \sqrt{1+\operatorname{csch}^2(x)\coth^2(x)}\, dx \approx 3.188$.

19. $\int_{0}^{4} \sqrt{1+(2t-10)^2}\, dt \approx 24.395$.

22. $\int_{0}^{4} \sqrt{(\frac{1}{1+t^2})^2+1}\, dt \approx 4.345$.

25. $\int_{\pi/4}^{\pi/3} 2\pi|\sin(x)|\sqrt{1+\cos^2(x)}\, dx \approx 1.521$.

28. $\int_{\pi/4}^{\pi/3} 2\pi|\sec(x)|\sqrt{1+\sec^2(x)\tan^2(x)}\, dx \approx 6.884$.

31. $\int_{1/4}^{1/3} 2\pi|\arcsin(x)|\sqrt{1+\frac{1}{1-x^2}}\, dx \approx 0.224$.

34. $\int_{1/4}^{1/3} 2\pi|\operatorname{arccot}(x)|\sqrt{1+(\frac{1}{1+x^2})^2}\, dx \approx 0.917$.

37. $\int_{1}^{4} 2\pi|\tanh(x)|\sqrt{1+\operatorname{sech}^4(x)}\, dx \approx 18.190$.

40. $\int_{1}^{4} 2\pi |\csch(x)| \sqrt{1 + \csch^2(x)\coth^2(x)}\, dx \approx 5.296.$

Section 7.6

1. $\int_{0}^{4} 3\sqrt{x}\, dx = 2x^{3/2} \Big]_{0}^{4} = 16.$

4. $k = 4,\quad W = \int_{3}^{8} 4x\, dx = 2x^2 \Big]_{3}^{8} = 110$ in-lbs.

7. $k = 15,\quad W = \int_{0}^{1/2} 15x\, dx = \tfrac{15}{2}x^2 \Big]_{0}^{1/2} = \tfrac{15}{8} = 1.875$ ft-lbs.

10. $k = \tfrac{10}{0.87} \approx 11.4943,$

$W \approx \int_{0}^{12} 11.4943x\, dx = 11.4943 \tfrac{x^2}{2} \Big]_{0}^{12} \approx 827.59$ in-lbs ≈ 68.97 ft-lbs.

13. $\Delta V = 2 \times 4 \Delta x,\ \Delta wt \approx 62.5(8)\Delta x,\ \Delta \text{work} \approx 62.5(8)(x+2)\Delta x.$

$W \approx \int_{1.5}^{3} 500(x+2)\, dx = 250x^2 + 1000x \Big]_{1.5}^{3} = 3187.5$ ft-lbs.

16. $\Delta V = \pi(15^2)\Delta x,\ \Delta wt \approx 0.0022(15^2)\pi \Delta x,\ \Delta \text{work} \approx 0.495\pi(x+20)\Delta x.$

$W \approx \int_{0}^{60} 0.495\pi(x+20)\, dx = 0.495\pi\left(\tfrac{x^2}{2} + 20x\right)\Big]_{0}^{60} = 1485\pi$ cm-lbs.

19. 62.5 lbs/ft$^3 \approx 0.0362$ lbs/in$^3.$

$\Delta V = \pi r^2 \Delta x,\ \Delta wt \approx 0.0362\pi r^2 \Delta x,\ \Delta \text{work} \approx 0.0362\pi r^2 x \Delta x.$ We also have the following relationship $\tfrac{40-x}{40} = \tfrac{r}{30} \Rightarrow$

$\Delta \text{work} \approx 0.0362\pi(\tfrac{3}{4}(40-x))^2 x \Delta x.$

$W \approx \int_{0}^{40} 0.0362\pi(\tfrac{9}{16})(40-x)^2 x\, dx$

$= (0.0362\pi)(\tfrac{9}{16})(800x^2 - \tfrac{80}{3}x^3 + \tfrac{x^4}{4})\Big]_{0}^{40} = 4344\pi$ in-lbs.

22. Half-filled $= \frac{1}{2}[\frac{1}{3}\pi(30)^2(40)] = 6000\pi$ in^3. Find the distance, x, from the top of the cone when it is half-filled.

$\frac{1}{3}\pi(r^2)(40-x) = 6000\pi$

$\frac{1}{3}(\frac{9}{16})(40-x)^2(40-x) = 6000$

$(40-x)^3 = 32,000$

$x \approx 8.2520$ in when the cone is half-filled. See #20 for additional information.

$W \approx 0.0362 \int_{8.2520}^{40} \frac{9}{16}\pi(40-x)^2 x \, dx$

$= 0.0362(\frac{9}{16}\pi)(800x^2 - \frac{80}{3}x^3 + \frac{x^4}{4})\Big]_{20}^{40} \approx 11,046.61$ in-lbs.

25. $\Delta V = 15(25)\Delta x$, $\Delta wt \approx 62.5(375)\Delta x$, $\Delta work \approx 62.5(375)(x+1)\Delta x$.

$W \approx \int_0^5 375(62.5)(x+1)\, dx = 11718.75x^2 + 23437.5x \Big]_0^5 \approx 410156$ ft-lbs.

Section 7.7

1. Integral converges.

$\lim_{b\to\infty} \int_1^b \frac{1}{x^3}\, dx = \lim_{b\to\infty} (-\frac{1}{2x^2})\Big]_1^b = \lim_{b\to\infty} (-\frac{1}{2b^2} + \frac{1}{2}) = \frac{1}{2}$.

4. Integral diverges.

$\lim_{b\to\infty} \int_8^b \frac{1}{x^{1/3}}\, dx = \lim_{b\to\infty} (\frac{3}{2}x^{2/3})\Big]_8^b = \lim_{b\to\infty} (\frac{3}{2}b^{2/3} - 6) = +\infty$.

7. Integral converges.

$\lim_{b\to\infty} \int_1^b \frac{1}{x^{1.01}}\, dx = \lim_{b\to\infty} (-\frac{100}{x^{0.01}})\Big]_1^b = \lim_{b\to\infty} (-\frac{100}{b^{0.01}} + 100) = 100$.

10. Integral diverges.

$\lim_{b\to\infty} \int_1^b \frac{1}{x-1}\, dx = \lim_{b\to\infty} (\frac{1}{2}x^2)\Big]_1^b = \lim_{b\to\infty} (\frac{1}{2}b^2 - \frac{1}{2}) = +\infty$.

13. Integral diverges.

$\lim_{b\to\infty} \int_0^b e^{2x}\, dx = \lim_{b\to\infty} (\frac{1}{2}e^{2x})\Big]_0^b = \lim_{b\to\infty} (\frac{1}{2}e^{2b} - \frac{1}{2}) = \infty$.

16. Integral diverges.

$\lim_{b\to-\infty} \int_b^0 e^{-2x}\, dx = \lim_{b\to-\infty} (-\frac{1}{2}e^{-2x})\Big]_b^0 = \lim_{b\to-\infty} (-\frac{1}{2} + \frac{1}{2}e^{-2b}) = \infty$.

19. Integral converges.
$$\lim_{b \to -\infty} \int_b^0 e^{x/2}\, dx = \lim_{b \to -\infty} (2e^{x/2})\Big]_b^0 = \lim_{b \to -\infty} (2 - 2e^{b/2}) = 2.$$

22. Integral diverges.
If $p = 0$, $\int_0^\infty e^{px}\, dx = \int_0^\infty 1\, dx = \infty.$

25. Integral diverges.
$$\lim_{a \to -\infty} \int_a^0 x\, dx + \lim_{b \to +\infty} \int_0^b x\, dx = \lim_{a \to -\infty} \left(-\tfrac{a^2}{2}\right) + \lim_{b \to +\infty} \left(\tfrac{b^2}{2}\right).$$
(Both integrals diverge.)

28. Integral converges.
$$\lim_{b \to 0^-} \int_{-3}^b \tfrac{1}{x^{2/3}}\, dx = \lim_{b \to 0^-} (3x^{1/3})\Big]_{-3}^b = \lim_{b \to 0^-} (3\sqrt[3]{b} - 3\sqrt[3]{-3}) = 3\sqrt[3]{3}.$$

31. Integral diverges.
$$\lim_{b \to 0^+} \int_b^2 \tfrac{1}{x^{3/2}}\, dx = \lim_{b \to 0^+} \left(-\tfrac{2}{\sqrt{x}}\right)\Big]_b^2 = \lim_{b \to 0^+} \left(-\tfrac{2}{\sqrt{2}} + \tfrac{2}{\sqrt{b}}\right) = \infty.$$

34. Integral converges.
$$\lim_{b \to 0^+} \int_b^1 \tfrac{1}{x^{0.99}}\, dx = \lim_{b \to 0^+} (100 x^{0.01})\Big]_b^1 = \lim_{b \to 0^+} (100 - 100 b^{0.01}) = 100.$$

37. If $p = 1$, the improper integral diverges:
$$\lim_{b \to 0^+} \int_b^1 \tfrac{1}{x}\, dx = \lim_{b \to 0^+} (-\ln(b)) = \infty.$$
Now, assume $p \neq 1$. Then
$$\lim_{b \to 0^+} \int_b^1 \tfrac{1}{x^p}\, dx = \lim_{b \to 0^+} \left(\tfrac{x^{1-p}}{1-p}\right)\Big]_b^1 = \lim_{b \to 0^+} \left(\tfrac{1}{1-p} - \tfrac{b^{1-p}}{1-p}\right).$$
This improper integral converges to $\tfrac{1}{1-p}$ for $p < 1$ and diverges for $p > 1$.

40. $\lim_{b \to 0^+} \int_b^1 \ln(x)\, dx = \lim_{b \to 0^+} (x \ln(x) - x)\Big]_b^1$
$= \lim_{b \to 0^+} (-1 + b - b \ln(b)) = -1.$

Note: $\lim_{b \to 0^+} -b \ln(b) = \lim_{b \to 0^+} -\tfrac{\ln(b)}{\tfrac{1}{b}} = \lim_{b \to 0^+} \tfrac{-\tfrac{1}{b}}{-\tfrac{1}{b^2}} = \lim_{b \to 0^+} b = 0.$ At this point students do not know L'Hopital's Rule. However, they can graphically or numerically investigate $\lim_{b \to 0^+} -b \ln(b)$.

41. Integral diverges.

Note that the graph of $y = \sec^2(x)$ is not continuous over $[0, \pi]$ so the Fundamental Theorem of Calculus does not apply.

$$\int_0^\pi \sec^2(x)\,dx = \lim_{b \to \frac{\pi}{2}^+} \int_0^b \sec^2(x)\,dx + \lim_{a \to \frac{\pi}{2}^-} \int_a^\pi \sec^2(x)\,dx$$
$$= \lim_{b \to \frac{\pi}{2}^+}(\tan(b) - \tan(0)) + \lim_{a \to \frac{\pi}{2}^-}(\tan(\pi) - \tan(a)). \text{ (Both integrals diverge.)}$$

44. We cannot use the property for integrals which says
$$\int_a^b f(x)\,dx = \int_a^c f(x)\,dx + \int_c^b f(x)\,dx \text{ unless both integrals converge.}$$
Since neither integral converges, we cannot justify the student's conclusion.

47. $\int_0^\infty e^{-sx}\sin(x)\,dx = \lim_{b \to \infty} \int_0^b e^{-sx}\sin(x)\,dx$

$= \lim_{b \to \infty} \left(\frac{e^{-sx}}{s^2+1}((-s)\sin(x) - \cos(x))\Big]_0^b\right)$

$= \lim_{b \to \infty} \left(\frac{e^{-sb}}{s^2+1}(-s\sin(b) - \cos(b)) - \frac{1}{s^2+1}(-1)\right) = \frac{1}{s^2+1}.$

ANSWERS TO SELECTED EXERCISES FROM CHAPTER 8

Section 8.1

1. $y(x) = 0.5 + \int_0^x e^{-t^2/2}\, dt.$

$y(3) = 0.5 + \int_0^3 e^{-t^2/2}\, dt \approx 1.750$ using a machine to compute.

4. $y(x) = 1 + \int_0^x (1+t)\, dt = \left(1 + t + \frac{t^2}{2}\right)\Big|_0^x = 1 + x + \frac{1}{2}x^2.$

$y(3) = 1 + 3 + \frac{9}{2} = \frac{17}{2}.$

7. $y(x) = 1 + \int_1^x \frac{1}{t^2+1}\, dt = 1 + (\arctan(t))\Big|_1^x = 1 + \arctan(x) - \frac{\pi}{4}.$

$y(3) = 1 + \arctan(3) - \frac{\pi}{4} \approx 1.464.$

10. Let $u = \text{arccot}(x)$ and $dv = dx$. We find that $du = -\frac{1}{1+x^2}dx$ and $v = x$.

$\int \text{arccot}(x)\, dx = x\,\text{arccot}(x) + \int \frac{x}{1+x^2}\, dx \qquad$ (Let $u = 1 + x^2$.)

$= x\,\text{arccot}(x) + \frac{1}{2}\int \frac{1}{u}\, du = x\,\text{arccot}(x) + \frac{1}{2}\ln(1+x^2) + C.$

13. Let $u = \arccos(x)$ and $dv = dx$. We find that $du = \frac{-1}{\sqrt{1-x^2}}dx$ and $v = x$.

$\int \arccos(x)\, dx = x\arccos(x) + \int x(1-x^2)^{-1/2}\, dx \qquad$ (Let $u = 1 - x^2$.)

$= x\arccos(x) - \frac{1}{2}\int u^{-1/2}\, du = x\arccos(x) - (1-x^2)^{1/2} + C.$

16. Let $u = x$ and $dv = \cos(5x)dx$. We find that $du = dx$ and $v = \frac{1}{5}\sin(5x)$.

$\int x\cos(5x)\, dx = \frac{1}{5}x\sin(5x) - \frac{1}{5}\int \sin(5x)\, dx \qquad$ (Let $u = 5x$.)

$= \frac{1}{5}x\sin(5x) - \frac{1}{25}\int \sin(u)\, du = \frac{1}{5}x\sin(5x) + \frac{1}{25}\cos(5x) + C.$

19. Let $u = x$ and $dv = \cos(x)dx$. We find that $du = dx$ and $v = \sin(x)$.

$\int x\cos(x)\, dx = x\sin(x) - \int \sin(x)\, dx = x\sin(x) + \cos(x) + C.$

22. Let $u = \ln\left(\frac{1}{x}\right)$ and $dv = x\,dx$. We find that

$du = \frac{1}{\frac{1}{x}}\left(-\frac{1}{x^2}\right)dx = -\frac{1}{x}dx$ and $v = \frac{x^2}{2}.$

$\int x\ln\left(\frac{1}{x}\right)dx = \frac{x^2}{2}\ln\left(\frac{1}{x}\right) + \int \frac{x}{2}dx = \frac{x^2}{2}\ln\left(\frac{1}{x}\right) + \frac{x^2}{4} + C.$

25. Let $u = (\ln(x))^3$ and $dv = dx$. We find that $du = \frac{3(\ln(x))^2}{x}dx$ and $v = x$.

$$\int (\ln(x))^3 dx = x(\ln(x))^3 - 3\int (\ln(x))^2 dx.$$

Applying integration by parts again, let $u = (\ln(x))^2$ and $dv = dx$. We find that $du = \frac{2\ln(x)}{x}dx$ and $v = x$.

$$\int (\ln(x))^3 dx = x(\ln(x))^3 - 3\left(x(\ln(x))^2 - 2\int \ln(x)dx\right)$$

$$= x(\ln(x))^3 - 3x(\ln(x))^2 + 6\int \ln(x)dx$$

$$= x(\ln(x))^3 - 3x(\ln(x))^2 + 6x\ln(x) - 6x + C$$

using result from example 5 in the text.

28. Let $u = (\ln(x))^2$ and $dv = xdx$. We find that $du = \frac{2\ln(x)}{x}dx$ and $v = \frac{x^2}{2}$.

$$\int x(\ln(x))^2 dx = \frac{x^2(\ln(x))^2}{2} - \int x\ln(x)dx = \frac{x^2(\ln(x))^2}{2} - \frac{x^2\ln(x)}{2} + \frac{x^2}{4} + C \text{ using}$$
result from #15.

31. Let $u = x$ and $dv = \sec(x)\tan(x)dx$. We find that

$du = dx$ and $v = \sec(x)$.

$$\int x\sec(x)\tan(x)dx = x\sec(x) - \int \sec(x)dx$$

$$= x\sec(x) - \ln|\sec(x) + \tan(x)| + C.$$

34. Let $u = \ln(\sin(x))$ and $dv = \cos(x)dx$. We find that

$du = \frac{\cos(x)}{\sin(x)}dx$ and $v = \sin(x)$.

$$\int \cos(x)\ln(\sin(x))dx = \sin(x)\ln(\sin(x)) - \int \cos(x)dx$$

$$= \sin(x)\ln(\sin(x)) - \sin(x) + C.$$

37. Let $u = \arctan(x)$ and $dv = xdx$. We find that $du = \frac{1}{1+x^2}dx$ and $v = \frac{x^2}{2}$.

$$\int x\arctan(x)dx = \frac{x^2(\arctan(x))}{2} - \frac{1}{2}\int \frac{x^2}{1+x^2}dx$$

and by using long divsion on the integrand

$$= \frac{x^2(\arctan(x))}{2} - \frac{1}{2}\int 1 - \frac{1}{x^2+1}dx = \frac{x^2(\arctan(x))}{2} - \frac{1}{2}x + \frac{1}{2}\arctan(x) + C.$$

40. Let $u = x$ and $dv = (x+1)^{10}dx$. We find that $du = dx$ and $v = \frac{(x+1)^{11}}{11}$.

$$\int x(x+1)^{10}dx = \frac{x(x+1)^{11}}{11} - \int \frac{(x+1)^{11}}{11}dx = \frac{x(x+1)^{11}}{11} - \frac{(x+1)^{12}}{132} + C.$$

43. Let $u = \sin(\ln(x))$ and $dv = dx$. We find that $du = \frac{\cos(\ln(x))}{x}dx$ and $v = x$.

$$\int \sin(\ln(x))dx = x\sin(\ln(x)) - \int \cos(\ln(x))dx.$$

Applying integration by parts again, let $u = \cos(\ln(x))$ and $dv = dx$. We find that $du = -\frac{\sin(\ln(x))}{x}dx$ and $v = x$.

$$\int \sin(\ln(x))dx = x\sin(\ln(x)) - x\cos(\ln(x)) - \int \sin(\ln(x))dx$$

$$2\int \sin(\ln(x))dx = x\sin(\ln(x)) - x\cos(\ln(x))$$

$$\int \sin(\ln(x))dx = \tfrac{1}{2}(x\sin(\ln(x)) - x\cos(\ln(x))) + C.$$

46. Integration by parts will not work on this problem. Rewrite
$\int e^x \sinh(x)dx$ as $\int e^x \left(\frac{e^x - e^{-x}}{2}\right) dx$.

$$\int e^x \sinh(x)dx = \tfrac{1}{2}\int e^{2x} - 1\, dx = \tfrac{1}{2}\left(\tfrac{1}{2}e^{2x} - x\right) + C = \tfrac{1}{4}e^{2x} - \tfrac{1}{2}x + C.$$

49. Let $u = e^{3x}$ and $dv = \cos(3x)$. We find that $du = 3e^{3x}$ and $v = \tfrac{1}{3}\sin(3x)$.

$$\int e^{3x}\cos(3x)dx = \tfrac{1}{3}e^{3x}\sin(x) - \int e^{3x}\sin(3x)dx.$$

Using integration by parts again, let $u = e^{3x}$ and $dv = \sin(3x)dx$. We find that $du = 3e^{3x}$ and $v = -\tfrac{1}{3}\cos(3x)$.

$$\int e^{3x}\cos(3x)dx = \tfrac{1}{3}e^{3x}\sin(3x) + \tfrac{1}{3}e^{3x}\cos(3x) - \int \cos(3x)e^{3x}dx$$

$$2\int e^{3x}\cos(3x)dx = \tfrac{1}{3}e^{3x}\sin(3x) + \tfrac{1}{3}e^{3x}\cos(3x)$$

$$\int e^{3x}\cos(3x)dx = \tfrac{1}{6}e^{3x}(\sin(3x) + \cos(3x)) + C.$$

Section 8.2

1. Exponential growth.

$y(0) = A = 4129$ and $y(1) = AC = 4501 \Rightarrow C \approx 1.09$.

$y(6) = 4129(1.09)^6 \approx 6925$.

4. Linear.

$y(t) = -68t + 4129$.

$y(6) = 3721$.

7. $\frac{y'}{y} = 4$

$\ln|y| = 4x + C$

$|y| = e^{4x}e^C$

$42 = e^4 e^C$

$e^C = \frac{42}{e^4}$

$y(x) = 42e^{4x-4}.\qquad D = (-\infty, \infty)$.

10. $\ln|y| = \ln|x| + C$

$|y| = |x|e^C$

$0.75 = e^C$

$|y| = 0.75|x|$

$y = -0.75x.$ $D = (-\infty, 0).$

13. $y = \frac{1}{2}\ln|x^2 + 1| + C$

$y = \frac{1}{2}\ln(x^2 + 1).$ $D = (-\infty, \infty).$

16. $y' = 6e^{2x},$ $y(0) = 3,$ $y(0) = 8.$

19. $y' = \frac{1}{x\ln(2)},$ $y(1) = 0,$ $y(1) = 5.$

Section 8.3

1. $i(t) = AC^t + 32.$ $i(1) = AC + 32 = 87.$ $i(2) = AC^2 + 32 = 76.2.$

$AC = 55$ and $AC^2 = 44.2$ so $\frac{55}{C} = \frac{44.2}{C^2} \Rightarrow C \approx 0.804$ and $A \approx 68.44.$

$i(3) = 68.44(0.804)^3 + 32 \approx 67.6°.$

3. $i(0) = AC^0 + 40 \approx 100.44 \Rightarrow A \approx 60.44.$

$i(1) = 60.44(0.804) + 40 \approx 88.59°.$

$i(2) = 60.44(0.804)^2 + 40 \approx 79.07°.$

7. $1.5 = 4.5\left(\frac{1}{2}\right)^{t/3.3}$

$\frac{1}{3} = \left(\frac{1}{2}\right)^{t/3.3}$

$-\ln(3) = -\left(\frac{t}{3.3}\right)\ln(2)$

$\left(\frac{\ln(3)}{\ln(2)}\right)(3.3) = t$ or $t \approx 5.2$ hours.

Since the minimal dose is half of the original dose and the half-life is 3.3 hours, the second dose should be administered after 3.3 hours. Immediately after the second dose there is 4.5 mg in the bloodstream. This will decrease to 1.5 mg after 5.2 hours. Hence, the third and later doses should be administered every 5.2 hours following the second dose.

10. $100\left(1 + \frac{0.0525}{4}\right)^4 = 100e^r$

$r = 4\ln\left(1 + \frac{.0525}{4}\right) \approx 0.052 \Rightarrow r \approx 5.2\%.$

13. Bank 1: $\quad y(1) = \left(1 + \frac{0.0525}{4}\right)^4 \approx 1.053542.$

Bank 2: $\quad z(1) = \left(1 + \frac{0.0528}{2}\right)^2 \approx 1.053497.$

Bank 1 has a better deal.

$\left(1 + \frac{0.0525}{4}\right)^4 = \left(1 + \frac{r}{2}\right)^2$

$r = 2\left((1 + \frac{0.0525}{4})^2 - 1\right) \approx 0.0528445 \Rightarrow r \approx 5.28445\%.$

16. $N(t) = Ce^{-kt}.$ Let $t = 0$ represent 950 A.D.

Find C: $N(0) = Ce^{-k(0)} = 210 \Rightarrow C = 210.$

Find k: $N(1000) = 210e^{-k(1000)} = 167$

$e^{-1000k} = \frac{167}{210}$

$-1000k = \ln\left(\frac{167}{210}\right)$

$k = -\frac{\ln\left(\frac{167}{210}\right)}{1000} \Rightarrow k \approx 0.000229.$

Find N when $t = 1575$: $N(1575) \approx 210e^{-0.000229(1575)} \approx 146$ words.

Section 8.4

1. The solution to the differential equation with the given initial condition does not lie in the region $[-3, 3] \times [0, 3]$.

2.

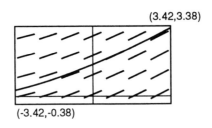

5.

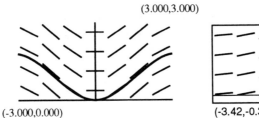

6.

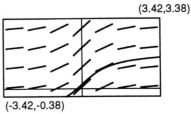

9.

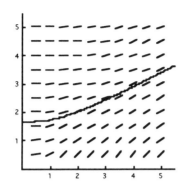

10.

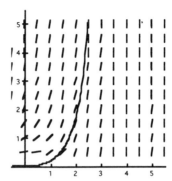

13.

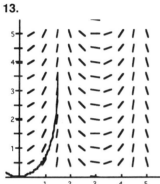

14.

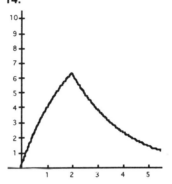

Amount peaks at 2 hours, and 1.4105 mg are left after 5 hours.

17. $\int v^{-2}(v')dt = -\int k^2 dt$

$-v^{-1} = -k^2 t + C$

$\frac{1}{v} = k^2 t + C$

$v = \frac{1}{k^2 t + C}.$

20. $f(x) = \arcsin(x).$

23. $f(x) = \mathrm{arcsec}(x).$

24. $f(x) = \arccos(x).$

26. $\int \operatorname{arcsec}(x)dx = x\operatorname{arcsec}(x) - \ln|x + \sqrt{x^2 - 1}| + C.$

$\int \operatorname{arccsc}(x)dx = x\operatorname{arccsc}(x) + \ln|x + \sqrt{x^2 - 1}| + C.$

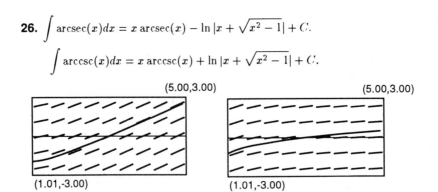

If you graph one of the antiderivatives found from a table of integrals, you should notice that for $x \geq 1$ your graph and the slopefield match up nicely. However, for $x \leq 1$, this is not the case. Try graphing $y = x\operatorname{arcsec}(x) + \ln|x + \sqrt{x^2 - 1}|$ together with the slopefield for $\operatorname{arcsec}(x)$ for $x \leq 1$ and $y = x\operatorname{arccsc}(x) - \ln|x + \sqrt{x^2 - 1}|$ together with the slopefield for $\operatorname{arccsc}(x)$ for $x \leq 1$.

29. $\int \sqrt{y}(y')dx = (1+x)dx$

$\frac{2}{3}y^{3/2} = x + \frac{x^2}{2} + C$

$\frac{2}{3}(9)^{3/2} = 2 + 2 + C$

$14 = C$

$\frac{2}{3}y^{3/2} = x + \frac{x^2}{2} + 14$

$y^{3/2} = \frac{3}{2}x + \frac{3}{4}x^2 + \frac{3}{2}(14)$

$y^{3/2} = \frac{3x^2 + 6x + 84}{4}$

$y = \left(\frac{3x^2 + 6x + 84}{4}\right)^{2/3}.$

32. $\int (y^3 + y)y'dx = \int dx$

$\frac{y^4}{4} + \frac{y^2}{2} = x + C$

$\frac{1}{4} + \frac{1}{2} = 1 + C$

$-\frac{1}{4} = C$

$\frac{y^4}{4} + \frac{y^2}{2} = x - \frac{1}{4}$

$y^4 + 2y^2 = 4x - 1$

$y^4 + 2y^2 + 1 = 4x$

$(y^2 + 1)^2 = 4x$

$y^2 + 1 = \sqrt{4x}$

$y^2 = -1 + 2\sqrt{x}$

$y = \sqrt{-1 + 2\sqrt{x}}.$

Section 8.5

1. (a)

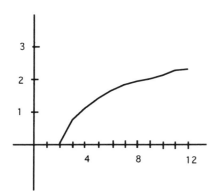

(b)

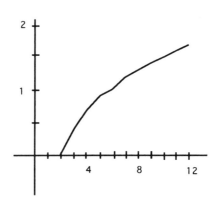

(c)

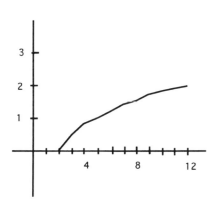

4. (a)

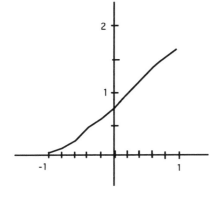

(b)

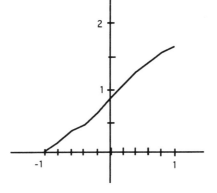

(c)

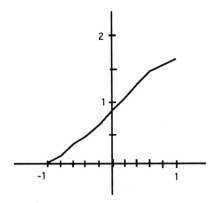

6. The trapezoidal approximation is

$\frac{1}{2}$ (left-end point approximation + right-end point approximation).

(For 9-13, results have been rounded at each step.)

9. $y(0.5) \approx 0 - 1(0.5) = -0.5.$

$y(1) \approx -0.5 - 1.429(0.5) \approx -1.215.$

$y(1.5)$ is undefined since the slope at $(1, -1.215)$ is undefined.

12. $y(0.5) \approx -1 + 1(0.5) = -0.5.$

$y(1) \approx -0.5 + 2(0.5) = 0.5.$

$y(1.5) \approx 0.5 + 0.8(0.5) = 0.9.$

$y(2) \approx 0.9 + 0.327(0.5) \approx 1.064.$

$y(2.5) \approx 1.064 + 0.195(0.5) \approx 1.162.$

$y(3) \approx 1.162 + 0.132(0.5) \approx 1.228.$

$y(1.2) \approx 0.5 + 0.4(0.4) = 0.66.$

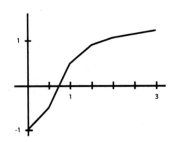

14.

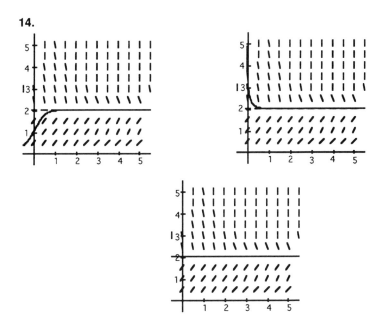

17. There are 2 families of solutions for the differential equation for $t \geq 0$. Using a large stepsize and the initial condition $u(0) = 3$ has caused the graph to cross over to another family of solutions. Using a smaller stepsize would eliminate this problem. A smaller stepsize should also keep the graph from crossing back and forth over the asymptote.

For problems 18-24, the students are expected to use various graphs to solve the problems. However, the algebraic solutions to these problems are also shown.

20. $\int \frac{dT}{90-T} = 0.75 \int dt$

$-\ln|90 - T| = 0.75t + C$

$-\ln|55| = C$

$\ln|90 - T| = -0.75t + \ln(55)$

$|90 - T| = 55e^{-0.75t}$

$90 - T = 55e^{-0.75t}$

(Initial condition determines the sign.)

$T = 90 - 55e^{-0.75t}$.

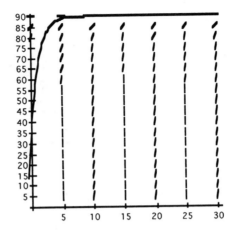

23. $T = 65 \Rightarrow t \approx 0.2$ hours using the expression from 18.

$T = 65 \Rightarrow t \approx 1.05$ hours and $T = 70 \Rightarrow t \approx 1.35$ using the expression from 20.

$1.35 - 1.05 = 0.3$ hours.

It takes approximately 0.2 hours for the temperature to go from 70° to 65° when the furnace is off and approximately 0.3 hours for the temperature to go from 65° to 70° when the furnace is on. This on/off cycle would be repeated over the 8 hour period.

26. For $T_{in} - T_{out} = 20°$: change is $0.5 \exp(-4(\frac{1}{6}))(20) \approx 5.13°$.

For $T_{in} - T_{out} = 3°$: change is $0.5 \exp(-4(\frac{1}{6}))(3) \approx 0.77°$.

ANSWERS TO SELECTED EXERCISES FROM CHAPTER 9

Section 9.1

Values for M may vary. Values for error estimate have been rounded.

1.

n	error bound($M = 1$)
2	0.021
4	0.00521
8	0.0013
16	0.000326
32	0.0000814

4.

n	error bound($M = 1.5$)
2	0.844
4	0.211
8	0.0527
16	0.0132
32	0.00330

7.

n	error bound($M = 3$)
1	3.47×10^{-4}
2	6.51×10^{-5}
4	4.07×10^{-6}
8	2.54×10^{-7}
16	1.59×10^{-8}

10.

n	error bound($M = 9$)
1	0.759
2	0.047
4	2.97×10^{-3}
8	1.85×10^{-4}
16	1.16×10^{-5}

13. $n = 128$. **16.** $n = 2048$.

19. $n = 8$.

22. $n = 64$.

25. The error bound for the trapezoidal rule estimate when f is any linear function over $[a, b]$ is 0 because for any partition of $[a, b]$, the geometric shapes formed are trapezoids.

28. $\frac{M^{(4)}(b-a)^5}{180(2n)^4} = \frac{M^{(4)}(b-a)(b-a)^4}{180(2n)^4} = \frac{M^{(4)}(b-a)}{180}(\Delta x)^4$ where $\Delta x = \frac{b-a}{2n}$.

30.. The size of the ratio in problem 29 is dependent on the size of Δx. When Δx is big which is the case for a small partition the value of the ratio is big. In particular, if the bound on the fourth derivative is larger than the bound on the second derivative, then for a small partition the trapezoidal estimate could be a better estimate than Simpson's rule.

32. Many answers. For example, using the function given we have $f(-1) = -15$, $f(0) = -7$, and $f(1) = 11$. Find a quadratic through $(-1, -15)$, $(0, -7)$, and $(1, 11)$.

$$\text{Solve} \begin{cases} a(-1)^2 + b(-1) + c = -15 \\ c = -7 \\ a(1)^2 + b(1) + c = 11. \end{cases}$$

$$a - b = -8$$
$$a + b = 18$$
$$2a = 10$$
$$a = 5.$$

We have $a = 5$ and $c = -7$. Substituting these values into the third equation, we get $5 + b - 7 = 11 \Rightarrow b = 13$. Our quadratic is $g(x) = 5x^2 + 13x - 7$.

34. $g(x) = ax^2 + bx + c$ where $a = \frac{2(y_0 - 2y_1 + y_2)}{(\Delta x)^2}$

$b = \frac{-4x_0(y_0 - 2y_1 + y_2) + \Delta x(-3y_0 + 4y_1 - y_2)}{(\Delta x)^2}$

$c = y_0 - ax_0^2 - bx_0$.

35. $\int_{x_0}^{x_2} ax^2 + bx + c\, dx = \frac{ax^3}{3} + \frac{bx^2}{2} + cx \Big]_{x_0}^{x_2}$

$= \frac{a}{3}(x_2^3 - x_0^3) + \frac{b}{2}(x_2^2 - x_0^2) + c(x_2 - x_0)$

$= \frac{a}{3}((x_0 + \Delta x)^3 - x_0^3) + \frac{b}{2}((x_0 + \Delta x)^2 - x_0^2) + c(x_2 - x_0)$

$= \frac{a}{3}(x_0^3 + 3\Delta x(x_0)^2 + 3(\Delta x)^2 x_0 + (\Delta x)^3 - x_0^3)$

$\quad + \frac{b}{2}(x_0^2 + 2\Delta x(x_0) + (\Delta x)^2 - x_0^2) + c(x_2 - x_0)$

$= \frac{a}{3}(3\Delta x(x_0)^2 + 3(\Delta x)^2 x_0 + (\Delta x)^3) + \frac{b}{2}(2\Delta x(x_0) + (\Delta x)^2) + c\Delta x$

$= \frac{a}{3}(3\Delta x(x_0)^2 + 3(\Delta x)^2 x_0 + (\Delta x)^3) + \frac{b}{2}(2\Delta x(x_0) + (\Delta x)^2) + (y_0 - ax_0^2 - bx_0)\Delta x$

$= \left(ax_0 + \frac{b}{2}\right)(\Delta x)^2 + y_0 \Delta x + \frac{a}{3}(\Delta x)^3$

$= (2y_0 - 4y_1 + 2y_2)x_0 - 2x_0 y_0 + 4x_0 y_1 - 2x_0 y_2$

$\quad - \frac{3}{2} y_0 \Delta x + 2y_1 \Delta x - \frac{y_2}{2}\Delta x + y_0 \Delta x + \left(\frac{2y_0}{3} - \frac{4y_1}{3} + \frac{2y_2}{3}\right)\Delta x$

(replacing a and b with the expressions found in problem 34)

$= \frac{1}{6} y_0 \Delta x + \frac{2}{3} y_1 \Delta x + \frac{1}{6} y_2 \Delta x = \frac{\Delta x}{6}(y_0 + 4y_1 + y_2).$

Section 9.2

1. $f(-3) = 9$, $f'(x) = 2x$, $f'(-3) = -6$, and $f''(x) = 2$.

$p(x) = 9 - 6(x + 3) + (x + 3)^2 = x^2.$

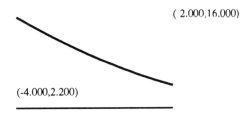

$p(-2) = 4,$	$f(-2) = 4,$	no error.
$p(-4) = 16,$	$f(-4) = 16,$	no error.
$p(-2.999) = 8.994001,$	$f(-2.999) = 8.994001,$	no error.
$p(-3.001) = 9.006001,$	$f(-3.001) = 9.006001,$	no error.

4. $f(-1.5) = 2.625$, $f'(x) = 3x^2 - 4$, $f'(-1.5) = 2.75$, $f''(x) = 6x$, $f''(-1.5) = -9$.
$p(x) = 2.625 + 2.75(x + 1.5) - 4.5(x + 1.5)^2$.

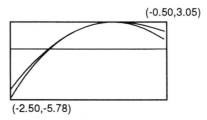
(-0.50, 3.05)
(-2.50, -5.78)

$p(-0.5) = 0.875$, $f(-0.5) = 1.875$,   error = 1.
$p(-2.5) = -4.625$, $f(-2.5) = -5.625$,   error = 1.
$p(-1.499) \approx 2.6277$, $f(-1.499) \approx 2.6277$,   error $\approx 1 \times 10^{-9}$.
$p(-1.501) \approx 2.6222$, $f(-1.501) \approx 2.622247$, error $\approx 1 \times 10^{-9}$.

7. $f\left(\frac{\pi}{4}\right) = 1$, $f'(x) = \sec^2(x)$, $f'\left(\frac{\pi}{4}\right) = 2$, $f''(x) = 2\sec^2(x)\tan(x)$, $f''\left(\frac{\pi}{4}\right) = 4$.
$p(x) = 1 + 2\left(x - \frac{\pi}{4}\right) + 2\left(x - \frac{\pi}{4}\right)^2$.

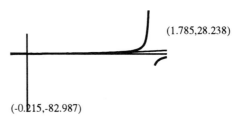
(1.785, 28.238)
(-0.215, -82.987)

$p(1 + \frac{\pi}{4}) = 5$, $f(1 + \frac{\pi}{4}) \approx -4.588$,   error ≈ 9.6.
$p(-1 + \frac{\pi}{4}) = 1$, $f(-1 + \frac{\pi}{4}) \approx -0.2180$,   error ≈ 1.2.
$p(0.001 + \frac{\pi}{4}) \approx 1.002$, $f(0.001 + \frac{\pi}{4}) \approx 1.002$,   error $\approx 2.7 \times 10^{-9}$.
$p(-0.001 + \frac{\pi}{4}) \approx 0.998$, $f(-0.001 + \frac{\pi}{4}) \approx 0.998$,   error $\approx 2.7 \times 10^{-9}$.

10. $f(132.78) = 5.42$, $f'(x) = 0$, $f''(x) = 0$. $p(x) = 5.42$.

(133.78, 5.96)
(131.78, 4.72)

$p(133.78) = 5.42$, $f(133.78) = 5.42$,   no error.
$p(131.78) = 5.42$, $f(131.78) = 5.42$,   no error.
$p(132.781) = 5.42$, $f(132.781) = 5.42$,   no error.
$p(131.779) = 5.42$, $f(131.779) = 5.42$,   no error.

11. a) $f'(x) = 2x + 1$, $f'(1.6) = 4.2$, $f''(x) = 2$, $\kappa = \frac{|2|}{(1+(4.2)^2)^{3/2}} \approx 0.025$.

b) $r = \frac{1}{\kappa} \approx 40.24$.

c) We need to find the center of the circle. It lies on a line passing through $(1.6, f(1.6))$ and perpendicular to the tangent line at $x = 1.6$. Let (a, b) be the center. The slope of this perpendicular line is $\frac{b - f(1.6)}{a - 1.6} = \frac{b - 5.16}{a - 1.6} = -\frac{1}{4.2}$. We also know the distance between (a, b) and $(1.6, 5.16)$ is approximately 40.24. So, $\sqrt{(a - 1.6)^2 + (b - 5.16)^2} \approx 40.24$. From above, $b - 5.16 = -\frac{(a - 1.6)}{4.2}$. So, $\sqrt{(a - 1.6)^2 + \left(\frac{(a-1.6)}{4.2}\right)^2} \approx 40.24 \Rightarrow \sqrt{1.06(a - 1.6)^2} \approx 40.24 \Rightarrow |(a - 1.6)| \approx 39.08 \Rightarrow a \approx 40.7$ or $a \approx -37.5$. We choose a so that the circle is on the same side of the tangent line as the graph of f. So, $a \approx -37.5$ and $b \approx 5.16 - \frac{(-37.5 - 1.6)}{4.2} \approx 14.47$. The equation of the osculating circle at $(1.6, 5.16)$ is $(x + 37.5)^2 + (y - 14.47)^2 = 40.24^2$.

d)

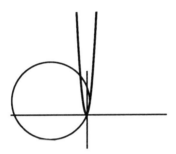

14. a) $f'(x) = \frac{3}{5}$, $f''(x) = 0$, $\kappa = 0$.

b) r does not exist.

c) Does not apply.

17. a) $f'(x) = 2\sin(x)\cos(x)$, $f'\left(\frac{\pi}{3}\right) = 2\left(\frac{\sqrt{3}}{2}\right)\left(\frac{1}{2}\right) = \frac{\sqrt{3}}{2}$,

$f''(x) = 2(\cos^2(x) - \sin^2(x))$, $f''\left(\frac{\pi}{3}\right) = 2\left(\frac{1}{4} - \frac{3}{4}\right) = -1$,

$\kappa = \frac{|-1|}{\left(1 + \left(\frac{\sqrt{3}}{2}\right)^2\right)^{3/2}} = \frac{1}{\left(\frac{7}{4}\right)^{3/2}} = \left(\frac{4}{7}\right)^{3/2} \approx 0.43$.

b) $r = \frac{1}{\kappa} = \left(\frac{7}{4}\right)^{3/2} \approx 2.32$.

c) Following the procedure of problem 11, part c, we need $f\left(\frac{\pi}{3}\right) = \frac{3}{4}$ and the slope of the line perpendicular to the tangent line which is $-\frac{2}{\sqrt{3}}$. Solve $\sqrt{\left(a - \frac{\pi}{3}\right)^2 + \left(b - \frac{3}{4}\right)^2} \approx 2.32$ and $\frac{b - \frac{3}{4}}{a - \frac{\pi}{3}} = -\frac{2}{\sqrt{3}}$ simultaneously. So, $\sqrt{\left(a - \frac{\pi}{3}\right)^2 + \left(\frac{2\left(a - \frac{\pi}{3}\right)}{\sqrt{3}}\right)^2} \approx 2.32 \Rightarrow \sqrt{\frac{7}{3}\left(a - \frac{\pi}{3}\right)^2} \approx 2.32 \Rightarrow \left|a - \frac{\pi}{3}\right| \approx 1.52$ $\Rightarrow a \approx 1.52 + \frac{\pi}{3} \approx 2.57$ or $a \approx -1.52 + \frac{\pi}{3} \approx -0.47$. From the graph, we see that $a \approx 2.57$. So, $b \approx 0.75 - \frac{2}{\sqrt{3}}\left(2.57 - \frac{\pi}{3}\right) \approx -1.01$. The equation of the osculating circle at $\left(\frac{\pi}{3}, \frac{3}{4}\right)$ is $(x - 2.57)^2 + (y + 1.01)^2 \approx 2.32^2$.

17. d)

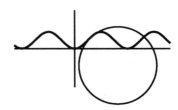

20. Notice, the graph of $f(x)$ is a half-circle of radius 5 and center at $(0,0)$.

a) $\kappa = \frac{1}{5}$.

b) $r = 5$.

c) $x^2 + y^2 = 25$.

d)

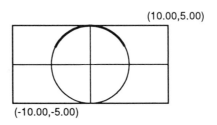

23. $\kappa = \frac{|f''(x_0)|}{(1+(f'(x_0))^2)^{3/2}} = 0$ since $f''(x_0) = 0$ (($x_0, f(x_0)$) is an inflection point). Also note that denominator is never 0. Since $f''(x_0) = 0$, we have

$$g(x) = f(x_0) + f'(x_0)(x-x_0) + (\tfrac{1}{2})f''(x_0)(x-x_0)^2 = f(x_0) + f'(x_0)(x-x_0).$$

26. $\kappa = \frac{|2a|}{(1+(2ax)^2)^{3/2}} = \frac{1}{2}$. At the point $(0,0)$, $\kappa = |2a|$ so $|2a| = \frac{1}{2} \Rightarrow a = \pm\frac{1}{4}$.

29. If q has no critical values, then q is always increasing or always decreasing. The graph of q then has minimum steepness only when $q''(x) = 0$, that is, when $x = -\frac{b}{3a}$. If q has one critical value, then, solving $q'(x) = 0$, we find $x = \frac{-2b \pm \sqrt{4b^2 - 12ac}}{6a} = -\frac{b}{3a}$ since $4b^2 - 12ac = 0$. If q has two critical values, then solving $q'(x) = 0$ we find those values are, as before, $x = \frac{-2b \pm \sqrt{4b^2 - 12ac}}{6a}$. Find the midpoint between the two roots: $x = -\frac{b}{3a}$.

Section 9.3

The Taylor polynomials for problems 1 - 10 can be found by machine.

1.

$f(x) = \sin(2x+1)$, $\qquad f(-1) \approx -0.8415$.
$f'(x) = 2\cos(2x+1)$, $\qquad f'(-1) \approx 1.081$.
$f''(x) = -4\sin(2x+1)$, $\qquad f''(-1) \approx 3.3659$.
$f'''(x) = -8\cos(2x+1)$, $\qquad f'''(-1) \approx -4.3224$.
$f^{(4)}(x) = 16\sin(2x+1)$, $\qquad f^{(4)}(-1) = 16\sin(-1) \approx -13.464$.
$f^{(5)}(x) = 32\cos(2x+1)$, $\qquad f^{(5)}(-1) = 32\cos(-1) \approx 17.290$.

The fifth degree Taylor polynomial at $a = -1$ is

$p_5(x) \approx -0.841+1.081(x+1)+3.366\frac{(x+1)^2}{2}-4.322\frac{(x+1)^3}{3!}-13.464\frac{(x+1)^4}{4!}+17.290\frac{(x+1)^5}{5!}$
$\approx -0.841 + 1.081(x+1) + 1.683(x+1)^2 - 0.720(x+1)^3 - 0.561(x+1)^4$
$\quad + 0.144(x+1)^5$.

4.

$f(x) = \sec(x)$, $\qquad f(0) = 1$.
$f'(x) = \sec(x)\tan(x)$, $\qquad f'(0) = 0$.
$f''(x) = \sec^3(x) + \tan^2(x)\sec(x)$, $\qquad f''(0) = 1$.
$f'''(x) = 5\sec^3(x)\tan(x) + \tan^3(x)\sec(x)$, $\qquad f'''(0) = 0$.
$f^{(4)}(0) = 5$, $f^{(5)}(0) = 0$ using a machine to compute.

The fifth degree Taylor polynomial at $a = 0$ is

$p_5(x) = 1 + \frac{1}{2}x^2 + \frac{5}{4!}x^4 = 1 + 0.5x^2 + 0.125x^4$.

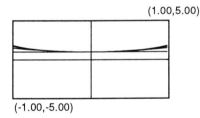

(1.00, 5.00)

(-1.00, -5.00)

6.

$f(x) = -\cot(x)$, $\qquad f(\frac{3\pi}{4}) = 1$.
$f'(x) = \csc^2(x)$, $\qquad f'(\frac{3\pi}{4}) = 2$.
$f''(x) = -2\csc^2(x)\cot(x)$, $\qquad f''(\frac{3\pi}{4}) = 4$.
$f'''(x) = 2\csc^4(x) + 4\csc^2(x)\cot^2(x)$, $\qquad f'''(\frac{3\pi}{4}) = 16$.
$f^{(4)}\left(\frac{3\pi}{4}\right) = 80$, $f^{(5)}\left(\frac{3\pi}{4}\right) = 512$ using a machine to compute.

The fifth degree Taylor polynomial at $a = \frac{3\pi}{4}$ is

$p_5(x) = 1 + 2\left(x - \frac{3\pi}{4}\right) + \frac{4}{2}\left(x - \frac{3\pi}{4}\right)^2 + \frac{16}{3!}\left(x - \frac{3\pi}{4}\right)^3 + \frac{80}{4!}\left(x - \frac{3\pi}{4}\right)^4 + \frac{512}{5!}\left(x - \frac{3\pi}{4}\right)^5$
$= 1 + 2\left(x - \frac{3\pi}{4}\right) + 2\left(x - \frac{3\pi}{4}\right)^2 + \frac{8}{3}\left(x - \frac{3\pi}{4}\right)^3 + \frac{10}{3}\left(x - \frac{3\pi}{4}\right)^4 + \frac{64}{15}\left(x - \frac{3\pi}{4}\right)^5$.

6. continued

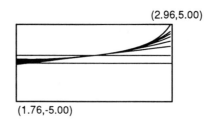

9.

$f(x) = (x^2 + 1)^{1/2}$, $\qquad f(0) = 1.$
$f'(x) = \frac{1}{2}(x^2 + 1)^{-1/2} \cdot 2x = \frac{x}{(x^2+1)^{1/2}}$, $\qquad f'(0) = 0.$
$f''(x) = \frac{(x^2+1)^{1/2} \cdot 1 - x \cdot \frac{1}{2}(x^2+1)^{-1/2} \cdot 2x}{x^2+1} = \frac{1}{(x^2+1)^{3/2}}$,
$\qquad\qquad\qquad\qquad\qquad\qquad\qquad\qquad f''(0) = 1.$
$f'''(x) = -\frac{3}{2}(x^2 + 1)^{-5/2} \cdot 2x$, $\qquad f'''(0) = 0.$
$f^{(4)}(0) = -3$, $f^{(5)}(0) = 0$ using a machine to compute.

The fifth degree Taylor polynomial with $a = 0$ is

$p_5(x) = 1 + \frac{1}{2}(x)^2 - \frac{3}{4!}(x)^4 = 1 + 0.5x^2 - 0.125x^4.$

Answers may vary for exercises 11 - 20.

12. 3, 1.01.

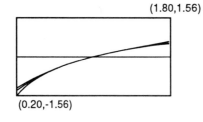

16. 2, 0.01.

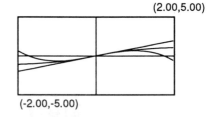

18. 3, 1.01.

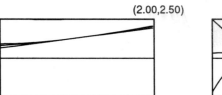

20. -2, 0.01.

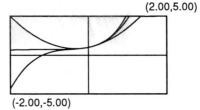

23. $\sqrt{f(0)} + \sqrt{f(0)} = 2$ so $\sqrt{f(0)} = 1$ or $f(0) = 1$.

Differentiating with respect to x, we get

$$\tfrac{1}{2}(f(x)+x)^{-1/2}(f'(x)+1) + \tfrac{1}{2}(f(x)+x^2)^{-1/2}(f'(x)+2x) = 0.$$

$$\tfrac{f'(x)+1}{2(f(x)+x)^{1/2}} + \tfrac{f'(x)+2x}{2(f(x)+x^2)^{1/2}} = 0.$$

Evaluating at $x = 0$, we get

$$\tfrac{f'(0)+1}{2(1)} + \tfrac{f'(0)}{2(1)} = 0 \text{ so } f'(0) + \tfrac{1}{2} = 0 \text{ or } f'(0) = -\tfrac{1}{2}.$$

Differentiating again with respect to x, we get

$$\tfrac{f''(x)(2)(f(x)+x)^{1/2} - (f'(x)+1)(f(x)+x)^{-1/2}(f'(x)+1)}{4(f(x)+x)}$$

$$+ \tfrac{(f''(x)+2)(2)(f(x)+x^2)^{1/2} - (f'(x)+2x)(f(x)+x^2)^{-1/2}(f'(x)+2x)}{4(f(x)+x^2)} = 0.$$

$$\tfrac{2f''(x)(f(x)+x) - (f'(x)+1)^2}{4(f(x)+x)^{3/2}} + \tfrac{2(f''(x)+2)(f(x)+x^2) - (f'(x)+2x)^2}{4(f(x)+x^2)^{3/2}} = 0.$$

Evaluating at $x = 0$, we get

$$\tfrac{2f''(0)(1) - \left(\tfrac{1}{4}\right)}{4} + \tfrac{2(f''(0)+2) - \left(\tfrac{1}{4}\right)}{4} = 0$$

$$f''(0) - \tfrac{1}{16} + 1 - \tfrac{1}{16} = 0 \text{ so } f''(0) = -\tfrac{7}{8}.$$

Differentiating again with respect to x, we get

$$\tfrac{\left(2f'''(x)(f(x)+x) + 2f''(x)(f'(x)+1) - 2(f'(x)+1)f''(x)\right)4(f(x)+x)^{3/2}}{16(f(x)+x)^3}$$

$$- \tfrac{\left(2f''(x)(f(x)+x) - (f'(x)+1)^2\right)6(f(x)+x)^{1/2}(f'(x)+1)}{16(f(x)+x)^3}$$

$$+ \tfrac{4(f(x)+x^2)^{3/2}(2(f'''(x))(f(x)+x^2) + 2(f''(x)+2)(f'(x)+2x) - 2(f'(x)+2x)(f''(x)+2))}{16(f(x)+x)^3}$$

$$- \tfrac{(2(f''(x)+2)(f(x)+x^2) - (f'(x)+2x)^2)6(f(x)+x^2)(f'(x)+2x)}{16(f(x)+x^2)^3} = 0.$$

Evaluating at $x = 0$, we get

$$\tfrac{\left(2f'''(0) + 2\left(-\tfrac{7}{8}\right)\left(\tfrac{1}{2}\right) - 2\left(\tfrac{1}{2}\right)\left(-\tfrac{7}{8}\right)\right)4}{16} - \tfrac{\left(2\left(-\tfrac{7}{8}\right) - \left(\tfrac{1}{4}\right)\right)6\left(\tfrac{1}{2}\right)}{16}$$

$$+ \tfrac{4\left(2f'''(0) + 2\left(-\tfrac{7}{8}+2\right)\left(-\tfrac{1}{2}\right) - 2\left(-\tfrac{1}{2}\right)\left(-\tfrac{7}{8}+2\right)\right)}{16} - \tfrac{\left(2\left(-\tfrac{7}{8}+2\right) - \left(\tfrac{1}{4}\right)\right)6\left(-\tfrac{1}{2}\right)}{16} = 0.$$

$$\tfrac{\left(2f'''(0) - \tfrac{7}{8} + \tfrac{7}{8}\right)4 - (-2)(3)}{16} + \tfrac{4\left(2f'''(0) - \tfrac{9}{8} + \tfrac{9}{8}\right) - (2)(-3)}{16} = 0.$$

$$f'''(0) + \tfrac{12}{16} = 0 \text{ so } f'''(0) = -\tfrac{3}{4}.$$

The third order Taylor polynomial for f about $a = 0$ is

$$p_3(x) = 1 - \tfrac{1}{2}x - \tfrac{7}{16}x^2 - \tfrac{1}{8}x^3.$$

24. $f(0) = 0$ by inspection.

Differentiating with respect to x, we get

$$\cos(f(x)) + x(-\sin(f(x)))f'(x) - f'(x) = 0.$$

Evaluating at $x = 0$, we get $\cos(0) + 0 - f'(0) = 0$ so $f'(0) = 1$.

Differentiating again with respect to x, we get

$$-\sin(f(x))f'(x) - x\sin(f(x))f''(x) - (\sin(f(x)))$$
$$+ x\cos(f(x)f'(x))(f'(x)) - f''(x) = 0.$$

Evaluating at $x = 0$, we get

$$-\sin(0)(1) - 0 - f''(0) = 0 \text{ so } f''(0) = 0.$$

Differentiating again with respect to x, we get

$$-\cos(f(x))f'(x)f'(x) - \sin(f(x))f''(x) - x\sin(f(x))f'''(x)$$
$$-(\sin(f(x)) + x\cos(f(x))f'(x))f''(x) - (\cos(f(x))f'(x) + x\cos(f(x))f''(x)$$
$$+(\cos(f(x)) - x\sin(f(x))f'(x))(f'(x)))f'(x) - f'''(x) = 0.$$

Evaluating at $x = 0$, we get

$$-\cos(0)(1)(1) - 0 - 0 - 0 - \cos(0)(1) - \cos(0) - f'''(0) = 0 \text{ or } f'''(0) = -3.$$

The third order Taylor polynomial about $a = 0$ is

$$p_3(x) = x - \tfrac{1}{2}x^3.$$

27. A fifth degree polynomial about $x = 0$ has the general form

$$y = c_0 + c_1 x + c_2 x^2 + c_3 x^3 + c_4 x^4 + c_5 x^5 \text{ with derivative}$$
$$y' = c_1 + 2c_2 x + 3c_3 x^2 + 4c_4 x^3 + 5c_5 x^4.$$

By substituting into the differential equation we get

$$c_1 + 2c_2 x + 3c_3 x^2 + 4c_4 x^3 + 5c_5 x^4$$
$$= x^2 - (0.1)(x + c_0 + c_1 x + c_2 x^2 + c_3 x^3 + c_4 x^4 + c_5 x^5).$$

The initial condition $y(0) = 0$ tells us that $c_0 = 0$. Now, if we equate coefficients of the two sides of the equation we get

$c_1 = -0.1 c_0 = 0$

$2c_2 = -0.1 - 0.1 c_1 \Rightarrow c_2 = -\frac{0.1}{2} = -0.05$

$3c_3 = 1 - 0.1 c_2 \Rightarrow c_3 = \frac{1 - 0.1(-0.05)}{3} \approx 0.335$

$4c_4 = -0.1 c_3 \Rightarrow c_4 = \frac{-0.1(0.335)}{4} = -0.008375$

$5c_5 = -0.1 c_4 \Rightarrow c_5 = \frac{-0.1(-0.008375)}{5} = 0.0001675.$

The fifth order polynomial approximation at $x = 0$

$$y(x) \approx -0.05 x^2 + 0.335 x^3 - 0.008375 x^4 + 0.0001675 x^5.$$

30. A fifth degree Taylor polynomial about $x = 0$ has the general form
$$y = c_0 + c_1 x + c_2 x^2 + c_3 x^3 + c_4 x^4 + c_5 x^5 \text{ with derivative}$$
$$y' = c_1 + 2c_2 x + 3c_3 x^2 + 4c_4 x^3 + 5c_5 x^4.$$

By substituting into the differential equation we get
$$c_1 + 2c_2 x + 3c_3 x^2 + 4c_4 x^3 + 5c_5 x^4 = x \cos(\tfrac{c_0}{2} + \tfrac{c_1}{2}x + \tfrac{c_2}{2}x^2 + \tfrac{c_3}{2}x^3 + \tfrac{c_4}{2}x^4 + \tfrac{c_5}{2}x^5).$$

Since $y(0) = 0$, we have $c_0 = 0$. Let $x = 0$ in the above equation to see that
$$c_1 = 0.$$

Differentiate the above equation to get
$$2c_2 + 6c_3 x + 12c_4 x^2 + 20c_5 x^3$$
$$= x(-\sin(\tfrac{c_0}{2} + \tfrac{c_1}{2}x + \tfrac{c_2}{2}x^2 + \tfrac{c_3}{2}x^3 + \tfrac{c_4}{2}x^4 + \tfrac{c_5}{2}x^5)$$
$$\cdot (\tfrac{c_1}{2} + c_2 x + \tfrac{3c_3}{2}x^2 + \tfrac{4c_4}{2}x^3 + \tfrac{5c_5}{2}x^4))$$
$$+ \cos(\tfrac{c_0}{2} + \tfrac{c_1}{2}x + \tfrac{c_2}{2}x^2 + \tfrac{c_3}{2}x^3 + \tfrac{c_4}{2}x^4 + \tfrac{c_5}{2}x^5).$$

Let $x = 0$ to see that $2c_2 = \cos\left(\tfrac{c_0}{2}\right) = \cos(0) = 1$ so $c_2 = \tfrac{1}{2}$.

Differentiating 3 more times and evaluating each result at $x = 0$, we find that $c_3 = c_4 = c_5 = 0$.

The fifth degree polynomial approximation about $x = 0$ is $y(x) = \tfrac{1}{2}x^2$.

33. $f(x) = x^2 e^{-x}$. Notice $f(0) = 0$, $f'(x) = 2xe^{-x} - x^2 e^{-x} \Rightarrow f'(0) = 0$, and $f''(x) = 2e^{-x} - 2xe^{-x} - 2xe^{-x} + x^2 e^{-x} \Rightarrow f''(0) = 2$. So, $p_1(x) = 0$ and $p_2(x) = 2$. Since $\lim_{x \to \infty} x^2 e^{-x} = 0$, $p_1(x)$ gives a better approximation for all large x than $p_2(x)$.

Section 9.4

1. Answers may vary, $[-0.1, 0.1] \times [-0.002, 0.002]$.

4. Answers may vary, $[-0.02, 0.02] \times [-1 \times 10^{-6}, 1 \times 10^{-6}]$.

6. f. **9.** f. **12.** f.

13. f. **16.** 3. **19.** 1.

22. 4. **25.** 2,3. **28.** 3,4.

31. 3,5. **34.** 3,4.

Section 9.5

1. $\lim_{x \to \frac{1}{2}} \frac{2x - \sin(\pi x)}{4x^2 - 1} = \lim_{x \to \frac{1}{2}} \frac{2 - \pi \cos(\pi x)}{8x} = \frac{1}{2}.$

4. $\exp(\lim_{x \to \infty} \ln((1 + \frac{1}{x})^x)) = \exp(\lim_{x \to \infty} x \ln(1 + \frac{1}{x})) = \exp(\lim_{x \to \infty} \frac{\ln(\frac{x+1}{x})}{\frac{1}{x}})$

$= \exp(\lim_{x \to \infty} \frac{\frac{x}{x+1}\left(\frac{x-(x+1)}{x^2}\right)}{-\frac{1}{x^2}}) = \exp(\lim_{x \to \infty} \frac{x}{x+1}) = \exp(1) = e.$

7. $\lim_{x \to \infty} x^{\frac{1}{x}} = \exp(\lim_{x \to \infty} \frac{1}{x} \ln(x)) = \exp(\lim_{x \to \infty} \frac{\frac{1}{x}}{1}) = \exp(0) = 1.$

10. $\lim_{x \to \infty} \frac{x + \cos(x)}{x} = \lim_{x \to \infty} (1 + \frac{\cos(x)}{x}).$

Since $-1 \leq \cos x \leq 1$ and $x > 0$, we have
$-\frac{1}{x} \leq \frac{\cos(x)}{x} \leq \frac{1}{x}.$

Adding 1 throughout the inequality gives us $1 + -\frac{1}{x} \leq 1 + \frac{\cos(x)}{x} \leq 1 + \frac{1}{x}.$
Now, $\lim_{x \to \infty} (1 + \frac{-1}{x}) = 1 = \lim_{x \to \infty} (1 + \frac{1}{x})$, so
$\lim_{x \to \infty} (1 + \frac{\cos(x)}{x}) = 1$ by the squeeze principle.

13. $\lim_{x \to \infty} \left(\sqrt{x + \sqrt{x}} - \sqrt{x} \right) = \lim_{x \to \infty} \left(\sqrt{x + \sqrt{x}} - \sqrt{x} \right) \left(\frac{\sqrt{x + \sqrt{x}} + \sqrt{x}}{\sqrt{x + \sqrt{x}} + \sqrt{x}} \right)$

$= \lim_{x \to \infty} \frac{x + \sqrt{x} - x}{\sqrt{x + \sqrt{x}} + \sqrt{x}} = \lim_{x \to \infty} \frac{\sqrt{x}}{\sqrt{x + \sqrt{x}} + \sqrt{x}} = \lim_{x \to \infty} \frac{\sqrt{x}}{\sqrt{x\left(1 + \frac{1}{\sqrt{x}}\right)} + \sqrt{x}}$

$= \lim_{x \to \infty} \frac{\sqrt{x}}{\sqrt{x}(\sqrt{1 + \frac{1}{\sqrt{x}}} + 1)} = \lim_{x \to \infty} \frac{1}{\sqrt{1 + \frac{1}{\sqrt{x}}} + 1} = \frac{1}{2}.$

16. $\lim_{x \to 0} \frac{a}{x} \ln(1 + x) = \lim_{x \to 0} \frac{a \ln(1+x)}{x} = \lim_{x \to 0} \frac{a}{1+x} = a.$

$\lim_{x \to 0} (1 + x)^{a/x} = e^{\lim_{x \to 0} a/x \ln(1 + x)} = e^a.$

19. $\lim_{x \to \infty} \frac{e^x}{x^n} = \cdots = \lim_{x \to \infty} \frac{e^x}{n!} = \infty.$

22. $\lim_{x \to \infty} \frac{\ln(x)}{x^{1/n}} = 0.$

Section 9.6

The cubic splines determined by the data in exercises 1-4 are shown below.

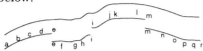

6. As can be seen below, none of the points f, g, or h can be ommitted without noticeably altering the curve, although leaving out g gives a "better" curve.

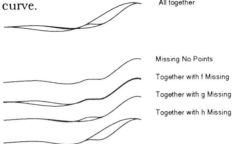

9. No, the curve and the spline approximation match very well.

12. In this region the spline approximation tracks the curve very well and there is no suggestion of other control points.

15. A cubic is determined by its value, together with the values of its first, second, and third derivatives, at a point. Since the point b together with the slope at b determine the value and first derivative at b for both the curve from a to b and the curve from b to c, these will automatically agree at b. For b to be superfluous, it must be that the curve from a to b and that from b to c must be part of one and the same cubic curve so the second and third derivatives at b of the curve from a to b must agree with the second and third derivatives at b of the curve from b to c, since they would be derivatives of one and the same cubic.

18.

$(1.25000, 9.23584)$ $(1.50000, 9.87109)$ $(1.75000, 10.14736)$

$(2.00000, 10.10625)$ $(2.25000, 9.78936)$ $(2.50000, 9.23828)$

$(2.75000, 8.49463)$ $(3.00000, 7.60000)$ $(3.25000, 6.59600)$

$(3.50000, 5.52422)$ $(3.75000, 4.42627)$ $(4.00000, 3.34375)$

$(4.25000, 2.31826)$ $(4.50000, 1.39141)$ $(4.75000, 0.60479)$

18. continued

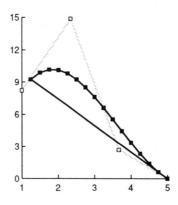

21.

$(2.18750, 2.28516)$ $(2.37500, 2.40625)$ $(2.56250, 2.38672)$

$(2.75000, 2.25000)$ $(2.93750, 2.01953)$ $(3.12500, 1.71875)$

$(3.31250, 1.37109)$ $(3.50000, 1.00000)$ $(3.68750, 0.62891)$

$(3.87500, 0.28125)$ $(4.06250, -0.01953)$ $(4.25000, -0.25000)$

$(4.43750, -0.38672)$ $(4.62500, -0.40625)$ $(4.81250, -0.28516)$

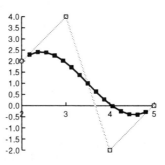

24. The instantaneous rate of change of y_n is 0.25; the instantaneous rate of change of x_n is 0; the direction is North.

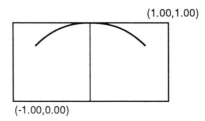

26. For $f(x) = \sin(x)$ $x_0 = 0$, $x_1 = \pi/2$, we have
spline approximation: $-0.110739816362x^3 - 0.05738534102x^2 + x$
Taylor approximation: x

Graphs of approximations: Graphs of errors

At $4\pi/9$ the spline approximation is better, and at $\pi/9$ the Taylor polynomial approximation is better.

29. For $f(x) = x^4$ $x_0 = 0$, $x_1 = 1$, we have
spline approximation: $2x^3 - x^2$ Taylor approximation: 0
Graphs of approximations: Graphs of errors

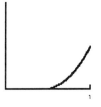
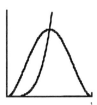

At 0.9 the spline is better, and at 0.1, the Taylor polynomial is better.

32. For

$$f(x) = \begin{cases} x & \text{for } x \leq 1 \\ 2-x & \text{otherwise} \end{cases} \quad x_0 = 0, \, x_1 = 2$$

spline approximation: $0.5x(2-x)$;   Taylor approximation: x.

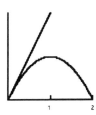

At 0.5 the spline is better, and at 1.8, the Taylor polynomial is better.

ANSWERS TO SELECTED EXERCISES FROM CHAPTER 10

Section 10.1

1. $\{3 - 5n\}_{n=0}^{\infty}$.

4. $\{0 + \pi n\}_{n=0}^{\infty}$.

7. $\{\pi(0)^n\}_{n=0}^{\infty}$.

10. False. A counterexample is the alternating harmonic sequence
$-1, \frac{1}{2}, -\frac{1}{3}, \frac{1}{4}, \ldots$ which converges to the limit $L = 0$.

13. $\frac{1}{2}, 2, \frac{1}{2}, 2, \frac{1}{2}, \ldots$. The sequence is bounded, non-monotonic, and divergent.

16. The sequence is bounded, non-monotonic, and converges to 2.

19. $a_{10} = \frac{11}{10} = 1.1$ is exactly 0.1 more than the limit. To be within 0.1 more than the limit look at $a_{12} = \frac{13}{12} = 1.08\overline{3}$. You may use trial and error method.

22. $\left\{\frac{(-1)^n}{n}\right\}_{n=1}^{\infty}$, $\left\{\left(\frac{-1}{2}\right)^n\right\}_{n=0}^{\infty}$, $\left\{\frac{\sin(n)}{n}\right\}_{n=1}^{\infty}$.

25. $\{(-1)^n\}_{n=1}^{\infty}$, $\{\cos(\frac{n\pi}{4})\}_{n=1}^{\infty}$, $\{\sin(n)\}_{n=1}^{\infty}$.

28. $L = 0$. $N = 9$ by trial and error.

The sequence is monotonic decreasing with
$a_9 = 0.001953125$ and $a_{10} = 0.0009765625$.

Section 10.2

1. Geometric series with $a = -\frac{1}{3}$, $r = -\frac{1}{3}$.
$$\sum_{n=1}^{\infty}\left(-\frac{1}{3}\right)^n = -\frac{1}{3}\sum_{n=0}^{\infty}\left(-\frac{1}{3}\right)^n = \frac{-\frac{1}{3}}{1-(-\frac{1}{3})} = \frac{-\frac{1}{3}}{\frac{4}{3}} = -\frac{1}{4}.$$

4. $\sum_{n=0}^{\infty} \frac{3^n + 2^n}{4^n} = \sum_{n=0}^{\infty}\left(\frac{3}{4}\right)^n + \sum_{n=0}^{\infty}\left(\frac{2}{4}\right)^n = \frac{1}{1-\frac{3}{4}} + \frac{1}{1-\frac{2}{4}} = 6.$

7. $a = \frac{1}{11}$, $r = -\frac{10}{11}$, $\sum_{n=0}^{\infty} \frac{(-10)^n}{11^{n+1}} = \frac{\frac{1}{11}}{1-(-\frac{10}{11})} = \frac{1}{21}$.

10. The series diverges since it's partial sums are twice that of the harmonic series which diverges.

13. $\sum_{n=3}^{N}\left(\frac{1}{n-1}-\frac{1}{n+1}\right)$

$=\frac{1}{2}-\frac{1}{4}+\frac{1}{3}-\frac{1}{5}+\frac{1}{4}-\frac{1}{6}+\cdots+\frac{1}{N-2}-\frac{1}{N}+\frac{1}{N-1}-\frac{1}{N+1}=\frac{1}{2}+\frac{1}{3}-\frac{1}{N}-\frac{1}{N+1}.$

$\sum_{n=3}^{\infty}\left(\frac{1}{n-1}-\frac{1}{n+1}\right)=\lim_{N\to\infty}\sum_{n=3}^{N}\left(\frac{1}{n-1}-\frac{1}{n+1}\right)$

$=\lim_{N\to\infty}\left(\frac{1}{2}+\frac{1}{3}-\frac{1}{N}-\frac{1}{N+1}\right)=\frac{1}{2}+\frac{1}{3}=\frac{5}{6}.$

16. $\sum_{n=0}^{\infty}(\sin(\pi n))=0+0+\cdots=0.$

19.

$S_0 = 1$ $S_5 = 2.71\overline{6}$

$S_1 = 2$ $S_6 = 2.7180\overline{5}$

$S_2 = 2.5$ $S_7 \approx 2.71825396826$

$S_3 = 2.\overline{6}$ $S_8 \approx 2.71827876985$

$S_4 = 2.708\overline{3}$ $S_9 \approx 2.71828152558.$

The limiting value is e.

22. $C_{40} \approx 0.58966358483,$ $C_{50} \approx 0.58718233289,$ $C_{60} \approx 0.58552585073.$

23. The sum of the lengths of the erased intervals is

$\frac{1}{3}+2\left(\frac{1}{9}\right)+4\left(\frac{1}{27}\right)+8\left(\frac{1}{81}\right)+\cdots=\sum_{n=0}^{\infty}\frac{2^n}{3^{n+1}}=\frac{\frac{1}{3}}{1-\frac{2}{3}}=1.$

Section 10.3

1. $\lim_{n\to\infty}\ln\left(1+\frac{1}{n}\right)=0$ and $\ln\left(1+\frac{1}{n}\right)>\ln(1+\frac{1}{n+1})$ for each n so the series converges.

4. $\lim_{n\to\infty}\frac{n}{2n^2-1}=0$ and $\frac{n}{2n^2-1}>\frac{n+1}{2(n+1)^2-1}$ for each n. So the series converges.

5. (1) Find N so that $\ln\left(1+\frac{1}{N}\right)<0.00005.$ $N = 20000.$

$S_{19999}=\sum_{n=1}^{19999}(-1)^n\ln\left(1+\frac{1}{n}\right)\approx -0.4516.$

8. $\frac{\ln(x)}{x} \geq 0$ for $x \geq 1$ and $f(x) = \frac{\ln(x)}{x}$ is a continuous, decreasing function for $x \geq 1$. Hence, the integral test applies.

$\int_1^{\infty}\frac{\ln(x)}{x}\,dx=\lim_{b\to\infty}\frac{1}{2}(\ln(b))^2-\frac{1}{2}(\ln(1))^2=\infty.$

The improper integral diverges so the series diverges.

10. (6) Find N so that the error $\leq \int_N^{N+1} 3e^{-x}\, dx \leq 0.00005$.

$$\int_N^{N+1} 3e^{-x}\, dx = -3e^{-x}\Big]_N^{N+1} = -3e^{-(N+1)} + 3e^{-N}.$$

When $N = 11$, we have $-3e^{-(N+1)} + 3e^{-N} \leq 0.00005$.

$$\sum_{n=1}^{11} 3e^{-n} \approx 1.7459 \text{ so } \sum_{n=1}^{\infty} 3e^{-n} \approx 1.7459.$$

13. Choose N so that the error, $\int_N^{N+1} \frac{1}{x^6}\, dx$ is less than 2×10^{-11}.

$$\int_N^{N+1} \frac{1}{x^6}\, dx = \frac{1}{-5x^5}\Big]_N^{N+1} = \frac{1}{-5(N+1)^5} - \frac{1}{-5N^5}.$$

When $N = 61$, the error $\approx 1.85 \times 10^{-11} < 2 \times 10^{-11}$.

$$S_{61} - \text{error} \leq \sum_{n=1}^{\infty} \frac{1}{n^6} \leq S_{61} + \text{error}.$$

$$1.01734306175 \leq \sum_{n=1}^{\infty} \frac{1}{n^6} \leq 1.01734306179.$$

$S_{61} \approx 1.017$.

15. Since $f(x) = \ln\left(1 + \frac{1}{x}\right)$ is positive, decreasing and continuous for $x \geq 1$, the integral test applies. The following computation uses integration by parts:

$$\int \ln\left(1 + \tfrac{1}{x}\right) dx = \left[\ln\left(1 + \tfrac{1}{x}\right)\right] x - \int x \left(\tfrac{1}{1+\frac{1}{x}}\right)\left(-\tfrac{1}{x^2}\right) dx$$

$$= x \ln\left(1 + \tfrac{1}{x}\right) + \int \tfrac{1}{x+1}\, dx$$

$$= x \ln\left(1 + \tfrac{1}{x}\right) + \ln|x+1| + C.$$

We now have

$$\int_1^{\infty} \ln\left(1 + \tfrac{1}{x}\right) dx = \lim_{b \to \infty} \tfrac{\ln(1+\frac{1}{b})}{\frac{1}{b}} - \ln(2) + (\ln|b+1| - \ln(2))$$

$$= 1 - \ln(2) + \infty - \ln(2) = \infty.$$

The improper integral diverges so the series must diverge.

18. $\sum_{n=1}^{\infty} \frac{1}{n}$ diverges, but

$\sum_{n=1}^{\infty} \frac{(-1)^{n+1}}{n}$ converges to $\ln(2)$.

Section 10.4

1. $\lim_{n\to\infty} \left| \frac{e^{n+1}}{e^{2(n+1)}+1} \cdot \frac{e^{2n}+1}{e^n} \right| = \lim_{n\to\infty} |e| \left| \frac{e^{2n}+1}{e^2 e^{2n}+1} \right|$ (Use L'Hôpital's Rule.)

$= \lim_{n\to\infty} |e| \left| \frac{2e^{2n}}{e^2 2e^{2n}} \right| = \frac{e}{e^2} < 1$ so the series converges.

4. $\lim_{n\to\infty} \left| \frac{(n+1)!(n+1)^2}{(2(n+1))!} \cdot \frac{(2n)!}{n!n^2} \right| = \lim_{n\to\infty} \left| \frac{(n+1)}{(2n+2)(2n+1)} \cdot \left(\frac{n+1}{n}\right)^2 \right| = 0 < 1$, so the series converges.

7. $\lim_{n\to\infty} \left| \left(\frac{n}{2n^2-1}\right) \right|^{1/n} = \exp\left(\lim_{n\to\infty} \left| \frac{\ln(\frac{n}{2n^2-1})}{n} \right| \right)$ (Use L'Hôpital's Rule.)

$= \exp\left(\lim_{n\to\infty} \left| \frac{2n^2-1}{n} \cdot \frac{(2n^2-1)-n(4n)}{(2n^2-1)^2} \right| \right) = \exp\left(\lim_{n\to\infty} \left| \frac{-(2n^2+1)}{n(2n^2-1)} \right| \right)$

$= \exp(0) = 1$, so the root test provides no information.

10. $\lim_{n\to\infty} \left| \frac{\pi^{1/(n+1)}}{n+1} \cdot \frac{n}{\pi^{1/n}} \right| = \lim_{n\to\infty} \left| \frac{n}{n+1} \cdot \pi^{1/(n+1)-1/n} \right| = 1 \cdot \left(\lim_{n\to\infty} \pi^{1/(n(n+1))} \right) = \pi^0 = 1$, so the ratio test is inconclusive.

$\lim_{n\to\infty} \frac{\pi^{1/n}}{n} = 0$, so the alternating series converges.

$\sum_{n=1}^{\infty} \left| \frac{(-1)^{n+1}\pi^{1/n}}{n} \right| = \sum_{n=1}^{\infty} \frac{\pi^{1/n}}{n}$ diverges by limit comparison with $\sum_{n=1}^{\infty} \frac{1}{n}$, using that

$\lim_{n\to\infty} \frac{\pi^{1/n}/n}{1/n} = \lim_{n\to\infty} \pi^{1/n} = \pi^0 = 1 \neq 0.$

So, the series converges conditionally.

13. $\sum_{n=1}^{\infty} |(-1)^{n+1} 3e^{-n}| = \sum_{n=1}^{\infty} 3e^{-n}$ converges since it is a geometric series with $r = \frac{1}{e} < 1$. So, the alternating series converges absolutely.

16. $\lim_{n\to\infty} \frac{e^{1/n}/n}{1/n} = \lim_{n\to\infty} e^{1/n} = 1 \neq 0.$

So the two series have the same behavior.

We know $\sum_{n=1}^{\infty} \frac{1}{n}$ diverges, so $\sum_{n=1}^{\infty} \frac{e^{1/n}}{n}$ diverges also.

19. $\sum_{n=1}^{\infty} \frac{1}{n+1000000}$ diverges by limit comparison test with $\sum_{n=1}^{\infty} \frac{1}{n}$.

22. $\lim_{n\to\infty} \frac{1}{2^{1/n}} = 1 \neq 0$ so $\sum_{n=1}^{\infty} \frac{1}{2^{1/n}}$ diverges by the nth term test.

25. $\frac{3^n}{n!} \leq \left(\frac{3}{4}\right)^n$ for $n \geq 9$. Since $\sum_{n=1}^{\infty} \left(\frac{3}{4}\right)^n$ is a convergent geometric series, $\sum_{n=1}^{\infty} \frac{3^n}{n!}$ converges by the comparison test.

26. $\sum_{n=1}^{\infty} \frac{n \ln(n)}{n^2+1}$ diverges by the limit comparison with $\sum_{n=1}^{\infty} \frac{1}{n}$.

$\frac{n \ln(n)}{n^2+1}$ is positive for $n \geq 1$ so the limit comparison test applies.

Note that $\lim_{n \to \infty} \frac{\frac{n \ln(n)}{n^2+1}}{\frac{1}{n}} = \lim_{n \to \infty} \frac{n^2}{n^2+1} \cdot \lim_{n \to \infty} \ln(n) = 1 \cdot \infty = \infty$ and $\sum_{n=1}^{\infty} \frac{1}{n}$ diverges, so $\sum_{n=1}^{\infty} \frac{n \ln(n)}{n^2+1}$ diverges.

28. $\sum_{n=1}^{\infty} \frac{(-1)^n}{\sqrt{n}}$ converges conditionally, since

$\sum_{n=1}^{\infty} \left|\frac{(-1)^n}{\sqrt{n}}\right|$ diverges, but

$\sum_{n=1}^{\infty} \frac{(-1)^n}{\sqrt{n}}$ converges by the alternating series test.

31. The partial sums oscillate between 1 and 0. Hence, the series diverges.

34. $\lim_{n \to \infty} \left(\frac{n-1}{n}\right)^n = e^{-1} \neq 0$ so the series diverges by the nth term test.

37. $\frac{5^n}{n!} \leq \left(\frac{5}{6}\right)^n$ for $n \geq 14$.

Hence, $\sum_{n=1}^{\infty} \left|\frac{(-5)^n}{n!}\right|$ converges by comparison with $\sum_{n=1}^{\infty} \left(\frac{5}{6}\right)^n$.

$\sum_{n=1}^{\infty} \frac{(-5)^n}{n!}$ converges absolutely.

Section 10.5

1. $\lim_{n \to \infty} \left|\frac{2^{n+1} x^{n+1}}{(n+1)!} \cdot \frac{n!}{2^n \cdot x^n}\right| = \lim_{n \to \infty} \left|\frac{2x}{n+1}\right| = 0$ for all x.

Hence, the radius of convergence is ∞ and the interval of convergence is the entire real line, $(-\infty, \infty)$.

4. $g(x) = f(-x) = \frac{1}{1-(-x)} = \sum_{n=0}^{\infty} (-x)^n = 1 - x + x^2 - x^3 + \ldots$.

The interval of convergence is $-1 < -x < 1$ or $1 > x > -1$.

7. $\ln(2) \approx \sum_{n=0}^{100} \frac{(-1)^n}{n+1} \approx 0.698073.$

$|R_{100}| < \frac{1}{102} \approx 0.0098$ is the maximum error in this result. (Note that $\ln(2) \approx 0.693147$.)

10. $1 - e^{-x} = 1 - \sum_{n=0}^{\infty} \frac{(-x)^n}{n!} = x - \frac{x^2}{2} + \frac{x^3}{6} - \frac{x^4}{24} + \cdots.$

13. S_N with $x = \frac{1}{2}$ approximates the integral to within 1×10^{-10} when $\frac{(\frac{1}{2})^{N+1}}{(N+1)(N+1)!} < 1 \times 10^{-10}.$

When $N = 9$, we have $\frac{(\frac{1}{2})^{10}}{10 \cdot 10!} \approx 2.6911 \times 10^{-11} < 1 \times 10^{-10}.$

S_9 with $x = \frac{1}{2}$ is approximately 0.443842079143. This approximates the value of the integral to within 1×10^{-10}.

16. $\lim_{x \to 0} \frac{x - \sin x}{x^3 \cos x} = \lim_{x \to 0} \frac{\frac{x^3}{3!} - \frac{x^5}{5!} + \frac{x^7}{7!} - \cdots}{x^3 - \frac{x^5}{2!} + \frac{x^7}{4!} - \cdots}$

$= \lim_{x \to 0} \frac{\frac{1}{6} - \frac{x^2}{5!} + \frac{x^4}{7!} - \cdots}{1 - \frac{x^2}{2} + \frac{x^4}{4!} - \cdots} = \frac{1}{6}.$

Section 10.6

1. $a_0 = 3,$ $\quad a_n = a_{n-1} - 5$ for $n = 1, 2, 3, \ldots.$

4. $a_0 = 0,$ $\quad a_n = a_{n-1} + \pi$ for $n = 1, 2, 3, \ldots.$

7. $a_0 = \pi,$ $\quad a_n = 2 \cdot a_{n-1}$ for $n = 1, 2, 3, \ldots.$

10. $a_1 = 1, \quad a_2 = \frac{1}{2}, \quad a_3 = \frac{1}{1 + \frac{1}{2}} = \frac{2}{3},$

$a_4 = \frac{1}{1 + \frac{2}{3}} = \frac{3}{5}, \quad a_5 = \frac{1}{1 + \frac{3}{5}} = \frac{5}{8}.$

14. $a_0 = 10, \quad a_1 = \frac{1}{10}, \quad a_2 = 10, \quad a_3 = \frac{1}{10}, \quad a_4 = 10.$

15. The denominators form the Fibonacci sequence. The numerators form the Fibonacci sequence starting with the second term. The Fibonacci sequence is defined recursively by $a_0 = 1, \ a_1 = 1, \ a_n = a_{n-1} + a_{n-2}$ for $n = 2, 3, 4, \ldots.$

18. $\frac{1597}{987} \approx 1.61803444782;$ $\quad \frac{2584}{1597} \approx 1.6180338134.$

20. $\frac{a_{n+1}}{a_n} = \dfrac{\frac{1}{\sqrt{5}}\left[\left(\frac{1+\sqrt{5}}{2}\right)^{n+1} - \left(\frac{1-\sqrt{5}}{2}\right)^{n+1}\right]}{\frac{1}{\sqrt{5}}\left[\left(\frac{1+\sqrt{5}}{2}\right)^{n} - \left(\frac{1-\sqrt{5}}{2}\right)^{n}\right]}$ for $n = 1, 2, 3, \ldots.$

23. $f(x) = \sin(x) - 0.7$; $f'(x) = \cos(x)$.

$g(x) = x - \frac{\sin(x)-0.7}{\cos(x)}$ is the iterating function.

$x_0 = 0.7$, $\quad x_1 \approx 0.773$, $\quad x_2 \approx 0.775$, $\quad x_3 \approx 0.775$.

26. $f(x) = x\exp(x^2) - \sqrt{x^2+1}$; $f'(x) = 2x^2 e^{x^2} + e^{x^2} - \frac{x}{\sqrt{x^2+1}}$.

$g(x) = x - \frac{xe^{x^2} - \sqrt{x^2+1}}{2x^2 e^{x^2} + e^{x^2} - \frac{x}{\sqrt{x^2+1}}}$ is the iterating function.

$x_0 = 0.4$, $\quad x_1 \approx 0.916$, $\quad x_2 \approx 0.778$, $\quad x_3 \approx 0.732$.

29. $f(x) = 3\cos(\cos(x)) - 2x$; $f'(x) = 3\sin(\cos(x))\sin(x) - 2$.

$g(x) = x - \frac{3\cos(\cos(x)) - 2x}{3\sin(\cos(x))\sin(x) - 2}$ is the iterating function.

$x_0 = 1.1$, $\quad x_1 \approx 1.699$, $\quad x_2 \approx 1.522$, $\quad x_3 \approx 1.496$.

ANSWERS TO SELECTED EXERCISES FROM CHAPTER 11

Section 11.1

1.

vector	initial point	terminal point
n	$(-1, 1)$	$(-1, -1)$
p	$(-1, 1)$	$(-1, 4)$
q	$(-1, 1)$	$(2, 1)$
u	$(2, 4)$	$(2, 1)$
v	$(2, 1)$	$(5, 1)$
w	$(-1, 1)$	$(5, -1)$

4. **q** and **v** are equal.

7. $A = 3(2) = 6$.

10. $(9, -6)$.

13. $\overrightarrow{PQ} = \langle -2.5, 3.4, 4.1 \rangle$.

16.
    (11) $S = (-4, -3, 4)$ (12) $S = (9, 0, 3)$
    (13) $S = (-3.5, 5.4, 7.1)$ (14) $S = (-1, 2, 3)$
    (15) $S = (2, 7, 2)$.

19. $\langle -2\sqrt{3}, 2 \rangle$.   **22.** $\langle 0, 84 \rangle$.

25. In addition to the North Pole, if you start 1 mile north of any circle of latitude of circumference $1/n$ miles (n a positive integer), then moving 1 mile south places you on this circle, moving 1 mile east brings you around the circles a whole number of times, and moving 1 mile north brings you back to your starting point.

Section 11.2

1. $\mathbf{p} + \mathbf{q} + \mathbf{u} = \mathbf{q}$.   **4.** $\mathbf{n} - \mathbf{q} = \mathbf{m}$.

7. Listing clockwise: $\mathbf{w} - \mathbf{v}, -\mathbf{v}, -\mathbf{w}, \mathbf{v} - \mathbf{w}$.

10. $-3\mathbf{w} = \langle -8.16, -9.42 \rangle$.   **13.** $\mathbf{w} - \mathbf{v} = \langle 6.72, 6.14 \rangle$.

16. $\mathbf{v} + (0.5)\langle 1, -1 \rangle = \langle -3.5, -3.5 \rangle$.   **19.** $\mathbf{v} + \mathbf{w} = \langle 2.5, -2, -6 \rangle$.

22. $2\mathbf{v} - 3\mathbf{w} = \langle 2.5, -9, 23 \rangle$.

25. (3) (using components) Let $\mathbf{u} = \langle u_1, u_2 \rangle$ be any vector in $\mathbb{R}^2$. Show $\mathbf{u} + \mathbf{0} = \mathbf{0} + \mathbf{u} = \mathbf{u}$.

$$\mathbf{u} + \mathbf{0} = \mathbf{0} + \mathbf{u} \text{ (by commutative law for addition property)}$$
$$= \langle 0, 0 \rangle + \langle u_1, u_2 \rangle$$
$$= \langle 0 + u_1, 0 + u_2 \rangle$$
$$= \langle u_1, u_2 \rangle$$
$$= \mathbf{u}.$$

(geometric illustration) If we represent $\mathbf{u}$ as a directed line segment, then the tail of $\mathbf{0}$ attaches to the head of $\mathbf{u}$. Since the tail of $\mathbf{0}$ coincides with the head of $\mathbf{0}$, the resultant vector is simply the original vector $\mathbf{u}$.

(4) (using components) Let $\mathbf{u} = \langle u_1, u_2 \rangle$ be any vector in $\mathbb{R}^2$. Show $\mathbf{u} + (-\mathbf{u}) = \mathbf{0}$.

$$\mathbf{u} + (-\mathbf{u}) = \langle u_1, u_2 \rangle + \langle -u_1, -u_2 \rangle$$
$$= \langle u_1 - u_1, u_2 - u_2 \rangle$$
$$= \langle 0, 0 \rangle$$
$$= \mathbf{0}.$$

(geometric illustration) If we represent $\mathbf{u}$ and $-\mathbf{u}$ as directed line segments, then the tail of $-\mathbf{u}$ attaches to the head of $\mathbf{u}$ and the head of $-\mathbf{u}$ then is located at the tail of $\mathbf{u}$. The resultant is the $\mathbf{0}$ vector.

(7) (using components) Let $\mathbf{u} = \langle u_1, u_2 \rangle$ be any vector in $\mathbb{R}^2$, and a and b any scalars. Show $(a + b)\mathbf{u} = a\mathbf{u} + b\mathbf{u}$.

$$(a + b)\mathbf{u} = (a + b)\langle u_1, u_2 \rangle$$
$$= \langle (a + b)u_1, (a + b)u_2 \rangle$$
$$= \langle au_1 + bu_1, au_2 + bu_2 \rangle$$
$$= \langle au_1, au_2 \rangle + \langle bu_1, bu_2 \rangle$$
$$= a\langle u_1, u_2 \rangle + b\langle u_1, u_2 \rangle$$
$$= a\mathbf{u} + b\mathbf{u}.$$

(continued on next page)

25. ((7) continued from previous page)

(geometric illustration) If we represent all vectors as directed line segments, $(a+b)\mathbf{u}$ is the vector with length $|a+b|$ times the length of $\mathbf{u}$ and the same direction as $\mathbf{u}$ if $a+b > 0$, opposite direction if $a+b < 0$. On the other hand, $a\mathbf{u}$ has length $|a|$ times the length of $\mathbf{u}$ while $b\mathbf{u}$ has length $|b|$ times the length of $\mathbf{u}$. If a and b are both positive then placing the tail of $b\mathbf{u}$ on the head of $a\mathbf{u}$ results in the new vector $a\mathbf{u}+b\mathbf{u}$ with the same direction as $\mathbf{u}$ but the length $a+b$ times the length of $\mathbf{u}$. If a and b are both negative, the result is similar, but the vector has opposite direction. So we still have $a\mathbf{u}+b\mathbf{u} = (a+b)\mathbf{u}$, both vectors of length $|a+b|$ but direction opposite that of $\mathbf{u}$. If one of a, b is negative, and the other positive, say $a > 0$, $b < 0$ (the reverse argument is identical) then placing $b\mathbf{u}$ on the head of $a\mathbf{u}$ will result in a vector having the same direction as $\mathbf{u}$ if $|b| < |a|$ but opposite direction if $|b| > |a|$. Thus, in either case we still have $(a+b)\mathbf{u} = a\mathbf{u}+b\mathbf{u}$.

(10) (using components) Let $\mathbf{u} = \langle u_1, u_2 \rangle$ be any vector in $\mathbb{R}^2$. Show $-1\mathbf{u} = -\mathbf{u}$.

$$-1\mathbf{u} = -1\langle u_1, u_2 \rangle$$
$$= \langle -u_1, -u_2 \rangle$$
$$= -\mathbf{u}.$$

(geometric illustration) By definition, $-\mathbf{u}$ is the vector having the same length as $\mathbf{u}$, but opposite direction. On the other hand, $(-1)\mathbf{u}$ is a vector of the form $a\mathbf{u}$. Thus, its length is $|a|$ times the length of $\mathbf{u}$, so equals the length of $\mathbf{u}$ in this case, and since $a = -1 < 0$, it has direction opposite that of $\mathbf{u}$. Thus, $(-1)\mathbf{u}$ is identical to $-\mathbf{u}$.

Section 11.3

1. $\mathbf{p} \cdot \mathbf{n} = -6$.

4. $\mathbf{w} \cdot \mathbf{n} = 4$.

7. $\mathbf{v} \cdot \mathbf{w} = \frac{1}{2}$.

11. $\|\mathbf{v}\| = 5$.

14. $\|\mathbf{v}\| + \|\mathbf{w}\| \approx 9.15428$.

17. $\langle -\frac{4}{5}, -\frac{3}{5} \rangle$.

20. $\mathbf{v} \cdot \mathbf{u} \approx -4.88653$ where $\mathbf{u}$ is the unit vector in the direction $\mathbf{w}$.

23. $\mathbf{v} \cdot \mathbf{w} = -9$.

26. $\|\mathbf{w}\| \approx 7.08872$.

29. $\|\mathbf{v} - \mathbf{w}\| \approx 9.06918$.

32. $\approx \langle 0.80160, 0.58116, 0.14028 \rangle$.

35. $(\mathbf{v} \cdot \mathbf{u})\mathbf{u} \approx \langle -8.95522, -0.17910, 1.25373 \rangle$.

38. Yes, $\|\mathbf{v} - \mathbf{w}\| = 0$ implies $\sqrt{(v_1 - w_1)^2 + (v_2 - w_2)^2} = 0$ which implies $(v_1 - w_1)^2 + (v_2 - w_2)^2 = 0$. The sum of the non-negative terms equals 0, so each term must equal 0. Thus, $v_1 = w_1$ and $v_2 = w_2$.

41. Assuming θ is expressed such that $0 \leq \theta < 2\pi$, we have

$$(r, \theta) = (\|\mathbf{v}\|, \arccos(\mathbf{v} \cdot \mathbf{i}/\|\mathbf{v}\|)), \qquad \text{if } 0 < \theta \leq \pi,$$

$$(r, \theta) = (\|\mathbf{v}\|, 2\pi - \arccos(\mathbf{v} \cdot \mathbf{i}/\|\mathbf{v}\|)), \qquad \text{if } \pi < \theta < 2\pi.$$

44. Let $\mathbf{u} = \langle u_1, u_2 \rangle$ be any vector in $\mathbb{R}^2$ and let α and β be the direction angles.

$$\cos^2 \alpha + \cos^2 \beta = \left(\frac{u_1}{\|\mathbf{u}\|}\right)^2 + \left(\frac{u_2}{\|\mathbf{u}\|}\right)^2$$

$$= \frac{u_1^2 + u_2^2}{\|\mathbf{u}\|^2}$$

$$= \frac{\|\mathbf{u}\|^2}{\|\mathbf{u}\|^2} = 1.$$

47. Let $\mathbf{u} = \langle u_1, u_2, u_3 \rangle$ be any vector in $\mathbb{R}^3$ and let α, β, and γ be the direction angles.

$$\cos^2 \alpha + \cos^2 \beta + \cos^2 \gamma = \left(\frac{u_1}{\|\mathbf{u}\|}\right)^2 + \left(\frac{u_2}{\|\mathbf{u}\|}\right)^2 + \left(\frac{u_3}{\|\mathbf{u}\|}\right)^2$$

$$= \frac{u_1^2 + u_2^2 + u_3^2}{\|\mathbf{u}\|^2}$$

$$= \frac{\|\mathbf{u}\|^2}{\|\mathbf{u}\|^2} = 1.$$

50. Let $\mathbf{v} = \langle 2, 2, 2 \rangle$ and $\mathbf{w} = \langle -1, -1, -1 \rangle$. Then $\mathbf{v} - \mathbf{w} = \langle 3, 3, 3 \rangle$ and $\|\mathbf{v} - \mathbf{w}\| = \sqrt{27}$ yet $\|\mathbf{v}\| = \sqrt{12}$ and $\|\mathbf{w}\| = \sqrt{3}$.

Section 11.4

3. $X = \begin{bmatrix} 4 & 2 \\ 9 & 2 \\ -5 & -11 \end{bmatrix}$.

6. $S + T = \begin{bmatrix} 2 & -1 \\ -3 & 4 \\ 0 & 1 \end{bmatrix} + \begin{bmatrix} 3 & -2 \\ 7 & 0 \\ 1 & -3 \end{bmatrix}$

$= \begin{bmatrix} 2+3 & -1+(-2) \\ -3+7 & 4+0 \\ 0+1 & 1+(-3) \end{bmatrix}$

$= \begin{bmatrix} 3+2 & (-2)+(-1) \\ 7+(-3) & 0+4 \\ 1+0 & (-3)+1 \end{bmatrix}$

$= \begin{bmatrix} 3 & -2 \\ 7 & 0 \\ 1 & -3 \end{bmatrix} + \begin{bmatrix} 2 & -1 \\ -3 & 4 \\ 0 & 1 \end{bmatrix}$

$= T + S.$

9. Distributive property of scalar multiplication over matrix addition.

12. $ET = \begin{bmatrix} -3 & 9 \\ 62 & -11 \\ 23 & 1 \end{bmatrix}; \quad SF = \begin{bmatrix} -3 & 9 \\ 62 & -11 \\ 23 & 1 \end{bmatrix}$, associativity for multiplication.

15. $\det(E) = 0.$

18. $\det(I_4) = 1; \quad \det(I_n) = 1$ for any n.

21. $\det(P) = -11.$

24. $\det(P) = 0.$

27. $\det(P) = 0.$

30. $\det(P) = 0.$

33. Let A be the original matrix.

Let $B = \begin{bmatrix} b_1 & b_2 & b_3 \\ a_1 & a_2 & a_3 \\ c_1 & c_2 & c_3 \end{bmatrix}$

$\det(B) = b_1(a_2 c_3 - a_3 c_2) - b_2(a_1 c_3 - a_3 c_1) + b_3(a_1 c_2 - c_1 a_2)$

$= b_1 a_2 c_3 - b_1 a_3 c_2 - b_2 a_1 c_3 + a_3 b_2 c_1 + b_3 a_1 c_2 - b_3 c_1 a_2$

$= -a_1(b_2 c_3 - b_3 c_2) + a_2(b_1 c_3 - b_3 c_1) - a_3(b_1 c_2 - b_2 c_1)$

$= -\det(A).$ Similar results occur if we interchange any 2 rows.

36. Let $B = \begin{bmatrix} db_1 + ec_1 & db_2 + ec_2 & db_3 + ec_3 \\ b_1 & b_2 & b_3 \\ c_1 & c_2 & c_3 \end{bmatrix}$ where d and e are scalars.

$\det(B) = (db_1 + ec_1)(b_2c_3 - b_3c_2) - (db_2 + ec_2)(b_1c_3 - b_3c_1) + (db_3 + ec_3)(b_1c_2 - b_2c_1)$

$= db_1b_2c_3 + ec_1b_2c_3 - db_1b_3c_2 - ec_1b_3c_2 - db_2b_1c_3 + db_2b_3c_1 - ec_2b_1c_3$

$ + ec_2b_3c_1 + db_3b_1c_2 - db_3b_2c_1 + ec_3b_1c_2 - ec_3b_2c_1$

$= 0.$

Similar results occur when any row is a linear combination of the other two.

Section 11.5

3. $(\mathbf{b} - \mathbf{a}) \times (\mathbf{c} - \mathbf{a}) = -10\mathbf{i} - 5\mathbf{k}$.

6. $\|\langle 1, 4, 7 \rangle \times \langle -2, 1, -6 \rangle\| = \|\langle -31, -8, 9 \rangle\| = \sqrt{1106}$.

9. Area ≈ 8.66025.

12. $\langle -2, 1, -6 \rangle \times \langle 3, 0, -5 \rangle = \langle -5, -28, -3 \rangle$;

$\langle 1, 4, 7 \rangle \cdot \langle -5, -28, -3 \rangle = -138$;

volume $= |\langle 1, 4, 7 \rangle \cdot (\langle -2, 1, -6 \rangle \times \langle 3, 0, -5 \rangle)| = |-138| = 138$.

15. $\approx 70.89339°$.

18. True. **21.** False. **24.** True.

Section 11.6

1. $\langle x(t), y(t), z(t) \rangle = \langle 2 + t, -1 - 3t, 1 - 2t \rangle$, other answers are possible.

4. $\langle x(t), y(t), z(t) \rangle = \langle 1 + 3t, 4 + 3t, 7 + 13t \rangle$, other answers are possible.

7. $\langle x(t), y(t), z(t) \rangle = \langle 1 - 2t, 3 + t, 5 + 3t \rangle$.

10. $\mathbf{v} = \langle 2, -3, 1 \rangle$; $\langle x(t), y(t), z(t) \rangle = \langle 1 + 2t, 2 - 3t, 3 + t \rangle$.

13. The lines cross when $1 + t = 2 - s$, $-2t = 1 - s$ and $1 - t = s$.

The first equation requires $t = 1 - s$. The second equation requires $-2t = 1 - s$, so $t = -2t \Rightarrow t = 0$. Then $s = 1$. When $t = 0$ the coordinates on the first line are $(1, 0, 1)$ and when $s = 1$ the coordinates on the second line are $(1, 0, 1)$.

The point of intersection is $(1, 0, 1)$.

16. $\mathbf{u} = \langle 1, 3, 5 \rangle$; $\mathbf{v} = \langle 2, 3, 4 \rangle$; $\mathbf{n} = \mathbf{u} \times \mathbf{v} = \langle -3, 6, -3 \rangle$.

Using the point $(1, 1, 1)$, we have

$$-3x + 6y - 3z + d = 0$$

$$-3 + 6 - 3 + d = 0 \Rightarrow d = 0.$$

Equation is: $x - 2y + z = 0$.

19. $3x - 6y + z = 3(1+t) - 6(2-t) + 5 - \frac{9}{2}t = 3 - 12 + 5 + 3t + 6t - \frac{9}{2}t = 1$ when $t = \frac{10}{9}$ so the line and the plane intersect at the point $(\frac{19}{9}, \frac{8}{9}, 0)$.

$3x - 6y + z = 3(3t) - 6(-6t) + t = 9t + 36t + t = 46t = 1$ when $t = \frac{1}{46}$ so the line and the plane intersect at the point $(\frac{3}{46}, -\frac{6}{46}, \frac{1}{46})$. Notice this line is perpendicular to the plane.

22. A normal vector is $\mathbf{u} \times \mathbf{v}$ where $\mathbf{u}$ and $\mathbf{v}$ are as in # 21.

$\mathbf{u} = \langle 3, 0, -4 \rangle$; $\mathbf{v} = \langle -1, 5, 2 \rangle$; $\mathbf{n} = \mathbf{u} \times \mathbf{v} = \langle 20, -2, 15 \rangle$.

A point on the line is given: $(1, 3, 4)$. We have

$$20x - 2y + 15z + d = 0$$

$$20 - 2 \cdot 3 + 15 \cdot 4 + d = 0$$

$$20 - 6 + 60 + d = 0$$

$$d = -74.$$

Equation is: $20x - 2y + 15z - 74 = 0$.

25. $\mathbf{n}_1 = \langle 1, 3, -1 \rangle$; $\mathbf{n}_2 = \langle 1, 2, 1 \rangle$;

$\mathbf{v} = \mathbf{n}_1 \times \mathbf{n}_2 = \langle 5, -2, -1 \rangle$ is parallel to the line; $(2, 0, 1)$ is a point on the line (it is on both planes).

$$\langle x(t), y(t), z(t) \rangle = \langle 2 + 5t, -2t, 1 - t \rangle.$$

ANSWERS TO SELECTED EXERCISES FROM CHAPTER 12

Section 12.1

4.

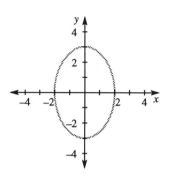

9. The orientation of the curves in problems 1,4,5, and 8 is counterclockwise. The orientation of the curves in problems 2,3,6, and 7 is clockwise. The conic section traced out by the position functions in problems 1-8 is an ellipse.

12.

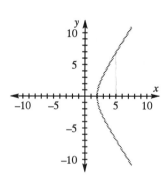

15.

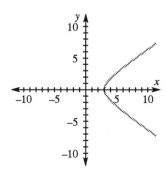

20. The curve traced out is one half of a hyperbola with x-axis intercept $(a, 0)$ and asymptotes $y = \pm \frac{b}{a} x$. When a and b have the same sign, the orientation of the curve is clockwise. When a and b are opposite in sign, the orientation of the curve is counterclockwise.

23. Three principal views:

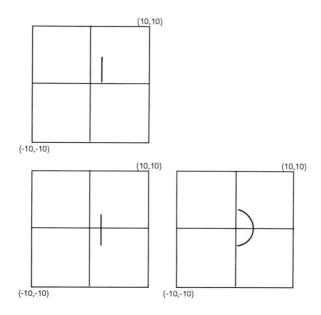

Curve:

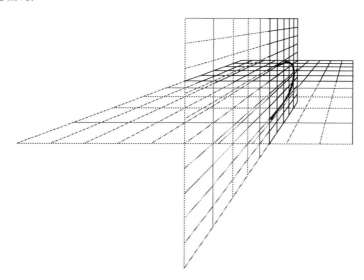

26. Three principal views:

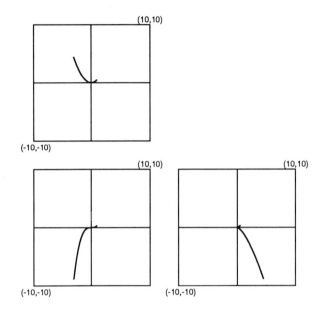

Curve:

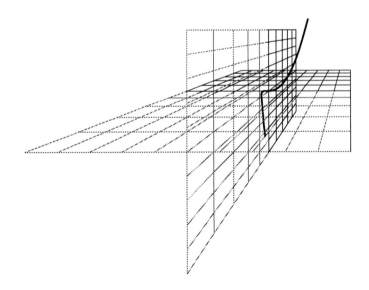

29. Three principal views:

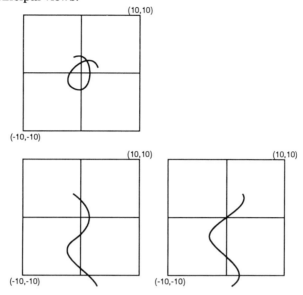

Curve:

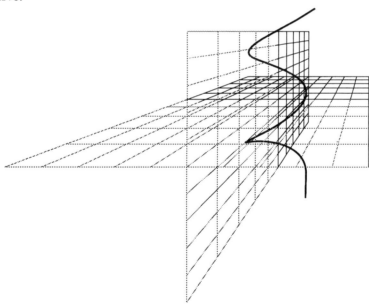

Section 12.2

2. $\mathbf{a}(t) = \langle 6, -\frac{1}{4}t^{-3/2}\rangle$; $\mathbf{a}(2) = \langle 6, -\frac{1}{8\sqrt{2}}\rangle$.

5. $\langle x(t), y(t)\rangle = \langle 12 + 12t, \sqrt{2} + \frac{1}{2\sqrt{2}}t\rangle$.

8. $\mathbf{T}(t) = \frac{\langle 3\cos(3t), -3\sin(3t), 3t^{1/2}\rangle}{3\sqrt{1+t}}$; $\mathbf{T}(1) \approx \langle -0.70003, 0.09979, 0.23570\rangle$.

11. $\mathbf{v}(t) = \langle t, 2t^{1/2}, 2\rangle$; $\mathbf{v}(1) = \langle 1, 2, 2\rangle$.

14. Speed = 3.

17. $\mathbf{a}(t) = \langle -\pi^2 \sin(\pi t), -\frac{1}{t^2}, e^t\rangle$; $\mathbf{a}(1) = \langle 0, -1, e\rangle$.

20. $t = e$. **23.** $t = 1$.

26. $\mathbf{v}(t) = \langle 2 + \frac{\pi}{2}\sin(\frac{\pi}{2}t), \frac{\pi}{2}\cos(\frac{\pi}{2}t), 2t\rangle$; $\mathbf{v}(3) = \langle 2 - \frac{\pi}{2}, 0, 6\rangle$.

29. $\|\mathbf{r}(t)\| = \sqrt{\left(2t - \cos(\frac{\pi}{2}t)\right)^2 + \sin^2(\frac{\pi}{2}t) + (t^2 + 16)^2}$. Each of the times in the sum increase as t increases between 3 and 4, so object is farthest from origin when $t = 4$.

32. $x = 0$ when $2t - \cos(\frac{\pi}{2}t) = 0$, i.e. when $t \approx 0.40308$ which is not in $[3, 4]$;

$y = 0$ when $\sin(\frac{\pi}{2}t) = 0$, i.e. when $t = 4$;

$z \neq 0$ for any t.

The object never crosses any of the coordinate axes.

35. $\mathbf{v}(t) = \langle -3\pi \sin(\frac{\pi}{2}t), \frac{\pi}{2}\cos(\frac{\pi}{2}t), 1\rangle$; $\mathbf{v}(1) = \langle -3\pi, 0, 1\rangle$.

38. $\|\mathbf{r}(t)\| = \sqrt{36\cos^2(\frac{\pi}{2}t) + \sin^2(\frac{\pi}{2}t) + t^2} = \sqrt{35\cos^2(\frac{\pi}{2}t) + 1 + t^2}$. This is greatest when $t = 2$.

41. $x = 0$ when $6\cos(\frac{\pi}{2}t) = 0$, i.e. when $t = 1$ and $t = 3$.

$y = 0$ when $\sin(\frac{\pi}{2}t) = 0$, i.e. when $t = 2$.

$z = 0$ when $t = 0$ (so never in $[1, 3]$).

The object never crosses any of the coordinate axes.

Section 12.3

1. $\mathbf{r}'(t) = \langle 6t, \frac{1}{2\sqrt{t}} \rangle; \quad \mathbf{r}'(\frac{1}{2}) = \langle 3, \frac{1}{\sqrt{2}} \rangle.$

4. $t \approx 0.15143.$

7. $\|\mathbf{r}'(t)\| = \sqrt{\frac{144t^3+1}{4t}}.$

10. If $\mathbf{r} \circ f = \langle 3(1-t^2)^2, \sqrt{1-t^2} \rangle$ then $\frac{d}{dt}(\mathbf{r} \circ f) = \langle -12t(1-t^2), -\frac{t}{\sqrt{1-t^2}} \rangle$ calculating directly. Now apply the chain rule to $\mathbf{r} \circ f$:
$$\mathbf{r}'(f(t))(f'(t)) = \langle 6(1-t^2), \frac{1}{2\sqrt{1-t^2}} \rangle (-2t) = \langle -12t(1-t^2), \frac{-t}{\sqrt{1-t^2}} \rangle.$$

13. $\mathbf{T}_1(t) = \frac{\langle \cos(3t), -\sin(3t), t^{1/2} \rangle}{\sqrt{1+t}}; \quad \mathbf{T}_2(t) = \frac{\langle t, 2t^{1/2}, 2 \rangle}{t+2}.$

16. $\mathbf{T}(t) = \frac{\langle 3t, 7t^{1/2}, 9 \rangle}{\sqrt{9t^2+49t+81}}.$

19. $\|\mathbf{r}_1\| = \sqrt{1+4t^3}; \quad \frac{d}{dt}(\|\mathbf{r}_1\|) = \frac{6t^2}{\sqrt{1+4t^3}}.$

22. Assume $\mathbf{r}(t)$ is non-zero and $\|\mathbf{r}(t)\| = C.$
$$\|\mathbf{r}(t)\| = \sqrt{\mathbf{r}(t) \cdot \mathbf{r}(t)} = C.$$
$$\frac{d}{dt}\|\mathbf{r}(t)\| = \frac{1}{2}(\mathbf{r}(t) \cdot \mathbf{r}(t))^{-1/2}[\mathbf{r}'(t) \cdot \mathbf{r}(t) + \mathbf{r}(t) \cdot \mathbf{r}'(t)] = 0 \Rightarrow$$
$$\frac{\mathbf{r}'(t) \cdot \mathbf{r}(t)}{\sqrt{\mathbf{r}(t) \cdot \mathbf{r}(t)}} = 0 \Rightarrow \mathbf{r}'(t) \cdot \mathbf{r}(t) = 0 \Rightarrow \mathbf{r}'(t) \text{ is orthogonal to } \mathbf{r}(t).$$

25. Let $\mathbf{r}(t) = \langle x(t), y(t), z(t) \rangle$, then
$$\frac{d\mathbf{T}}{dt} = \left(\frac{y'(y'x''-x'y'')+z'(z'x''-x'z'')}{((x')^2+(y')^2+(z')^2)^{3/2}}\right)\mathbf{i}$$
$$+ \left(\frac{x'(x'y''-y'x'')+z'(z'y''-y'z'')}{((x')^2+(y')^2+(z')^2)^{3/2}}\right)\mathbf{j} + \left(\frac{x'(x'z''-z'x'')+y'(y'z''-z'y'')}{((x')^2+(y')^2+(z')^2)^{3/2}}\right)\mathbf{k},$$

and
$$\frac{d\mathbf{T}}{dt} \times \mathbf{T} = \frac{[x'(x'y''-y'x'')+z'(z'y''-y'z'')]z' - [x'(x'z''-z'x'')+y'(y'z''-z'y'')]y'}{((x')^2+(y')^2+(z')^2)^2} \mathbf{i}$$
$$- \frac{[(y'(y'x''-x'y'')+z'(z'x''-x'z'')]z' + [(x'(x'z''-z'x'')+y'(y'z''-z'y'')]x'}{((x')^2+(y')^2+(z')^2)^2} \mathbf{j}$$
$$+ \frac{[y'(y'x''-x'y'')+z'(z'x''-x'z'')]y' - [(x'(x'y''-y'x'')+z'(z'y''-y'z'')]x'}{((x')^2+(y')^2+(z')^2)^2} \mathbf{k}$$
$$= \frac{[(x')^2+(y')^2+(z')^2]}{[(x')^2+(y')^2+(z')^2]^2} \langle y''z'-z''y', z''x'-x''z', x''y'-y''x' \rangle$$
$$= \frac{\langle y''z'-z''y', z''x'-x''z', x''y'-y''x' \rangle}{(x')^2+(y')^2+(z')^2}$$
$$= \frac{\mathbf{r}''(t)}{\|\mathbf{r}'(t)\|} \times \mathbf{T}(t).$$

Section 12.4

3. $\mathbf{r}(4) - \mathbf{r}(1) = \langle 64, \frac{16}{3}\rangle - \langle 1, \frac{2}{3}\rangle = \langle 63, \frac{14}{3}\rangle$.

Therefore, the net distance is $\|\langle 63, \frac{14}{3}\rangle\| \approx 63.17620$.

6. $\mathbf{r}(t) = \langle t^3 - 3, \frac{2}{3}t^{3/2} + \frac{7}{3}\rangle$.

9. $\int \mathbf{r}(t)\,dt = \langle -\frac{\cos(3t)}{3}, \frac{\sin(3t)}{3}, \frac{4}{5}t^{5/2}\rangle + \mathbf{C}$.

12. Total distance $= \int_0^2 \|\mathbf{v}(t)\|\,dt = 2(3^{3/2} - 1) \approx 8.39230$.

15. $\mathbf{v}(t) = \langle \frac{t^3}{3} + 1, \frac{t^4}{4} + 2, t^2 + 3\rangle$; terminal velocity $= \langle \frac{11}{3}, 6, 7\rangle$.

18. Total distance $= \int_0^2 \|\mathbf{v}(t)\|\,dt \approx 10.86686$.

Section 12.5

1. $\mathbf{T}(t) = \frac{\langle 12t^{3/2}, 1\rangle}{\sqrt{144t^3 + 1}}$.

4. $\mathbf{N}(\frac{3}{2}) = \frac{\langle 1, -18\sqrt{\frac{3}{2}}\rangle}{\sqrt{487}} \approx \langle 0.04531, -0.99897\rangle$;

$\mathbf{N}(2) = \frac{\langle 1, -24\sqrt{2}\rangle}{\sqrt{1153}} \approx \langle 0.02945, -0.99957\rangle$;

$\mathbf{N}(\frac{5}{2}) = \frac{\langle 1, -30\sqrt{\frac{5}{2}}\rangle}{\sqrt{2251}} \approx \langle 0.02108, -0.99978\rangle$;

7. $\mathbf{B} \cdot \mathbf{B} = \|\mathbf{B}\|^2 = \|1\|^2 = 1$.

10. $\mathbf{N}$ is orthogonal to both $\mathbf{T}$ and $\mathbf{B}$ so if $\frac{d\mathbf{B}}{ds}$ is also orthogonal to $\mathbf{T}$ and $\mathbf{B}$, then $\frac{d\mathbf{B}}{ds}$ must have the same or opposite direction as $\mathbf{N}$.

13. Approximate equation: $0.71174x - 0.01677y + 0.70225z - 1.52154 = 0$.

16. $\mathbf{T}(t) = \frac{\langle t, 2t^{1/2}, 2\rangle}{t+2}$; $\mathbf{T}(1) = \langle \frac{1}{3}, \frac{2}{3}, \frac{2}{3}\rangle$.

19. $\mathbf{B}(1) = \langle \frac{2}{3}, -\frac{2}{3}, \frac{1}{3}\rangle$.

22. $\mathbf{N}(1) \approx \langle 0.16837, -0.95880, 0.22884\rangle$.

25. $\tau = -0.23868$.

Section 12.6

3. $\kappa(1) = \frac{1}{9}; \quad \rho(1) = 9.$

6. $v(t)\mathbf{T}(t) = v(t)\frac{\mathbf{v}(t)}{v(t)} = \mathbf{v}(t).$

9. $\mathbf{a}(t) = \dfrac{dv}{dt}\mathbf{T}(t) + v(t)[\kappa(t)v(t)\mathbf{N}(t)] = \dfrac{dv}{dt}\mathbf{T}(t) + \kappa(t)v^2(t)\mathbf{N}(t).$

12. $\dfrac{\mathbf{a}(t)\times\mathbf{v}(t)}{v^3(t)} = \dfrac{\kappa(t)v^3(t)\mathbf{N}(t)\times\mathbf{T}(t)}{v^3(t)} = \kappa(t)\mathbf{N}(t)\times\mathbf{T}(t) = \kappa(t)\mathbf{B}(t).$

15. $f'(x) = 3x^2 - 2; \quad f'(1) = 1; \quad f''(x) = 6x; \quad f''(1) = 6;$
$\kappa = \dfrac{6}{2^{3/2}} \approx 2.12132; \quad \rho(1) \approx 0.47140.$

ANSWERS TO SELECTED EXERCISES FROM CHAPTER 13

Section 13.1

1. A vector parallel to the line of intersection is $\langle 1, 0, 1 \rangle$. The line of intersection can be described by $\langle x(t), y(t), z(t) \rangle = \langle 0, 0, 0 \rangle + t \langle 1, 0, 1 \rangle$.

4. Hyperbolic cylinder running parallel to the z-axis.

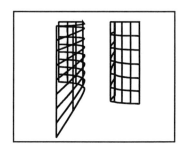

7. Circular cylinder with the y-axis running down its center.

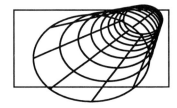

10. When $z = f(x)$ or $z = f(y)$, the graph will be a cylindrical surface. Two examples are $z = x^2$ and $z = y^3$.

13. When $a = b = c$.

16. It is not the graph of a function.

19. It is the graph of a function.

22. $\frac{x^2}{c^2} = \frac{y^2}{a^2} + \frac{z^2}{b^2}$; $\frac{y^2}{c^2} = \frac{x^2}{a^2} + \frac{z^2}{b^2}$.

25. $cx = \frac{y^2}{a^2} + \frac{z^2}{b^2}$; $cy = \frac{x^2}{a^2} + \frac{z^2}{b^2}$.

Section 13.2

2. Minima at $\approx (0.9, 3)$, $(3, 3)$;   maxima at $\approx (2.5, 3)$, $(3, 2.4)$.

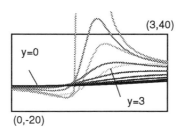

5. Minimum at $(0, 3)$;   maximum at $(3, 0)$.

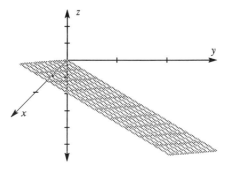

8. Saddle.

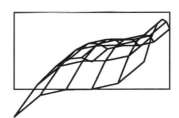

11. Saddle.

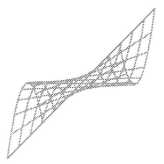

14. Uphill.

17. Uphill.

20. Downhill.

23. N, S.

26.

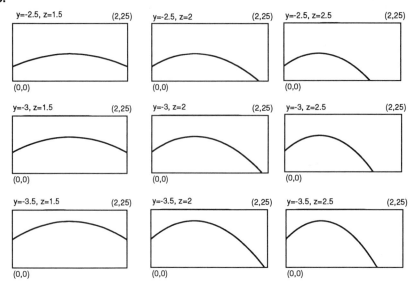

From this graph, it is clear that neither a minimum nor a maximum can occur at $(1, -3)$.

Section 13.3

3. $z = 2x + \frac{1}{2}y - \frac{3}{2}$.

6. $\frac{\partial z}{\partial x} = 2$; $\frac{\partial z}{\partial y} = -5$; z-intercept: $(0, 0, 1)$; $z = 2(x + 1) - 5(y - 4) - 21$.

9. Slopes undefined, plane contains entire z-axis, plane is not the graph of a function.

12. Find the discriminant:

$A = -1$, $B = 3$, $C = -2$, $D = (-1)(-2) - (3)^2 = -7$.

Since $D < 0$, a saddle point occurs at $(0, 0)$.

15. Simplifies to $k(x, y) = 0$. Graph is the xy plane.

18. $z = -3 + 3(x + 1) - (y - 3) + (x + 1)(y - 3)$.

21. Find the discriminant: $A = -1$, $B = 4$, $C = -4$, $D = (-1)(-4) - 4^2 = -12$. Since $D < 0$, a saddle point occurs at $(-1, 3, 17)$.

(Answers may vary for exercises 24-30.)

24. $z = 10(x-1)^2 + 2(y+2)^2 + 3$.

27. $z = (2x - y + 3)^2 + 5$.

30. $z = xy$.

ANSWERS TO SELECTED EXERCISES FROM CHAPTER 14

Section 14.1

1. $\dfrac{\partial f}{\partial x} = \dfrac{(x-y)-(x+y)}{(x-y)^2} = \dfrac{-2y}{(x-y)^2}$; $\left.\dfrac{\partial f}{\partial x}\right|_{(1,2)} = \dfrac{-2(2)}{(1-2)^2} = -4$;

   $\dfrac{\partial f}{\partial y} = \dfrac{(x-y)-(x+y)(-1)}{(x-y)^2} = \dfrac{2x}{(x-y)^2}$; $\left.\dfrac{\partial f}{\partial y}\right|_{(1,2)} = \dfrac{2(1)}{(1-2)^2} = 2$.

4. $\dfrac{\partial f}{\partial x} = 2$; $\left.\dfrac{\partial f}{\partial x}\right|_{(0,0)} = 2$; $\dfrac{\partial f}{\partial y} = -3$; $\left.\dfrac{\partial f}{\partial y}\right|_{(0,0)} = -3$.

7. $\dfrac{\partial f}{\partial x} = -\left(\dfrac{y}{x^2}\right)\sec^2\left(\dfrac{y}{x}\right)$; $\left.\dfrac{\partial f}{\partial x}\right|_{(2,\frac{\pi}{2})} = -\dfrac{\frac{\pi}{2}}{2^2}\sec^2\left(\dfrac{\pi}{2}{2}\right) = -\dfrac{\pi}{8}(2) = -\dfrac{\pi}{4}$;

   $\dfrac{\partial f}{\partial y} = \left(\dfrac{1}{x}\right)\sec^2\left(\dfrac{y}{x}\right) + \csc^2(y)$; $\left.\dfrac{\partial f}{\partial y}\right|_{(2,\frac{\pi}{2})} = \left(\dfrac{1}{2}\right)\sec^2\left(\dfrac{\pi}{4}\right) + \csc^2\left(\dfrac{\pi}{2}\right) = 2$.

10. $\dfrac{\partial f}{\partial x} = yx^{y-1}$; $\left.\dfrac{\partial f}{\partial x}\right|_{(e,2)} = 2e^{2-1} = 2e$; (Rewrite $f(x,y) = x^y = e^{\ln x^y} = e^{y\ln x}$.)

    $\dfrac{\partial f}{\partial y} = \ln x(e^{y\ln x}) = \ln x(e^{\ln x^y}) = \ln x(x^y)$; $\left.\dfrac{\partial f}{\partial y}\right|_{(e,2)} = \ln e(e^2) = e^2$.

13. $f_x = -2y\cos(xy)\sin(xy)$; $f_x\left(\dfrac{1}{2}, \dfrac{\pi}{2}\right) = -\dfrac{\pi}{2}$;

    $f_y = -2x\cos(xy)\sin(xy)$; $f_y\left(\dfrac{1}{2}, \dfrac{\pi}{2}\right) = -\dfrac{1}{2}$.

16. $f_x = -y^2\sin x e^{\cos x}$; $f_y = 2y\, e^{\cos x}$; $f_x(0,-2) = 0$; $f_y(0,-2) = -4e$.

19. $f_x = \dfrac{(x+y)(0)-z(1)}{(x+y)^2} = -\dfrac{z}{(x+y)^2}$; $f_x(1,2,3) = -\dfrac{3}{(1+2)^2} = -\dfrac{1}{3}$;

    $f_y = \dfrac{(x+y)(0)-z(1)}{(x+y)^2} = -\dfrac{z}{(x+y)^2}$; $f_y(1,2,3) = -\dfrac{3}{(x+2)^2} = -\dfrac{1}{3}$;

    $f_z = \dfrac{1}{x+y}$; $f_z(1,2,3) = \dfrac{1}{1+2} = \dfrac{1}{3}$.

22. $f_x = -2x$; $f_x(0,1,-1) = -2(0) = 0$; $f_y = -2y$;

    $f_y(0,1,-1) = -2(1) = -2$; $f_z = -2z$; $f_z(0,1,-1) = -2(-1) = 2$.

25. $f_x = yz\cos(xyz)$; $f_y = xz\cos(xyz)$; $f_z = xy\cos(xyz)$;

    $f_x(1,-\pi,0) = 0$; $f_y(1,-\pi,0) = 0$; $f_z(1,-\pi,0) = -\pi$.

28. $f_x = ye^{xy}\cos(yz)$; $f_y = -ze^{xy}\sin(yz) + xe^{xy}\cos(yz)$;

    $f_z = -ye^{xy}\sin(yz)$;

    $f_x\left(1,1,\dfrac{\pi}{4}\right) = \dfrac{e}{\sqrt{2}}$; $f_y\left(1,1,\dfrac{\pi}{4}\right) = \dfrac{e}{\sqrt{2}}\left(-\dfrac{\pi}{4}+1\right)$; $f_z\left(1,1,\dfrac{\pi}{4}\right) = -\dfrac{e}{\sqrt{2}}$.

31. $f_x = \frac{-2x^3+y^2+z-3x^2y}{(x^3+y^2+z)^2}$; $f_y = \frac{x^3-y^2+z-2xy}{(x^3+y^2+z)^2}$; $f_z = -\frac{(x+y)}{(x^3+y^2+z)^2}$;
$f_x(1,-1,1) = \frac{1}{3}$; $f_y(1,-1,1) = \frac{1}{3}$; $f_z(1,-1,1) = 0$.

34. $\frac{\partial H}{\partial x} = 4x^3 - 8xy^2$; $\frac{\partial H}{\partial x}\Big|_{(3,-2)} = 4(3)^3 - 8(3)(-2)^2 = 12$ (uphill).

37. $\frac{\partial H}{\partial y} = -\frac{x}{y^2}$; $-\frac{\partial H}{\partial y}\Big|_{(-1,2)} = -\left(\frac{-(-1)}{(2)^2}\right) = -0.25$ (downhill).

40. $\frac{\partial P}{\partial x} = 2(-\frac{1}{2})((x-0.2)^2+(y-.01)^2)^{-3/2}(2)(x-0.2)$
$\quad -1(-\frac{1}{2})((x-0.8)^2+(y-0.01)^2)^{-3/2}2(x-0.8)$
$\quad = \frac{-2(x-0.2)}{((x-0.2)^2+(y-0.01)^2)^{3/2}} + \frac{(x-0.8)}{((x-0.8)^2+(y-0.01)^2)^{3/2}}$;
$\frac{\partial P}{\partial x}\Big|_{(0,0)} = \frac{0.4}{((-0.2)^2+(-0.01)^2)^{3/2}} + \frac{-0.8}{((-0.8)^2+(-0.01)^2)^{3/2}} \approx 48.25095$.

43. $\frac{\partial P}{\partial y}\Big|_{(0.5,-0.12)}$
$\quad = \frac{-2(-0.12-0.01)}{((0.5-0.2)^2+(-0.12-0.01)^2)^{3/2}} + \frac{(-0.12-0.01)}{((0.5-0.8)^2+(-0.12-0.01)^2)^{3/2}} \approx 3.71943$.

46. $\frac{\partial P}{\partial x}\Big|_{(-0.35,-0.42)}$
$\quad = \frac{-2(-0.35-0.2)}{((-0.35-0.2)^2+(-0.42-0.01)^2)^{3/2}} - \frac{(-0.35-0.8)}{((-0.35-0.8)^2+(-0.42-0.01)^2)^{3/2}} \approx 3.85407$.

49. $\frac{\partial T}{\partial y} = -\frac{600y}{(1+x^2+y^2+z^2)^2}$; $\frac{\partial T}{\partial y}\Big|_{(0,-3,0)} = \frac{-600(-3)}{(1+(-3)^2)^2} = 18$.

Section 14.2

1. $Df = \begin{bmatrix} -\frac{2y}{(x-y)^2} & \frac{2x}{(x-y)^2} \end{bmatrix}$;
$Df(1,2) = [-4 \quad 2]$; $f(x,y) \approx f(1,2) + [-4 \quad 2]\begin{bmatrix} x-1 \\ y-2 \end{bmatrix}$ or
$f(x,y) \approx -3 - 4(x-1) + 2(y-2)$.

4. $Df = [2 \quad -3]$; $Df(0,0) = [2 \quad -3]$;
$f(x,y) = f(0,0) + [2-3]\begin{bmatrix} x \\ y \end{bmatrix}$ or $f(x,y) = 2x - 3y$. **(The function is linear.)**

7. $Df = \begin{bmatrix} -\frac{y}{x^2}\sec^2\left(\frac{y}{x}\right) & \left(\frac{1}{x}\right)\sec^2\left(\frac{y}{x}\right) + \csc^2(y) \end{bmatrix}$; $Df(2,\frac{\pi}{2}) = [-\frac{\pi}{4} \quad 2]$;
$f(x,y) \approx f(2,\frac{\pi}{2}) + [-\frac{\pi}{4} \quad 2]\begin{bmatrix} x-2 \\ y-\frac{\pi}{2} \end{bmatrix}$ or
$f(x,y) \approx 1 - \frac{\pi}{4}(x-2) + 2(y-\frac{\pi}{2})$.

10. $Df = [yx^{y-1} \quad \ln x (x^y)]; \quad Df(e, 2) = [2e \quad e^2];$

$f(x, y) \approx f(e, 2) + [2e \quad e^2] \begin{bmatrix} x - e \\ y - 2 \end{bmatrix}$ or

$f(x, y) \approx e^2 + 2e(x - e) + e^2(y - 2).$

13. $Df = [5 \quad -4 \quad 3]; \quad Df(-3, 2, 4) = [5 \quad -4 \quad 3];$

$f(x, y, z) \approx f(-3, 2, 4) + [5 \quad -4 \quad 3] \begin{bmatrix} x + 3 \\ y - 2 \\ z - 4 \end{bmatrix}$

or $f(x, y, z) \approx -11 + 5(x + 3) - 4(y - 2) + 3(z - 4)$. (Note, this is $f(x, y, z)$ since $f(x, y, z)$ is linear.)

16. $H(2.1, 3.1) - H(2, 3) = -57.7202 - (-47) = -10.7202.$

19. $DH = [4x^3 - 8xy^2 \quad 4y^3 - 8x^2y]; \quad DH(-1, 2) = [28 \quad 16];$

$H(x, y) \approx H(-1, 2) + [28 \quad 16] \begin{bmatrix} x + 1 \\ y - 2 \end{bmatrix}$ or $H(x, y) \approx 1 + 28(x + 1) + 16(y - 2);$

$H(-1.3, 1.8) - 1 \approx 28(-1.3 + 1) + 16(1.8 - 2) = -11.6.$

22. $H(-2.2, -0.9) - H(-2, -1) = \frac{4}{9}.$

25. $DT = [-\frac{600x}{(1+x^2+y^2+z^2)^2} \quad -\frac{600y}{(1+x^2+y^2+z^2)^2} \quad -\frac{600z}{(1+x^2+y^2+z^2)^2}];$

$DT(1, 2, 3) = [-\frac{8}{3} \quad -\frac{16}{3} \quad -8];$

$T(x, y, z) \approx T(1, 2, 3) + [-\frac{8}{3} \quad -\frac{16}{3} \quad -8] \begin{bmatrix} x - 1 \\ y - 2 \\ z - 3 \end{bmatrix}$ or

$T(x, y, z) \approx 20 - \frac{8}{3}(x - 1) - \frac{16}{3}(y - 2) - 8(z - 3);$

$T(1.1, 1.9, 3.2) - 20 \approx -\frac{8}{3}(1.1 - 1) - \frac{16}{3}(1.9 - 2) - 8(3.2 - 3) = -\frac{4}{3}.$

28. (a) $f_x = 2xe^{x^2+y^2}; \quad f_x(1, 2) = 2e^5;$

$f_y = 2ye^{x^2+y^2}; \quad f_y(1, 2) = 4e^5; \quad f(1, 2) = e^5;$

$z = e^5 + 2e^5(x - 1) + 4e^5(y - 2).$

(b)

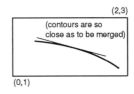

(continued on next page)

28. (c)

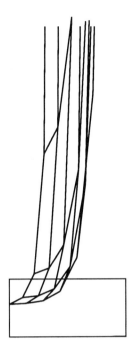

Section 14.3

1. $\nabla f = \langle -\frac{2y}{(x-y)^2}, \frac{2x}{(x-y)^2}\rangle$; $\nabla f(1,2) = \langle -4, 2\rangle$; $\mathbf{u} = \langle \frac{1}{\sqrt{2}}, \frac{1}{\sqrt{2}}\rangle$;
$D_{\mathbf{u}}f(1,2) = \langle -4, 2\rangle \cdot \langle \frac{1}{\sqrt{2}}, \frac{1}{\sqrt{2}}\rangle = -\sqrt{2}.$

4. $\nabla f = \langle -\frac{y}{x^2+y^2}, \frac{x}{x^2+y^2}\rangle$; $\nabla f(-3, 3) = \langle \frac{1}{6}, -\frac{1}{6}\rangle$; $\mathbf{u} = \langle 1, 0\rangle$;
$D_{\mathbf{u}}f(-3, -3) = \langle \frac{1}{6}, -\frac{1}{6}\rangle \cdot \langle 1, 0\rangle = \frac{1}{6}.$

7. $\nabla f = \langle -\frac{z}{(x+y)^2}, -\frac{z}{(x+y)^2}, \frac{1}{x+y}\rangle$; $\nabla f(1,2,3) = \langle -\frac{1}{3}, -\frac{1}{3}, \frac{1}{3}\rangle$;
$\mathbf{u} = \frac{\langle -\frac{1}{3}, -\frac{1}{3}, \frac{1}{3}\rangle}{\sqrt{(-\frac{1}{3})^2+(-\frac{1}{3})^2+(\frac{1}{3})^2}} \approx \langle -0.57735, -0.57735, 0.57735\rangle.$

10. $\nabla f = \langle -2xe^{1-x^2-y^2-z^2}, -2ye^{1-x^2-y^2-z^2}, -2ze^{1-x^2-y^2-z^2}\rangle$;
$\nabla f(0, 1, -1) = \langle 0, -2e^{-1}, 2e^{-1}\rangle$; $\mathbf{u} \approx \langle 0, -0.70711, 0.70711\rangle.$

13. $\nabla T(-1, -3, 2) = \langle -6, -12, -4\rangle$; $\mathbf{u} = \frac{-\langle -1, -3, 2\rangle}{\sqrt{(1)^2+3^2+(-2)^2}} = \langle \frac{1}{\sqrt{14}}, \frac{3}{\sqrt{14}}, -\frac{2}{\sqrt{14}}\rangle$;
$D_{\mathbf{u}}T(-1, -3, 2) \approx \langle -6, -12, -4\rangle \cdot \langle \frac{1}{\sqrt{14}}, \frac{3}{\sqrt{14}}, -\frac{2}{\sqrt{14}}\rangle \approx -9.08688.$

16. $\nabla T(\mathbf{r}(0)) = \nabla T(\langle 1, 0, 0\rangle) = \langle 6, 0, -4\rangle$;
$\mathbf{r}'(t) = \langle -\sin(t), \cos(t), 1\rangle$; $\mathbf{r}'(0) = \langle 0, 1, 1\rangle$;
$\nabla T(\mathbf{r}(0)) \cdot \mathbf{r}'(0) = \langle 6, 0, -4\rangle \cdot \langle 0, 1, 1\rangle = -4$ **degrees per second.**

19. $D_{\mathbf{i}}f = \left\langle \frac{2x}{1+(x^2+y^2+z^2)^2}, \frac{2y}{1+(x^2+y^2+z^2)^2}, \frac{2z}{1+(x^2+y^2+z^2)^2} \right\rangle \cdot \langle 1,0,0 \rangle$

$= \frac{2x}{1+(x^2+y^2+z^2)^2} = f_x;$

$D_{\mathbf{j}}f = \left\langle \frac{2x}{1+(x^2+y^2+z^2)^2}, \frac{2y}{1+(x^2+y^2+z^2)^2}, \frac{2z}{1+(x^2+y^2+z^2)^2} \right\rangle \cdot \langle 0,1,0 \rangle$

$= \frac{2y}{1+(x^2+y^2+z^2)^2} = f_y;$

$D_{\mathbf{k}}f = \left\langle \frac{2x}{1+(x^2+y^2+z^2)^2}, \frac{2y}{1+(x^2+y^2+z^2)^2}, \frac{2z}{1+(x^2+y^2+z^2)^2} \right\rangle \cdot \langle 0,0,1 \rangle$

$= \frac{2z}{1+(x^2+y^2+z^2)^2} = f_z.$

22. $\nabla H = \langle 4x^3 - 8xy^2, 4y^3 - 8x^2y \rangle;\quad \nabla H(2,3) = \langle -112, 12 \rangle;$

$\mathbf{u} = \langle \frac{1}{\sqrt{2}}, \frac{1}{\sqrt{2}} \rangle;\quad D_{\mathbf{u}}H(2,3) = \langle -112, 12 \rangle \cdot \langle \frac{1}{\sqrt{2}}, \frac{1}{\sqrt{2}} \rangle \approx -70.71068;$

$\mathbf{u} = \frac{\langle -112, 12 \rangle}{\sqrt{(-112)^2 + 12^2}} \approx \langle -0.99431, 0.10653 \rangle.$

25. $\nabla H(-2,-1) = \langle -16, 28 \rangle;\quad \mathbf{u} = \langle \frac{1}{\sqrt{2}}, -\frac{1}{\sqrt{2}} \rangle;$

$D_{\mathbf{u}}H(-2,-1) = \langle -16, 28 \rangle \cdot \langle \frac{1}{\sqrt{2}}, -\frac{1}{\sqrt{2}} \rangle \approx -31.11270;$

$\mathbf{u} = \frac{-\langle -16, 28 \rangle}{\sqrt{(-16)^2 + (28)^2}} \approx \langle 0.49614, -0.86824 \rangle.$

28. $\nabla H(3,-2) = \langle -0.5, -0.75 \rangle;\quad \mathbf{u} = \langle \frac{1}{\sqrt{2}}, \frac{1}{\sqrt{2}} \rangle;$

$D_{\mathbf{u}}H(3,-2) = \langle -0.5, -0.75 \rangle \cdot \langle \frac{1}{\sqrt{2}}, \frac{1}{\sqrt{2}} \rangle \approx -0.88388;$

$\mathbf{u} = \frac{-\langle -0.5, -0.75 \rangle}{\sqrt{(-0.5)^2 + (-0.75)^2}} \approx \langle 0.55470, 0.83205 \rangle.$

Section 14.4

2. $f_x = 3x^2;\quad f_x(0,0) = 0;\quad f_y = 3y^2;\quad f_y(0,0) = 0;\quad f_{xx} = 6x;$
$f_{xx}(0,0) = 0;\quad f_{xy} = 0;\quad f_{xy}(0,0) = 0;\quad f_{yy} = 6y;\quad f_{yy}(0,0) = 0;\quad f(x,y) \approx 0.$

The quadratic approximation is the same as the linear approximation (the horizontal plane $z = 0$). Hence, the graph of the quadratic approximation is flat at the origin.

5. $f(x,y) = xy$ is quadratic.

$A = C = 0,\ B = \frac{1}{2},\ D = -\frac{1}{4};\ D < 0 \Rightarrow$ a saddle point at $(0,0)$.

8. $f_x = 2xy^2;\quad f_x(0,0) = 0;\quad f_y = 2x^2y;\quad f_y(0,0) = 0;\quad f_{xx} = 2y^2;$
$f_{xx}(0,0) = 0;\quad f_{xy} = 4xy;\quad f_{xy}(0,0) = 0;\quad f_{yy} = 2x^2;\quad f_{yy}(0,0) = 0;$
$f(x,y) \approx 0.$

The quadratic approximation is the same as the linear approximation (the horizontal plane $z = 0$). Hence, the graph of the quadratic approximation is flat at the origin.

11. $f_x(x,y) = -\frac{z}{(x+y)^2}$; $\quad f_y(x,y) = -\frac{z}{(x+y)^2}$;
$f_z(x,y) = \frac{1}{x+y}$; $\quad f_{xy}(x,y) = \frac{2z}{(x+y)^3} = f_{yx}(x,y)$;
$f_{xz}(x,y) = -\frac{1}{(x+y)^2} = f_{zx}(x,y)$;
$f_{yz}(x,y) = -\frac{1}{(x+y)^2} = f_{zy}(x,y)$;
$f_{xx}(x,y) = \frac{2z}{(x+y)^3} = f_{yy}(x,y)$;
$f_{zz}(x,y) = 0$.

14. $f_x(x,y) = -2x$; $\quad f_y(x,y) = -2y$; $\quad f_z(x,y) = -2z$;
$f_{xx}(x,y) = f_{yy}(x,y) = f_{zz}(x,y) = -2$.
All second order mixed partial derivatives are 0.

17. $f_x(x,y) = y + \frac{1}{y}$; $\quad f_y(x,y) = x - \frac{x}{y^2}$; $\quad f_{xx}(x,y) = 0$;
$f_{xy}(x,y) = 1 - \frac{1}{y^2}$; $\quad f_{yy}(x,y) = \frac{2x}{y^3}$;
$f(x,y) \approx -\frac{10}{3} + \frac{10}{3}(x+1) - \frac{8}{9}(y-3) + \frac{1}{2}(2(\frac{8}{9})(x+1)(y-3) - \frac{2}{27}(y-3)^2)$
or
$f(x,y) \approx -\frac{10}{3} + \frac{10}{3}(x+1) - \frac{8}{9}(y-3) + \frac{8}{9}(x+1)(y-3) - \frac{1}{27}(y-3)^2$.

20. $f_x(x,y) = -\frac{y}{x^2+y^2}$; $\quad f_y(x,y) = \frac{x}{x^2+y^2}$; $\quad f_{xx}(x,y) = \frac{2xy}{(x^2+y^2)^2}$;
$f_{yy}(x,y) = -\frac{2xy}{(x^2+y^2)^2}$; $\quad f_{xy}(x,y) = \frac{(x^2+y^2)(-1)+y(2y)}{(x^2+y^2)^2} = \frac{-x^2+y^2}{(x^2+y^2)^2}$;
$f(x,y) \approx \frac{\pi}{4} + \frac{1}{6}(x+3) - \frac{1}{6}(y+3) + \frac{1}{2}(\frac{1}{18}(x+3)^2 - \frac{1}{18}(y+3)^2)$ or
$f(x,y) \approx \frac{\pi}{4} + \frac{1}{6}(x+3) - \frac{1}{6}(y+3) + \frac{1}{36}(x+3)^3 - \frac{1}{36}(y+3)^2$.

23. $f_x = -2y \cos(xy) \sin(xy)$; $\quad f_y = -2x \cos(xy) \sin(xy)$;
$f_{xx} = -2y^2 \cos^2(xy) + 2y^2 \sin^2(xy)$;
$f_{xy} = 2xy \sin^2(xy) - 2xy \cos^2(xy) - 2 \sin(xy) \cos(xy)$;
$f_{yy} = -2x^2 \cos^2(xy) + 2x^2 \sin^2(xy)$.

26. $f_x = 2xe^{x^2+y^2}$; $\quad f_y = 2ye^{x^2+y^2}$; $\quad f_{xx} = (4x^2+2)e^{x^2+y^2}$;
$f_{xy} = 4xye^{x^2+y^2}$; $\quad f_{yy} = (4y^2+2)e^{x^2+y^2}$.

29. $f_x = \frac{2x}{x^2+y^2}$; $\quad f_y = \frac{2y}{x^2+y^2}$; $\quad f_{xx} = \frac{2(x^2+y^2)-4x^2}{(x^2+y^2)^2} = \frac{-2x^2+2y^2}{(x^2+y^2)^2}$;
$f_{xy} = -\frac{4xy}{(x^2+y^2)^2}$; $\quad f_{yy} = \frac{2(x^2+y^2)-4y^2}{(x^2+y^2)^2} = \frac{2x^2-2y^2}{(x^2+y^2)^2}$.

32. $f_x = \frac{1}{y^2} - \frac{2y}{x^3}$; $\quad f_y = -\frac{2x}{y^3} + \frac{1}{x^2}$; $\quad f_{xx} = \frac{6y}{x^4}$; $\quad f_{xy} = -\frac{2}{y^3} - \frac{2}{x^3}$; $\quad f_{yy} = \frac{6x}{y^4}$.

35. $f_x = 2e^{2x+y^2+z}$; $\quad f_y = 2ye^{2x+y^2+z}$; $\quad f_z = e^{2x+y^2+z}$;
$f_{xx} = 4e^{2x+y^2+z}$; $\quad f_{xy} = 4ye^{2x+y^2+z}$; $\quad f_{xz} = 2e^{2x+y^2+z}$;
$f_{yy} = (4y^2+2)e^{2x+y^2+z}$; $\quad f_{yz} = 2ye^{2x+y^2+z}$; $\quad f_{zz} = e^{2x+y^2+z}$.

38. $f_x = \frac{x}{\sqrt{x^2+y^2+z^2}};$ $f_y = \frac{y}{\sqrt{x^2+y^2+z^2}};$ $f_z = \frac{z}{\sqrt{x^2+y^2+z^2}};$

$f_{xx} = \frac{\sqrt{x^2+y^2+z^2} - x \cdot \frac{x}{\sqrt{x^2+y^2+z^2}}}{x^2+y^2+z^2} = \frac{x^2+y^2+z^2-x^2}{(x^2+y^2+z^2)^{3/2}} = \frac{y^2+z^2}{(x^2+y^2+z^2)^{3/2}};$

$f_{xy} = \frac{-x \frac{y}{\sqrt{x^2+y^2+z^2}}}{x^2+y^2+z^2} = -\frac{xy}{(x^2+y^2+z^2)^{3/2}};$ $f_{xz} = -\frac{xz}{(x^2+y^2+z^2)^{3/2}};$

$f_{yy} = \frac{x^2+z^2}{(x^2+y^2+z^2)^{3/2}};$ $f_{yz} = -\frac{yz}{(x^2+y^2+z^2)^{3/2}};$ $f_{zz} = \frac{x^2+y^2}{(x^2+y^2+z^2)^{3/2}}.$

Section 14.5

1. $f_x(x,y) = 6x + 3 = 0 \Rightarrow x = -0.5;$ $f_y(x,y) = -4y - 2 = 0 \Rightarrow y = -0.5;$

$(-0.5, -0.5)$ is stationary critical point;

$f_{xx}(x,y) = 6,$ $f_{xy}(x,y) = 0,$ $f_{yy}(x,y) = -4;$

$A = 6, B = 0, C = -4, \Delta = -24 \Rightarrow$ a saddle point at $(-0.5, -0.5).$

4. $f_x(x,y) = -2x = 0 \Rightarrow x = 0;$ $f_y(x,y) = 2y = 0 \Rightarrow y = 0;$

$(0,0)$ is a stationary critical point;

$f_{xx}(x,y) = -2,$ $f_{xy}(x,y) = 0,$ $f_{yy}(x,y) = 2;$

$A = -2, B = 0, C = 2, \Delta = -4 \Rightarrow$ a saddle point at $(0,0).$

7. $f_x(x,y) = 4x^3 - 16x = 0 \Rightarrow 4x(x^2 - 4) = 0 \Rightarrow x = 0, \pm 2;$
$f_y(x,y) = 3y^2 - 3 = 0 \Rightarrow 3(y^2 - 1) = 0 \Rightarrow y = \pm 1;$

$(0,1), (0,-1), (2,1), (2,-1), (-2,1), (-2,-1)$ are the stationary critical points;

$f_{xx}(x,y) = 12x^2 - 16,$ $f_{xy}(x,y) = 0;$ $f_{yy}(x,y) = 6y;$

point	A	B	C	Δ	classification
$(0,1)$	-16	0	6	-96	saddle point
$(0,-1)$	-16	0	-6	96	relative maximum
$(2,1)$	32	0	6	192	relative minimum
$(2,-1)$	32	0	-6	-192	saddle point
$(-2,1)$	32	0	6	192	relative minimum
$(-2,-1)$	32	0	-6	-192	saddle point

10. $f_x(x,y) = 2xe^{-y} = 0 \Rightarrow x = 0$;

$f_y(x,y) = \frac{e^y(2y)(-(x^2+y^2)e^y}{e^{2y}} = \frac{-x^2-y^2+2y}{e^y} = 0$;

$-y^2 + 2y = 0 \Rightarrow y(-y+2) = 0 \Rightarrow y = 0, 2$;

$(0,0), (0,2)$ are stationary critical points;

$f_{xx}(x,y) = 2e^{-y}, \quad f_{xy}(x,y) = -2xe^{-y}$,

$f_{yy}(x,y) = \frac{e^y(-2y+2)-(-x^2-y^2+2y)e^y}{e^{2y}} = \frac{-2y+2+x^2+y^2-2y}{e^y} = \frac{x^2+y^2-4y+2}{e^y}$;

point	A	B	C	Δ	classification
$(0,0)$	2	0	2	4	relative minimum
$(0,2)$	$2e^{-2}$	0	$-\frac{2}{e^2}$	$-4e^{-4}$	saddle point

13. $f_x(x,y) = -\sin(x)\sinh(y) = 0 \Rightarrow x = n\pi$, n an integer, $y = 0$;

$f_y(x,y) = \cos(x)\cosh(y) = 0 \Rightarrow x = \frac{\pi}{2} + n\pi$, n an integer;

$(\frac{\pi}{2} + n\pi, 0)$, n an integer, are stationary critical points;

$f_{xx}(x,y) = -\cos(x)\sinh(y), \quad f_{xy}(x,y) = -\sin(x)\cosh(y)$,

$f_{yy}(x,y) = \cos(x)\sinh(y)$;

$A = 0$, $B = -1$(n odd), $B = 1$(n even), $C = 0$, $\Delta = -1 \Rightarrow (\frac{\pi}{2} + n\pi, 0)$ are saddle points.

16. $f_x(x,y) = 6x + 3 = 0 \Rightarrow x = -\frac{1}{2}$; $\quad f_y(x,y) = -4y - 2 = 0$.

Hence, $y = -\frac{1}{2}$ and $(-\frac{1}{2}, -\frac{1}{2})$ is critical point within the domain.

Along the edge $y = -1$, we have $z = f(x, -1) = 3x^2 + 3x + 5$.

$\frac{dz}{dx} = 6x + 3 = 0 \Rightarrow x = -\frac{1}{2} \Rightarrow (-\frac{1}{2}, -1)$ is a critical point.

Along the edge $x = 0$, we have $z = f(0, y) = -2y^2 - 2y + 5$.

$\frac{dz}{dy} = -4y - 2 = 0 \Rightarrow y = -\frac{1}{2} \Rightarrow (0, -\frac{1}{2})$ is a critical point.

Along the edge $y = 0$, we have $z = f(x, 0) = 3x^2 + 3x + 5$.

$\frac{dz}{dx} = 6x + 3 = 0 \Rightarrow x = -\frac{1}{2} \Rightarrow (-\frac{1}{2}, 0)$ is a critical point.

Along the edge $x = -1$, we have $z = f(-1, y) = -2y^2 - 2y + 5$.

$\frac{dz}{dy} = -4y - 2 = 0 \Rightarrow y = -\frac{1}{2} \Rightarrow (-1, -\frac{1}{2})$ is a critical point.

(continued on next page)

16. (continued from previous page)

Evaluate all critical points and endpoints of each edge.

(a,b)	$f(a,b)$
$(-0.5,-0.5)$	4.75
$(-0.5,-1)$	4.25
$(0,-0.5)$	5.5
$(-0.5,0)$	4.25
$(-1,-0.5)$	5.5
$(0,0)$	5
$(-1,0)$	5
$(-1,-1)$	5
$(0,-1)$	5

Absolute minimum value is 4.25 and absolute maximum value is 5.5.

19. $f_x(x,y) = -2x = 0 \Rightarrow x = 0; \quad f_y(x,y) = 2y = 0 \Rightarrow y = 0.$

$(0,0)$ is critical point occurring on the boundary.

Along the edge $y = -1$, we have $z = f(x,-1) = -x^2 + 2$.

$\dfrac{dz}{dx} = -2x = 0 \Rightarrow x = 0 \Rightarrow (0,-1)$ is a critical point.

Along the edge $x = 0$, we have $z = f(0,y) = 1 + y^2$.

$\dfrac{dz}{dy} = 2y = 0 \Rightarrow y = 0 \Rightarrow (0,0)$ is a critical point.

Along the edge $y = 0$, we have $z = f(x,0) = 1 - x^2$.

$\dfrac{dz}{dx} = -2x = 0 \Rightarrow x = 0 \Rightarrow (0,0)$ is a critical point.

Along the edge $x = -1$, we have $z = f(-1,y) = y^2$.

$\dfrac{dz}{dy} = 2y = 0 \Rightarrow x = 0 \Rightarrow (-1,0)$ is a critical point.

Evaluate all critical points and endpoints of each edge.

(a,b)	$f(a,b)$
$(0,0)$	1
$(0,-1)$	2
$(-1,0)$	0
$(-1,-1)$	1

Absolute minimum value is 0 and absolute maximum value is 2.

22. $f_x(x,y) = 4x^3 - 16x = 0 \Rightarrow 4x(x^2 - 4) = 0 \Rightarrow x = 0, 2, -2$;
$f_y(x,y) = 3y^2 - 3 = 0 \Rightarrow 3(y^2 - 1) = 0 \Rightarrow y = 1, -1$;

$(0,1)$ is critical point occurring on the boundary;

$(0,-1)$, $(2,1)$, $(-2,1)$, $(2,-1)$ and $(-2,-1)$ are critical points occurring outside the domain.

Along the edge $y = 0$, we have $z = f(x,0) = x^4 - 8x^2 + 5$.

$\frac{dz}{dx} = 4x^3 - 16x = 0 \Rightarrow 4x(x^2 - 4) \Rightarrow x = 0, 2, -2 \Rightarrow (0,0)$ is a critical point; $(2,0)$ and $(-2,0)$ are outside the domain.

Along the edge $y = 1 - x$, we have $z = f(x, 1-x) = x^4 - x^3 - 5x^2 + 3$.

$\frac{dz}{dx} = 4x^3 - 3x^2 - 10x = 0 \Rightarrow x(4x^2 - 3x - 10) = 0 \Rightarrow x(x-2)(4x+5) = 0$.

This implies $x = 0, 2, -\frac{5}{4}$, and so $(0,1)$ is a critical point; $(2,-1)$ and $(-\frac{5}{4}, 1)$ are outside the domain.

Along the edge $x = 0$, we have $z = f(0,y) = y^3 - 3y + 5$.

$\frac{dz}{dy} = 3y^2 - 3 = 0 \Rightarrow 3(y^2 - 1) = 0 \Rightarrow y = 1, -1 \Rightarrow (0,1)$ is a critical point.

Evaluate all critical points and endpoints of each edge.

(a,b)	$f(a,b)$
$(0,1)$	3
$(0,0)$	5
$(1,0)$	-2

Absolute minimum value is -2 and absolute maximum value is 5.

25. $f_x(x,y) = 2xe^{-y} = 0 \Rightarrow x = 0$;

$f_y(x,y) = \frac{e^y(2y) - (x^2+y^2)e^y}{e^{2y}} = (-x^2 - y^2 + 2y)e^{-y} = 0$;

$x = 0 \Rightarrow -y^2 + 2y = 0 \Rightarrow -y(y-2) = 0 \Rightarrow y = 0, 2$.

$(0,0)$ is critical point on the boundary; $(0,2)$ is critical point outside the domain.

Along the edge $y = 0$, we have $z = f(x,0) = x^2$.

$\frac{dz}{dx} = 2x = 0 \Rightarrow x = 0 \Rightarrow (0,0)$ is a critical point.

Along the edge $y = 1 - x$,

we have $z = f(x, 1-x) = \frac{x^2 + (1-x)^2}{e^{1-x}} = \frac{2x^2 - 2x + 1}{e^{1-x}}$.

$\frac{dz}{dx} = 2x^2 - 2x + 1 = 0 \Rightarrow$ no critical points along this edge.

Along the edge $x = 0$, we have $z = f(0,y) = y^2 e^{-y}$.

$\frac{dz}{dy} = 2ye^{-y} - y^2 e^{-y} = 0 \Rightarrow ye^{-y}(2-y) = 0 \Rightarrow y = 0, 2$ so

$(0,0)$ is a critical point; $(0,2)$ is outside the domain.

(continued on next page)

25. (continued from previous page)

Evaluate all critical points and endpoints of each edge.

(a,b)	$f(a,b)$
$(0,0)$	0
$(0,1)$	e^{-1}
$(1,0)$	1

Absolute minimum value is 0 and absolute maximum value is 1.

Section 14.6

3. Minimize $f(x,y,z) = x^2 + y^2 + z^2$; $\nabla f = \langle 2x, 2y, 2z \rangle$; $\nabla g = \langle yz, xz, xy \rangle$.

Solve: $2x = \lambda yz$, $2y = \lambda xz$, $2z = \lambda xy$, $xyz = 8$;

$yz = \frac{8}{x} \Rightarrow 2x = \lambda(\frac{8}{x}) \Rightarrow \frac{x^2}{4} = \lambda$;

$xz = \frac{8}{y} \Rightarrow 2y = \lambda(\frac{8}{y}) \Rightarrow \frac{y^2}{4} = \lambda$;

$xy = \frac{8}{z} \Rightarrow 2z = \lambda(\frac{8}{z}) \Rightarrow \frac{z^2}{4} = \lambda$;

$\frac{x^2}{4} = \frac{y^2}{4} = \frac{z^2}{4} \Rightarrow |x| = |y| = |z| \Rightarrow -x^3 = 8$ or $x^3 = 8$ since $xyz = 8$.

$(2,2,2)$, $(2,-2,-2)$, $(-2,-2,2)$, $(-2,2,-2)$ are the points closest to the origin.

6. Maximize $V = xyz$ subject to $2xy + 2yz + 2xz = 1200$; $\nabla V = \langle yz, xz, xy \rangle$, $\nabla g = \langle 2y + 2z, 2x + 2z, 2x + 2y \rangle$.

Solve: $yz = \lambda(2y + 2z)$, $xz = \lambda(2x + 2z)$, $xy = \lambda(2x + 2y)$, and $2xy + 2yz + 2xz = 1200$;

$\frac{yz}{2y+2z} = \frac{xz}{2x+2z} = \frac{xy}{2x+2y}$, which implies

$yz(2x+2z) = xz(2y+2z)$, so $2xyz+2yz^2 = 2xyz+2xz^2 \Rightarrow y = x$ (assuming $z \neq 0$).

Also, $xz(2x+2y) = xy(2x+2z)$, which implies $2x^2z+2xyz = 2x^2y+2xyz \Rightarrow y = z$ (assuming $x \neq 0$).

Together, this means $x = y = z$, so $6x^2 = 1200$ (from the constraint), which implies $x = \sqrt{200}$. Hence, the dimensions of the box are

$\sqrt{200} \times \sqrt{200} \times \sqrt{200} \approx 14.14214$ cm $\times$ 14.14214 cm $\times$ 14.14214 cm, yielding a maximum volume of $(\sqrt{200})^3 \approx 2828.42712$ cm^3.

9. Minimize $C = 2xy + 2xz + 2yz$ subject to $xyz = 6000$. This is the same problem as 7.

ANSWERS TO SELECTED EXERCISES FROM CHAPTER 15

Section 15.1

1. $\int_1^2 \int_1^4 x+y\,dx\,dy = \int_1^2 \frac{x^2}{2} + xy\Big]_{x=1}^{x=4} dy = \int_1^2 (8+4y) - (\frac{1}{2}+y)\,dy$

$= \int_1^2 \frac{15}{2} + 3y\,dy = \frac{15}{2}y + \frac{3y^2}{2}\Big]_{y=1}^{y=2} = 21 - 9 = 12.$

4. $\int_1^e \int_1^2 \frac{x}{y}\,dx\,dy = \int_1^e \frac{x^2}{2y}\Big]_{x=1}^{x=2} dy = \int_1^e \frac{2}{y} - \frac{1}{2y}\,dy = 2\ln|y| - \frac{1}{2}\ln|y|\Big]_{y=1}^{y=e} = \frac{3}{2}.$

7. $\int_{-1}^1 \int_0^{\pi/2} x\sin(y) - ye^x\,dy\,dx = \int_{-1}^1 x(-\cos(y)) - \frac{y^2}{2}e^x\Big]_{y=0}^{y=\pi/2}$

$= \int_{-1}^1 -\frac{\pi^2}{8}e^x - (x(-1))\,dx = -\frac{\pi^2}{8}e^x + \frac{x^2}{2}\Big]_{x=-1}^{x=1}$

$= \left(-\frac{\pi^2}{8}e + \frac{1}{2}\right) - \left(-\frac{\pi^2}{8}e^{-1} + \frac{1}{2}\right) = \frac{\pi^2}{8}(e^{-1} - e).$

10. $\int_{-2}^{-1} \int_0^1 \frac{y}{1+x^2}\,dx\,dy = \int_{-2}^{-1} y\arctan(x)\Big]_{x=0}^{x=1} dy$

$= \int_{-2}^{-1} \frac{\pi}{4}y\,dy = \frac{\pi}{4}(\frac{y^2}{2})\Big]_{y=-2}^{y=-1} = \frac{\pi}{4}(\frac{1}{2} - 2) = -\frac{3}{8}\pi.$

13. $\int_{-1}^2 \int_1^4 2x + 6x^2y\,dx\,dy = \int_{-1}^2 x^2 + 2x^3y\Big]_{x=1}^{x=4} dy = \int_{-1}^2 15 + 126y\,dy$

$= 15y + 63y^2\Big]_{y=-1}^{y=2} = 234.$

16. $\int_1^2 \int_{-1}^2 4xy^3 + y\,dy\,dx = \int_1^2 xy^4 + \frac{1}{2}y^2\Big]_{y=-1}^{y=2} dx = \int_1^2 15x + \frac{3}{2}\,dx$

$= \frac{15}{2}x^2 + \frac{3}{2}x\Big]_{x=1}^{x=2} = 24.$

$\int_{-1}^2 \int_1^2 4xy^3 + y\,dx\,dy = \int_{-1}^2 2x^2y^3 + yx\Big]_{x=1}^{x=2} dy = \int_{-1}^2 6y^3 + y\,dy$

$= \frac{3}{2}y^4 + \frac{1}{2}y^2\Big]_{y=-1}^{y=2} = 24.$

19. $\int_{-1}^{1}\int_{-1}^{1}\int_{-1}^{1}\frac{y^2 e^z}{1+x^2}\,dx\,dy\,dz = \int_{-1}^{1}\int_{-1}^{1}y^2 e^z \arctan(x)\Big]_{x=-1}^{x=1}\,dy\,dz$

$= \int_{-1}^{1}\int_{-1}^{1}y^2 e^z(\frac{\pi}{4}-(-\frac{\pi}{4}))\,dy\,dz = \frac{\pi}{2}\int_{-1}^{1}\int_{-1}^{1}y^2 e^z\,dy\,dz$

$= \frac{\pi}{2}\int_{-1}^{1}\frac{y^3}{3}e^z\Big]_{y=-1}^{y=1}\,dz = \frac{\pi}{2}\int_{-1}^{1}\frac{e^z}{3}+\frac{e^z}{3}\,dz$

$= \frac{\pi}{3}\int_{-1}^{1}e^z\,dz = \frac{\pi}{3}e^z\Big]_{z=-1}^{z=1} = \frac{\pi}{3}(e-e^{-1}).$

22. $\int_{0}^{2}\int_{-1}^{3}\int_{1}^{4}3xy^3 z^2\,dy\,dx\,dz = \int_{0}^{2}\int_{-1}^{3}\frac{3}{4}xy^4 z^2\Big]_{y=1}^{y=4}\,dx\,dz$

$= \frac{3}{4}\int_{0}^{2}\int_{-1}^{3}xz^2(255)\,dx\,dz = \frac{765}{4}\int_{0}^{2}\frac{1}{2}x^2 z^2\Big]_{x=-1}^{x=3}\,dz$

$= \frac{765}{8}\int_{0}^{2}8z^2\,dz = \frac{765}{3}z^3\Big]_{z=0}^{z=2} = 2040.$

(There are 5 other ways to set up the integral.)

Section 15.2

2. $\int_{0}^{1}\int_{y^2}^{y}(x^2+y^3)\,dx\,dy = \int_{0}^{1}\frac{x^3}{3}+y^3 x\Big]_{x=y^2}^{x=y}\,dy = \int_{0}^{1}(\frac{y^3}{3}+y^4)-(\frac{y^6}{3}+y^5)\,dy$

$= \frac{y^4}{12}+\frac{y^5}{5}-\frac{y^7}{21}-\frac{y^6}{6}\Big]_{y=0}^{y=1} = \frac{1}{12}+\frac{1}{5}-\frac{1}{21}-\frac{1}{6} = \frac{29}{420}.$

5. $\int_{1}^{2}\int_{-y}^{y}(x+y)\,dx\,dy = \int_{1}^{2}\frac{x^2}{2}+xy\Big]_{x=-y}^{x=y}\,dy$

$= \int_{1}^{2}(\frac{y^2}{2}+y^2)-(\frac{y^2}{2}-y^2)\,dy = \int_{1}^{2}2y^2\,dy$

$= \frac{2y^3}{3}\Big]_{y=1}^{y=2} = \frac{16}{3}-\frac{2}{3} = \frac{14}{3}.$

8. $\int_{0}^{1}\int_{0}^{2}e^{x-y}\,dy\,dx = \int_{0}^{1}-e^{x-y}\Big]_{y=0}^{y=2}\,dx = \int_{0}^{1}-e^{x-2}+e^x\,dx$

$= -e^{x-2}+e^x\Big]_{x=0}^{x=1} = (-e^{-1}+e)-(-e^{-2}+1) = -e^{-1}+e^{-2}+e-1.$

11. $\int_{0}^{1}\int_{e^y}^{e}xe^y\,dx\,dy.$

14. $\int_0^1 \int_{1-y}^1 2xy\, dx\, dy = \int_0^1 x^2 y \Big]_{x=1-y}^{x=1} dy = \int_0^1 y - (1-y)^2 y\, dy$

$= \int_0^1 y - y + 2y^2 - y^3\, dy = \frac{2y^3}{3} - \frac{y^4}{4} \Big]_{y=0}^{y=1} = \frac{2}{3} - \frac{1}{4} = \frac{5}{12}.$

17. $\int_1^2 \int_{x^3}^x e^{y/x} dy\, dx = \int_1^2 xe^{y/x}\Big]_{y=x^3}^{y=x} dx = \int_1^2 xe - xe^{x^2} dx$

$= \frac{1}{2}x^2 e - \frac{1}{2}e^{x^2}\Big]_{x=1}^{x=2} = 2e - 0.5e^4.$

20. $\int_0^1 \int_y^1 \frac{1}{1+y^2} dx\, dy = \int_0^1 \frac{x}{1+y^2}\Big]_{x=y}^{x=1} dy = \int_0^1 \frac{1}{1+y^2} - \frac{y}{1+y^2} dy$

$= \arctan(y) - \frac{1}{2}\ln(1+y^2)\Big]_{y=0}^{y=1} = \frac{\pi}{4} - \frac{1}{2}\ln(2).$

23. $\int_0^1 \int_{x^3}^{\sqrt{x}} f(x,y)\, dy\, dx.$

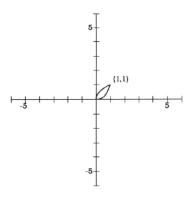

26. $\int_{-1}^2 \int_{-y^2}^{4+y} dx\, dy = \int_{-1}^2 4 + y + y^2\, dy = 4y + \frac{1}{2}y^2 + \frac{1}{3}y^3\Big]_{x=-1}^{x=2} = 16.5.$

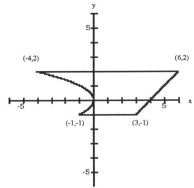

29. $2\int_0^1 \int_0^{e^y} dx\, dy = 2\int_0^1 e^y\, dy = 2e^y\Big]_{y=0}^{y=1} = 2(e-1)$.

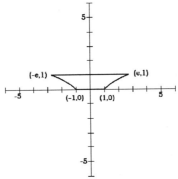

32. $\int_0^3 \int_0^{2x} \sqrt{9-x^2}\, dy\, dx = \int_0^3 \sqrt{9-x^2}\, y\Big]_{y=0}^{y=2x} dx = \int_0^3 2x\sqrt{9-x^2}\, dx$

$= -\frac{2}{3}(9-x^2)^{3/2}\Big]_{x=0}^{x=3} = 18$.

Section 15.3

1. $\int_D\int x^2+y^2\, dx\, dy = \int_0^{\pi/2}\int_0^2 r^2(r\, dr\, d\theta) = \int_0^{\pi/2} \frac{r^4}{4}\Big]_{r=0}^{r=2} d\theta$

$= 4\int_0^{\pi/2} d\theta = 4\theta\Big]_{\theta=0}^{\theta=\pi/2} = 4(\frac{\pi}{2}) = 2\pi$.

4. $\int_{-1}^1\int_{-\sqrt{1-y^2}}^{\sqrt{1-y^2}} x^2+y^2\, dx\, dy = \int_0^{2\pi}\int_0^1 r^2(r\, dr\, d\theta)$

$= \int_0^{2\pi}\int_0^1 r^3\, dr\, d\theta = \int_0^{2\pi} \frac{r^4}{4}\Big]_{r=0}^{r=1} d\theta = \int_0^{2\pi} \frac{1}{4}\, d\theta = \frac{\theta}{4}\Big]_{\theta=0}^{\theta=2\pi} = \frac{\pi}{2}$.

7. $\int_{7\pi/6}^{11\pi/6}\int_0^{1+2\sin(\theta)} r\, dr\, d\theta = \int_{7\pi/6}^{11\pi/6} \frac{r^2}{2}\Big]_{r=0}^{r=1+2\sin(\theta)} d\theta = \frac{1}{2}\int_{7\pi/6}^{11\pi/6} (1+2\sin(\theta))^2\, d\theta$

$= \frac{1}{2}\int_{7\pi/6}^{11\pi/6} 1+4\sin(\theta)+4\sin^2(\theta)\, d\theta$

$= \frac{1}{2}\int_{7\pi/6}^{11\pi/6} 1+4\sin(\theta)+\frac{4(1-\cos(2\theta))}{2}\, d\theta$

$= \frac{1}{2}(\theta - 4\cos(\theta) + 2\theta - \sin(2\theta))\Big]_{\theta=7\pi/6}^{\theta=11\pi/6}$

$= \frac{1}{2}(((\frac{11\pi}{6} - 4(\frac{\sqrt{3}}{2}) + 2(\frac{11\pi}{6}) - (-\frac{\sqrt{3}}{2})) - (\frac{7\pi}{6} - 4(-\frac{\sqrt{3}}{2}) + 2(\frac{7\pi}{6}) - (\frac{\sqrt{3}}{2})))$

$= \frac{1}{2}(\frac{12\pi}{6} - 3\sqrt{3}) = \pi - \frac{3\sqrt{3}}{2}$.

10. $\displaystyle\int_{\pi/4}^{3\pi/4}\int_0^{\cos(2\theta)} r\,dr\,d\theta = \int_{\pi/4}^{3\pi/4} \frac{r^2}{2}\Big]_{r=0}^{r=\cos(2\theta)} d\theta = \frac{1}{2}\int_{\pi/4}^{3\pi/4} \cos^2(2\theta)\,d\theta$

$= \frac{1}{2}\int_{\pi/4}^{3\pi/4} \frac{1+\cos(4\theta)}{2}\,d\theta = \frac{1}{4}\left(\theta + \frac{\sin 4\theta}{4}\right)\Big]_{\theta=\pi/4}^{\theta=3\pi/4} = \frac{1}{4}\left(\frac{3\pi}{4} - \frac{\pi}{4}\right) = \frac{\pi}{8}.$

13. $\displaystyle\int_0^{2\pi}\int_0^{3+2\sin(\theta)} r\,dr\,d\theta = \int_0^{2\pi} \frac{1}{2}r^2\Big]_{r=0}^{r=3+2\sin(\theta)} d\theta$

$= \frac{1}{2}\int_0^{2\pi} (3+2\sin(\theta))^2\,d\theta = \frac{1}{2}\int_0^{2\pi} 9 + 12\sin(\theta) + 4\sin^2(\theta)\,d\theta$

$= \frac{1}{2}\int_0^{2\pi} 9 + 12\sin(\theta) + 4\frac{(1-\cos(2\theta))}{2}\,d\theta$

$= \frac{9}{2}\theta - 6\cos(\theta) + \theta - \frac{1}{2}\sin(2\theta)\Big]_{\theta=0}^{\theta=2\pi} = 11\pi.$

16. $\displaystyle\int_0^{\pi/2}\int_0^1 e^r\,r\,dr\,d\theta \approx 1.57080.$

Section 15.4

1. $\displaystyle\int_0^1\int_{x^2}^x\int_0^{xy} (x+2y+3z)\,dz\,dy\,dx = \int_0^1\int_{x^2}^x xz + 2yz + \frac{3z^2}{2}\Big]_{z=0}^{z=xy} dy\,dx$

$= \int_0^1\int_{x^2}^x x^2y + 2xy^2 + \frac{3x^2y^2}{2}\,dy\,dx = \int_0^1 \frac{x^2y^2}{2} + \frac{2xy^3}{3} + \frac{3x^2y^3}{6}\Big]_{y=x^2}^{y=x} dx$

$= \int_0^1 \left(\frac{x^4}{2} + \frac{2x^4}{3} + \frac{3x^5}{6}\right) - \left(\frac{x^6}{2} + \frac{2x^7}{3} + \frac{3x^8}{6}\right) dx = \int_0^1 \frac{7x^4}{6} + \frac{x^5}{2} - \frac{x^6}{2} - \frac{2x^7}{3} - \frac{x^8}{2}\,dx$

$= \frac{7x^5}{30} + \frac{x^6}{12} - \frac{x^7}{14} - \frac{x^8}{12} - \frac{x^9}{18}\Big]_{x=0}^{x=1} = \frac{7}{30} + \frac{1}{12} - \frac{1}{14} - \frac{1}{12} - \frac{1}{18} = \frac{67}{630}.$

4. $\displaystyle\int_0^{2\pi}\int_0^1\int_r^1 r\,dz\,dr\,d\theta = \int_0^{2\pi}\int_0^1 rz\Big]_{z=r}^{z=1} dr\,d\theta = \int_0^{2\pi}\int_0^1 r(1-r)\,dr\,d\theta$

$= \int_0^{2\pi}\int_0^1 r - r^2\,dr\,d\theta = \int_0^{2\pi} \frac{r^2}{2} - \frac{r^3}{3}\Big]_{r=0}^{r=1} d\theta = \int_0^{2\pi} \frac{1}{6}\,d\theta = \frac{\theta}{6}\Big]_{\theta=0}^{\theta=2\pi} = \frac{\pi}{3}.$

7. $\displaystyle\int_0^1 \int_0^{\sqrt{1-x^2}} \int_0^{\sqrt{1-x^2-y^2}} dz\, dy\, dx$

$\displaystyle = \int_0^1 \int_0^{\sqrt{1-x^2}} z \Big]_{z=0}^{z=\sqrt{1-x^2-y^2}} dy\, dx$

$\displaystyle = \int_0^1 \int_0^{\sqrt{1-x^2}} \sqrt{1-x^2-y^2}\, dy\, dx$

$\displaystyle = \int_0^1 \tfrac{y}{2}\sqrt{1-x^2-y^2} + \tfrac{1-x^2}{2}\arcsin\left(\tfrac{y}{\sqrt{1-x^2}}\right)\Big]_{y=0}^{y=\sqrt{1-x^2}} dx$

$\displaystyle = \int_0^1 \left(\tfrac{\sqrt{1-x^2}}{2}\sqrt{1-x^2-(1-x^2)} + \tfrac{1-x^2}{2}\arcsin(1)\right) dx$

$\displaystyle = \tfrac{\pi}{4}\int_0^1 1-x^2\, dx = \tfrac{\pi}{4}(x - \tfrac{x^3}{3})\Big]_{x=0}^{x=1}$

$\displaystyle = \tfrac{\pi}{4}(1 - \tfrac{1}{3}) = \tfrac{\pi}{6}.$

(The second integration was performed using a formula from an integral table:

$\displaystyle \int \sqrt{a^2 - u^2}\, du$ with $a^2 = 1 - x^2$ and $u^2 = y^2$).

10. $\displaystyle I_{yz} = \int_0^1 \int_0^{\sqrt{1-x^2}} \int_0^{\sqrt{1-x^2-y^2}} x^3 yz\, dz\, dy\, dx$

$\displaystyle = \int_0^1 \int_0^{\sqrt{1-x^2}} \tfrac{1}{2}x^3 yz^2\Big]_{z=0}^{z=\sqrt{1-x^2-y^2}} dy\, dz$

$\displaystyle = \int_0^1 \int_0^{\sqrt{1-x^2}} \tfrac{1}{2}x^3 y(1-x^2-y^2)\, dy\, dz = \int_0^1 \int_0^{\sqrt{1-x^2}} \tfrac{1}{2}x^3 y - \tfrac{1}{2}x^5 y - \tfrac{1}{2}x^3 y^3\, dy\, dz$

$\displaystyle = \int_0^1 \tfrac{1}{4}x^3 y^2 - \tfrac{1}{4}x^5 y^2 - \tfrac{1}{8}x^3 y^4\Big]_{y=0}^{y=\sqrt{1-x^2}} dx$

$\displaystyle = \int_0^1 \tfrac{1}{4}x^3(1-x^2) - \tfrac{1}{4}x^5(1-x^2) - \tfrac{1}{8}x^3(1-x^2)^2\, dx$

$\displaystyle = \int_0^1 \tfrac{1}{4}x^3 - \tfrac{1}{4}x^5 - \tfrac{1}{4}x^5 + \tfrac{1}{4}x^7 - \tfrac{1}{8}x^3 + \tfrac{1}{4}x^5 - \tfrac{1}{8}x^7\, dx = \int_0^1 \tfrac{1}{8}x^3 - \tfrac{1}{4}x^5 + \tfrac{1}{8}x^7\, dx$

$\displaystyle = \tfrac{1}{32}x^4 - \tfrac{1}{24}x^6 + \tfrac{1}{64}x^8\Big]_{x=0}^{x=1} = \tfrac{1}{192};$

$\displaystyle I_{xz} = \int_0^1 \int_0^{\sqrt{1-x^2}} \int_0^{\sqrt{1-x^2-y^2}} xy^3 z\, dz\, dy\, dx = \tfrac{1}{192};$

$\displaystyle I_{xy} = \int_0^1 \int_0^{\sqrt{1-x^2}} \int_0^{\sqrt{1-x^2-y^2}} xyz^3\, dz\, dy\, dx = \tfrac{1}{192}.$

13. $I_{xy} = \int_0^1 \int_0^{1-z} \int_0^{1-y-z} x^2 z^2 + y^2 z^2 + z^4 \, dx \, dy \, dz$

$= \int_0^1 \int_0^{1-z} \frac{x^3}{3} z^2 + x y^2 z^2 + x z^4 \Big]_{x=0}^{x=1-y-z} dy \, dz$

$= \int_0^1 \int_0^{1-z} \frac{(1-y-z)^3}{3} z^2 + (1-y-z) y^2 z^2 + (1-y-z) z^4 \, dy \, dz$

$= \int_0^1 \int_0^{1-z} \frac{(1-y-z)^3}{3} z^2 + y^2 z^2 - y^3 z^2 - y^2 z^3 + z^4 - y z^4 - z^5 \, dy \, dz$

$= \int_0^1 \frac{-(1-y-z)^4 z^2}{12} + \frac{y^3 z^2}{3} - \frac{y^4 z^2}{4} - \frac{y^3 z^3}{3} + z^4 y - \frac{y^2 z^4}{2} - y z^5 \Big]_{y=0}^{y=1-z} dz$

$= \int_0^1 \frac{(1-z)^3 z^2}{3} - \frac{(1-z)^4 z^2}{4} - \frac{(1-z)^3 z^3}{3} + z^4 (1-z)$
$\quad - \frac{(1-z)^2 z^4}{2} - (1-z) z^5 + \frac{(1-z)^4 z^2}{12} \, dz$

$= \int_0^1 \frac{(1-z)^3 z^2}{3} - \frac{(1-z)^3 z^3}{3} - \frac{(1-z)^4 z^2}{4}$
$\quad + \frac{(1-z)^4 z^2}{12} + z^4 - z^5 - \frac{z^4}{2} + z^5 - \frac{z^6}{2} - z^5 + z^6 \, dz$

$= \int_0^1 \frac{z^2 (1-z)^4}{3} - \frac{z^2 (1-z)^2}{6} + \frac{z^4}{2} - z^5 + \frac{z^6}{2} \, dz$

$= \int_0^1 \frac{z^2 (1-z)^4}{6} + \frac{z^4}{2} - z^5 + \frac{z^6}{2} \, dz$

$= \int_0^1 \frac{1}{6}(z^2 - 4z^3 + 6z^4 - 4z^5 + z^6) + \frac{z^4}{2} - z^5 + \frac{z^6}{2} \, dz$

$= \int_0^1 \frac{z^2}{6} - \frac{2}{3} z^3 + \frac{3}{2} z^4 - \frac{5}{3} z^5 + \frac{2 z^6}{3} \, dz$

$= \frac{z^3}{18} - \frac{z^4}{6} + \frac{3 z^5}{10} - \frac{5 z^6}{18} + \frac{2 z^7}{21} \Big]_{z=0}^{z=1}$

$= \frac{1}{18} - \frac{1}{6} + \frac{3}{10} - \frac{5}{18} + \frac{2}{21}$

$= \frac{4}{630} = \frac{2}{315}$ so $I_{xy} = \frac{2}{315}$; $I_{xz} = I_{yz} = \frac{2}{315}$

since all these integrals are essentially the same.

16. $\int_1^2 \int_0^{z^2} \int_{x+z}^{x-z} 4 \, dy \, dx \, dz = \int_1^2 \int_0^{z^2} 4y \Big]_{y=x+z}^{y=x-z} dx \, dz$

$= \int_1^2 \int_0^{z^2} 4(x-z) - 4(x+z) \, dx \, dz = \int_1^2 \int_0^{z^2} -8z \, dx \, dz$

$= \int_1^2 -8zx \Big]_{x=0}^{x=z^2} dz = \int_1^2 -8z^3 \, dz = -2z^4 \Big]_{z=1}^{z=2} = -30.$

19. $4\int_0^{3/2}\int_0^{\sqrt{9-4x^2}}\int_0^{9-4x^2-y^2} dz\,dy\,dx = 4\int_0^{3/2}\int_0^{\sqrt{9-4x^2}} 9-4x^2-y^2\,dy\,dx$

$= 4\int_0^{3/2}\left[(9-4x^2)y-\tfrac{1}{3}y^3\right]_{y=0}^{y=\sqrt{9-4x^2}} dx = 4\int_0^{3/2}(9-4x^2)^{3/2}-\tfrac{1}{3}(9-4x^2)^{3/2}\,dx$

$= \tfrac{8}{3}\int_0^{3/2}(9-4x^2)^{3/2}\,dx = \tfrac{4}{3}\left(-\tfrac{2x}{8}(8x^2-45)\sqrt{9-4x^2}+\tfrac{243}{8}\arcsin\left(\tfrac{2x}{3}\right)\right)\Big]_{x=0}^{x=3/2}$

$= \tfrac{81\pi}{4}$ using # 38 from Table of Integrals.

Other ways to set up the integral are:

$4\int_0^3\int_0^{\sqrt{9-y^2}/2}\int_0^{9-4x^2-y^2} dz\,dx\,dy;\quad 4\int_0^{3/2}\int_0^{9-4x^2}\int_0^{\sqrt{9-4x^2-z}} dy\,dz\,dx;$

$4\int_0^9\int_0^{\sqrt{9-z}/2}\int_0^{\sqrt{9-4x^2-z}} dy\,dx\,dz;\quad 4\int_0^3\int_0^{9-y^2}\int_0^{\sqrt{9-y^2-z}/2} dx\,dz\,dy;$

$4\int_0^9\int_0^{\sqrt{9-z}}\int_0^{\sqrt{9-y^2-z}/2} dx\,dy\,dz.$

Section 15.5

2. $(-2\sqrt{2},-2\sqrt{2},-3)\approx(-2.82843,-2.82843,-3);$
$(5,\tfrac{5\pi}{4},\pi+\arctan(-\tfrac{4}{3}))\approx(5,3.92699,2.21430).$

5. $(-\tfrac{7}{\sqrt{2}},\tfrac{7}{\sqrt{2}},6)\approx(-4.94975,4.94975,6);$
$(\sqrt{85},\tfrac{3\pi}{4},\arctan(\tfrac{7}{6}))\approx(9.21954,2.35619,0.86217).$

8. $(2,0,0);\quad (2,0,\tfrac{\pi}{2}).$

11. $(\sqrt{3},-1,-1)\approx(1.73205,-1,-1);$
$(\sqrt{5},330°,\pi+\arctan(-2))\approx(2.23607,330°,116.56505°).$

14. $(0,0,0).$

17. Points on the z-axis have more than one representation in the cylindrical and spherical coordinate systems. Since $r=0$, θ can be any angle.

20. $r^2+z^2=16;\quad \rho^2=16$ or $\rho=4.$

23. $r^2=9$ or $r=3;\quad \rho^2\sin^2\phi=9$ or $\rho\sin\phi=3.$

26. $z=2r^2\cos\theta\sin\theta$ or $z=r^2\sin 2\theta;\quad \rho\cos\phi=2\rho^2\sin^2\phi\cos\theta\sin\theta$
or $1=\rho\tan\phi\sin\phi\sin 2\theta.$

29. $z^2 = r^2 + 1$; $\rho^2 \cos^2 \phi = \rho^2 \sin^2 \phi + 1$ or $\rho^2 \cos 2\phi = 1$.

32. $\displaystyle\int_0^{\pi/2}\int_0^{2\pi}\int_0^1 \rho^2 \sin\phi\, d\rho\, d\theta\, d\phi = \int_0^{\pi/2}\int_0^{2\pi} \tfrac{1}{3}\rho^3 \sin\phi \Big]_{\rho=0}^{\rho=1} d\theta\, d\phi$

$= \displaystyle\int_0^{\pi/2}\int_0^{2\pi} \tfrac{1}{3}\sin\phi\, d\theta\, d\phi = \int_0^{\pi/2} \tfrac{1}{3}\sin\phi(\theta)\Big]_{\theta=0}^{\theta=2\pi} d\phi = \int_0^{\pi/2} \tfrac{2\pi}{3}\sin\phi\, d\phi$

$= \tfrac{2\pi}{3}(-\cos\phi)\Big]_{\phi=0}^{\phi=\pi/2} = \tfrac{2\pi}{3}.$

Section 15.6

Exercises 1-3 use the following tabulated data:

(x,y)	$e^{x^2+y^2}$	(x,y)	$e^{x^2+y^2}$	(x,y)	$e^{x^2+y^2}$
$(0,0)$	1	$(0.2,0.2)$	1.08	$(0.1,0.1)$	1.02
$(0.2,0)$	1.04	$(0.4,0.2)$	1.22	$(0.1,0.3)$	1.11
$(0.4,0)$	1.17	$(0.6,0.2)$	1.49	$(0.1,0.5)$	1.30
$(0.6,0)$	1.43	$(0.8,0.2)$	1.97	$(0.1,0.7)$	1.65
$(0.8,0)$	1.90	$(1,0.2)$	2.83	$(0.1,0.9)$	2.27
$(0,0.2)$	1.04	$(0.2,0.4)$	1.22	$(0.3,0.1)$	1.11
$(0.2,0.2)$	1.08	$(0.4,0.4)$	1.38	$(0.3,0.3)$	1.20
$(0.4,0.2)$	1.22	$(0.6,0.4)$	1.68	$(0.3,0.5)$	1.40
$(0.6,0.2)$	1.49	$(0.8,0.4)$	2.23	$(0.3,0.7)$	1.79
$(0.8,0.2)$	1.97	$(1,0.4)$	3.19	$(0.3,0.9)$	2.46
$(0,0.4)$	1.17	$(0.2,0.6)$	1.49	$(0.5,0.1)$	1.30
$(0.2,0.4)$	1.22	$(0.4,0.6)$	1.68	$(0.5,0.3)$	1.40
$(0.4,0.4)$	1.38	$(0.6,0.6)$	2.05	$(0.5,0.5)$	1.65
$(0.6,0.4)$	1.68	$(0.8,0.6)$	2.72	$(0.5,0.7)$	2.10
$(0.8,0.4)$	2.23	$(1,0.6)$	3.90	$(0.5,0.9)$	2.89
$(0,0.6)$	1.43	$(0.2,0.8)$	1.97	$(0.7,0.1)$	1.65
$(0.2,0.6)$	1.49	$(0.4,0.8)$	2.23	$(0.7,0.3)$	1.79
$(0.4,0.6)$	1.68	$(0.6,0.8)$	2.72	$(0.7,0.5)$	2.10
$(0.6,0.6)$	2.05	$(0.8,0.8)$	3.60	$(0.7,0.7)$	2.66
$(0.8,0.6)$	2.72	$(1,0.8)$	5.16	$(0.7,0.9)$	3.67
$(0,0.8)$	1.90	$(0.2,1)$	2.83	$(0.9,0.1)$	2.27
$(0.2,0.8)$	1.97	$(0.4,1)$	3.19	$(0.9,0.3)$	2.46
$(0.4,0.8)$	2.23	$(0.6,1)$	3.90	$(0.9,0.5)$	2.89
$(0.6,0.8)$	2.72	$(0.8,1)$	5.16	$(0.9,0.7)$	3.67
$(0.8,0.8)$	3.60	$(1,1)$	7.40	$(0.9,0.9)$	5.05

2. $\displaystyle\int_0^1\int_0^1 e^{x^2+y^2}\, dx\, dy \approx 68.29(0.04) = 2.7316.$

5. From example 26, omit calculations for the points $(0.8, 1), (1, 0.8)$, and $(1, 1)$; $22 - (1.64 + 1.64 + 2.00) = 16.72$;
$$\iint_D x^2 + y^2 \, dA \approx 16.72(0.04) \approx 0.67.$$

8. $\int_0^{\pi/2} \int_0^1 e^{r^2}(r \, dr \, d\theta) = \int_0^{\pi/2} \frac{e^{r^2}}{2}\Big]_{r=0}^{r=1} d\theta = \int_0^{\pi/2} \frac{e}{2} - \frac{1}{2} \, d\theta$

$= \left(\frac{e-1}{2}\right) \theta \Big]_{\theta=0}^{\theta=\pi/2} = \left(\frac{e-1}{2}\right)\left(\frac{\pi}{2}\right) = \frac{\pi(e-1)}{4} \approx 1.34954.$

(Note: For the first integration, let $u = r^2$ and $du = 2r \, dr$.)

ANSWERS TO SELECTED EXERCISES FROM CHAPTER 16

Section 16.1

1.
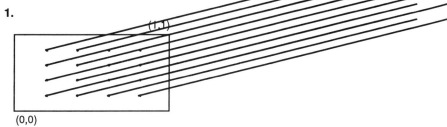

4.
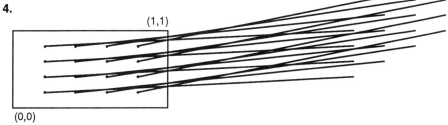

7.
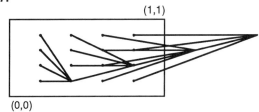

10.
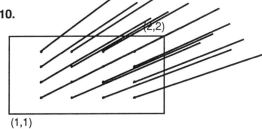

13. $\nabla h = \langle yze^{xyz}, xze^{xyz}, xye^{xyz} \rangle$.

16. $\nabla \cdot \mathbf{G} = yz + 2xyz^3 + x^3y^2$.

19. $\nabla \times \mathbf{F} = \langle 2y - 2z, 2z - 2x, 2x - 2y \rangle$.

22. $\nabla \times \mathbf{P} = \langle -2y - 3z^2, 1 - 3x^2, 0 \rangle.$

25. $\nabla h = \langle yze^{xyz}, xze^{xyz}, xye^{xyz} \rangle \Rightarrow \nabla^2 y = y^2z^2 e^{xyz} + x^2z^2 e^{xyz} + x^2y^2 e^{xyz}.$

28. $\nabla^2 G = \langle 0, 2xz^3 + 6xy^2z, 6xy^2z + 2x^3z \rangle.$

31. $\nabla f \times \nabla g = \langle f_y g_z - f_z g_y, \quad g_x f_z - f_x g_z, \quad f_x g_y - f_y g_x \rangle;$
$$\nabla \cdot (\nabla f \times \nabla g) = \frac{\partial}{\partial x}(f_y g_z - f_z g_y) + \frac{\partial}{\partial y}(g_x f_z - f_x g_z) + \frac{\partial}{\partial z}(f_x g_y - f_y g_x)$$
$$= f_{yx} g_z + f_y g_{zx} - f_{zx} g_y - f_z g_{yx} + g_{xy} f_z + g_x f_{zy}$$
$$- f_{xy} g_z - f_x g_{zy} + f_{xz} g_y + f_x g_{yz} - f_{yz} g_x - f_y g_{xz} = 0.$$

34. Let $\mathbf{G} = \langle G_1, G_2, G_3 \rangle$.
$$\nabla \times \mathbf{G} = \left\langle \frac{\partial G_3}{\partial y} - \frac{\partial G_2}{\partial z}, \frac{\partial G_1}{\partial z} - \frac{\partial G_3}{\partial x}, \frac{\partial G_2}{\partial x} - \frac{\partial G_1}{\partial y} \right\rangle,$$
so
$$\nabla \times (\nabla \times \mathbf{G}) = \left\langle \frac{\partial}{\partial y}\left(\frac{\partial G_2}{\partial x} - \frac{\partial G_1}{\partial y}\right) - \frac{\partial}{\partial z}\left(\frac{\partial G_1}{\partial z} - \frac{\partial G_3}{\partial x}\right), \right.$$
$$\frac{\partial}{\partial z}\left(\frac{\partial G_3}{\partial y} - \frac{\partial G_2}{\partial z}\right) - \frac{\partial}{\partial x}\left(\frac{\partial G_2}{\partial x} - \frac{\partial G_1}{\partial y}\right),$$
$$\left. \frac{\partial}{\partial x}\left(\frac{\partial G_1}{\partial z} - \frac{\partial G_3}{\partial x}\right) - \frac{\partial}{\partial y}\left(\frac{\partial G_3}{\partial y} - \frac{\partial G_2}{\partial z}\right) \right\rangle.$$

In terms of the standard basis components, we can write
$$\nabla \times (\nabla \times \mathbf{G}) = \left(\frac{\partial^2 G_2}{\partial y \partial x} - \frac{\partial^2 G_1}{\partial y^2} - \frac{\partial^2 G_1}{\partial z^2} + \frac{\partial^2 G_3}{\partial z \partial x}\right) \mathbf{i}$$
$$+ \left(\frac{\partial^2 G_3}{\partial z \partial y} - \frac{\partial^2 G_2}{\partial z^2} - \frac{\partial^2 G_2}{\partial x^2} + \frac{\partial^2 G_1}{\partial x \partial y}\right) \mathbf{j}$$
$$+ \left(\frac{\partial^2 G_1}{\partial x \partial z} - \frac{\partial^2 G_3}{\partial x^2} - \frac{\partial^2 G_3}{\partial y^2} + \frac{\partial^2 G_2}{\partial y \partial z}\right) \mathbf{k}$$
$$= \left(\frac{\partial^2 G_1}{\partial x^2} + \frac{\partial^2 G_2}{\partial y \partial x} + \frac{\partial^2 G_3}{\partial z \partial x}\right) \mathbf{i} + \left(\frac{\partial^2 G_1}{\partial x \partial y} + \frac{\partial^2 G_2}{\partial y^2} + \frac{\partial^2 G_3}{\partial z \partial y}\right) \mathbf{j}$$
$$+ \left(\frac{\partial^2 G_1}{\partial x \partial z} + \frac{\partial^2 G_2}{\partial y \partial z} + \frac{\partial^2 G_3}{\partial z^2}\right) \mathbf{k}$$
$$- \left(\left(\frac{\partial^2 G_1}{\partial x^2} + \frac{\partial^2 G_1}{\partial y^2} + \frac{\partial^2 G_1}{\partial z^2}\right) \mathbf{i} + \left(\frac{\partial^2 G_2}{\partial x^2} + \frac{\partial^2 G_2}{\partial y^2} + \frac{\partial^2 G_2}{\partial z^2}\right) \mathbf{j}\right.$$
$$\left. + \left(\frac{\partial^2 G_3}{\partial x^2} + \frac{\partial^2 G_3}{\partial y^2} + \frac{\partial^2 G_3}{\partial z^2}\right) \mathbf{k}\right)$$
$$= \nabla(\nabla \cdot \mathbf{G}) - \langle \nabla \cdot \nabla G_1, \nabla \cdot \nabla G_2, \nabla \cdot \nabla G_3 \rangle$$
$$= \nabla(\nabla \cdot \mathbf{G}) - \nabla^2 \mathbf{G}.$$

37. $p.$

40. Yes. The first statement says the divergence of the curl of a vector field **F** is 0 which was shown in the text. The second statement says the curl of the gradient of a function f is $\langle 0,0,0\rangle$. To show this calculate
$$\nabla \times (\nabla f)$$
$$= \nabla \times \left\langle \frac{\partial f}{\partial x}, \frac{\partial f}{\partial y}, \frac{\partial f}{\partial z}\right\rangle$$
$$= \left\langle \frac{\partial}{\partial y}\left(\frac{\partial f}{\partial z}\right) - \frac{\partial}{\partial z}\left(\frac{\partial f}{\partial y}\right), \frac{\partial}{\partial z}\left(\frac{\partial f}{\partial x}\right) - \frac{\partial}{\partial x}\left(\frac{\partial f}{\partial z}\right), \frac{\partial}{\partial x}\left(\frac{\partial f}{\partial y}\right) - \frac{\partial}{\partial y}\left(\frac{\partial f}{\partial x}\right)\right\rangle$$
$$= \langle 0,\ 0,\ 0\rangle.$$

Section 16.2

2. $D\mathbf{G}(u,v) = \begin{bmatrix} v & u \\ \frac{1}{v} & -\frac{u}{v^2} \\ -\frac{v}{u^2} & \frac{1}{u} \end{bmatrix}$.

5. Does not make sense because $\mathbf{G}:\mathbb{R}^2 \to \mathbb{R}^3$ and $\mathbf{F}:\mathbb{R}^2 \to \mathbb{R}^2$.

8. Does not make sense because $\mathbf{P}:\mathbb{R}^3 \to \mathbb{R}^3$ and $\mathbf{H}:\mathbb{R}^3 \to \mathbb{R}^2$.

11. $\mathbf{H}\circ\mathbf{G}(u,v) = \mathbf{H}(uv,\frac{u}{v},\frac{v}{u}) = \langle uv, v^2\rangle$;

$D(\mathbf{H}\circ\mathbf{G})(u,v) = \begin{bmatrix} v & u \\ 0 & 2v \end{bmatrix}$;

$D\mathbf{H}(x,y,z) = \begin{bmatrix} yz & xz & xy \\ y^2z^3 & 2xyz^3 & 3xy^2z^2 \end{bmatrix}$ and $D\mathbf{G}(u,v) = \begin{bmatrix} v & u \\ \frac{1}{v} & -\frac{u}{v^2} \\ -\frac{v}{u^2} & \frac{1}{u} \end{bmatrix}$;

$(D\mathbf{H}\circ\mathbf{G})D\mathbf{G} = \begin{bmatrix} 1 & v^2 & u^2 \\ \frac{v}{u} & \frac{2v^3}{u} & 3uv \end{bmatrix}\begin{bmatrix} v & u \\ \frac{1}{v} & -\frac{u}{v^2} \\ -\frac{v}{u^2} & \frac{1}{u} \end{bmatrix} = \begin{bmatrix} v & u \\ 0 & 2v \end{bmatrix}$.

14. $\mathbf{G}\circ\mathbf{H}(x,y,z) = \mathbf{G}(xyz, xy^2z^3) = \langle x^2y^3z^4, \frac{1}{yz^2}, yz^2\rangle$;

$D(\mathbf{G}\circ\mathbf{H})(x,y,z) = \begin{bmatrix} 2xy^3z^4 & 3x^2y^2z^4 & 4x^2y^3z^3 \\ 0 & -\frac{1}{y^2z^2} & -\frac{2}{yz^3} \\ 0 & z^2 & 2yz \end{bmatrix}$;

$D\mathbf{G}(u,v) = \begin{bmatrix} v & u \\ \frac{1}{v} & -\frac{u}{v^2} \\ -\frac{v}{u^2} & \frac{1}{u} \end{bmatrix}$ and $D\mathbf{H}(x,y,z) = \begin{bmatrix} yz & xz & xy \\ y^2z^3 & 2xyz^3 & 3xy^2z^2 \end{bmatrix}$;

(continued on next page)

14. (continued from previous page)

$(DG \circ \mathbf{H})(D\mathbf{H})$

$$= \begin{bmatrix} xy^2z^3 & xyz \\ \frac{1}{xy^2z^3} & -\frac{1}{xy^3z^5} \\ -\frac{z}{x} & \frac{1}{xyz} \end{bmatrix} \begin{bmatrix} yz & xz & xy \\ y^2z^3 & 2xyz^3 & 3xy^2z^2 \end{bmatrix}$$

$$= \begin{bmatrix} xy^3z^4 + xy^3z^4 & x^2y^2z^4 + 2x^2y^2z^4 & x^2y^3z^3 + 3x^2y^3z^3 \\ \frac{1}{xyz^2} - \frac{1}{xyz^2} & \frac{1}{y^2z^2} - \frac{2}{y^2z^2} & \frac{1}{yz^3} - \frac{3}{yz^3} \\ -\frac{z^2y}{x} + \frac{yz^2}{x} & -z^2 + 2z^2 & -zy + 3yz \end{bmatrix}$$

$$= \begin{bmatrix} 2xy^3z^4 & 3x^2y^2z^4 & 4x^2y^3z^3 \\ 0 & -\frac{1}{y^2z^2} & -\frac{2}{yz^3} \\ 0 & z^2 & 2yz \end{bmatrix}$$

17. $x = \frac{u}{y}$ and $x = \frac{y}{v} \Rightarrow \frac{u}{y} = \frac{y}{v} \Rightarrow y^2 = uv \Rightarrow y = \sqrt{uv}$;

$y = \frac{u}{x}$ and $y = vx \Rightarrow \frac{u}{x} = vx \Rightarrow x^2 = \frac{u}{v} \Rightarrow x = \sqrt{\frac{u}{v}}$;

$$J(u,v) = \begin{vmatrix} \frac{1}{2}\left(\frac{u}{v}\right)^{-1/2}\left(\frac{1}{v}\right) & \frac{1}{2}\left(\frac{u}{v}\right)^{-1/2}\left(-\frac{u}{v^2}\right) \\ \frac{1}{2}(uv)^{-1/2}(v) & \frac{1}{2}(uv)^{-1/2}(u) \end{vmatrix}$$

$= \frac{1}{4}(u^2)^{-1/2}\left(\frac{u}{v}\right) - \frac{1}{4}(u^2)^{-1/2}\left(-\frac{u}{v}\right)$

$= \frac{1}{4}\left(\frac{u}{v}\right)\left(\frac{1}{u} + \frac{1}{u}\right) = \frac{1}{2v}$.

Borderlines of the new region lie on $v = 1$ (since $\frac{y}{x} = 1$) and $v = 2$ (since $\frac{y}{x} = 2$) and $u = 1$ and $u = 2$. We have

$$\int_1^2 \int_1^2 \frac{1}{2v} \, du \, dv = \int_1^2 \frac{u}{2v}\Big]_1^2 dv = \int_1^2 \frac{2}{2v} - \frac{1}{2v} \, dv = \int_1^2 \frac{1}{2v} \, dv$$

$= \frac{1}{2}\ln(2v)\Big]_1^2 = \frac{1}{2}\ln(4) - \frac{1}{2}\ln(2) = \frac{1}{2}\ln(2)$.

20. $J(\rho, \theta, \varphi) = \begin{vmatrix} \cos\theta\sin\varphi & -\rho\sin\theta\sin\varphi & \rho\cos\theta\cos\varphi \\ \sin\theta\sin\varphi & \rho\cos\theta\sin\varphi & \rho\sin\theta\cos\varphi \\ \cos\varphi & 0 & -\rho\sin\varphi \end{vmatrix}$

$= \cos\varphi(-\rho^2\sin^2\theta\sin\varphi\cos\varphi - \rho^2\cos^2\theta\sin\varphi\cos\varphi)$

$+ 0 - \rho\sin\varphi(\rho\cos^2\theta\sin^2\varphi + \rho\sin^2\theta\sin^2\varphi)$

$= \cos\varphi(-\rho^2\sin\varphi\cos\varphi) - \rho\sin\varphi(\rho\sin^2\varphi)$

$= -\rho^2\sin\varphi(\cos^2\varphi + \sin^2\varphi) = -\rho^2\sin\varphi$.

Section 16.3

3. Parametrization of line segment: $\mathbf{r}(t) = \langle 0, -1+2t, 1-2t \rangle$, $0 \le t \le 1$;
$\mathbf{r}'(t) = \langle 0, 2, -2 \rangle$ and $\|\mathbf{r}'(t)\| = \sqrt{8}$.

$$\oint_C f\, ds = \int_0^1 \sqrt{8}(1-2t)^2\, dt = \sqrt{8}\int_0^1 (1-4t+4t^2)\, dt$$
$$= \sqrt{8}\left(t - 2t^2 + \tfrac{4t^3}{3}\right)\Big]_0^1 = \tfrac{\sqrt{8}}{3}.$$

$$\oint_C \mathbf{F}\cdot d\mathbf{s} = \int_0^1 \langle 0, (-1+2t)^3, (1-2t)\rangle \cdot \langle 0, 2, -2\rangle\, dt$$
$$= \int_0^1 2(-1+2t)^3 - 2(1-2t)\, dt = \tfrac{(1-2t)^4}{4} - 2t + 2t^2\Big]_0^1 = 0.$$

6. Parametrization of line segment: $\mathbf{r}(t) = \langle 1-t, -1+t, t\rangle$, $0 \le t \le 1$;
$\mathbf{r}'(t) = \langle -1, 1, 1\rangle$ and $\|\mathbf{r}'(t)\| = \sqrt{3}$.

$$\oint_C f\, ds = \int_0^1 \sqrt{3}((1-t)^2 + (-1+t)^2)\, dt$$
$$= \int_0^1 \sqrt{3}(2 - 4t + 2t^2)\, dt = \sqrt{3}\left(2t - 2t^2 + \tfrac{2t^3}{3}\right)\Big]_0^1 = \tfrac{2\sqrt{3}}{3}.$$

$$\oint_C \mathbf{F}\cdot d\mathbf{s} = \int_0^1 \langle t-t^2, -t+t^2, t^2\rangle \cdot \langle -1, 1, 1\rangle\, dt$$
$$= \int_0^1 3t^2 - 2t\, dt = t^3 - t^2\Big]_0^1 = 0.$$

9. $\mathbf{r}'(t) = \langle -2\sin 2t, 2\cos 2t, 1\rangle$, so

$$\oint_C \mathbf{F}\cdot \mathbf{T}\, ds = \int_0^\pi \langle 1, t, 2t\rangle \cdot \langle -2\sin 2t, 2\cos 2t, 1\rangle\, dt$$
$$= \int_0^\pi (-2\sin 2t + 2t\cos 2t + 2t)\, dt = \cos(2t)\Big]_0^\pi + t^2\Big]_0^\pi + \int_0^\pi 2t\cos t\, dt$$
$$= \pi^2 + \int_0^\pi 2t\cos t\, dt = \pi^2 + t\sin 2t\Big]_0^\pi - \int_0^\pi \sin 2t\, dt = \pi^2 \approx 9.86960.$$

12. A particle traveling along the path $\mathbf{q}(t)$ moves the slowest. A particle traveling along the path $\mathbf{p}(t)$ moves the fastest. A particle travelling along the path $\mathbf{q}(t)$ will collide with a particle moving along the path $\mathbf{r}(t)$ at $t = \tfrac{\pi}{3}$ and also with a particle moving along the path $\mathbf{p}(t)$ at $t \approx 0.43135$.

15. $\mathbf{q}'(t) = \langle \tfrac{1}{2}\cos(\tfrac{t}{2}), -\tfrac{1}{2}\sin(\tfrac{t}{2})\rangle$; $\|\mathbf{q}'(t)\| = \tfrac{1}{2}$.

$$\oint_C f\, ds = \tfrac{1}{2}\int_0^\pi \sin^2(\tfrac{t}{2})\cos^3(\tfrac{t}{2})\, dt = \tfrac{2}{15}.$$

18. $\mathbf{p}'(t) = \langle -\pi\sin(\pi t), \pi\cos(\pi t)\rangle$.

$$\oint_C \mathbf{F}\cdot d\mathbf{s} = \int_0^{1/2} \langle \sin(\pi t), -\cos(\pi t)\rangle \cdot \langle -\pi\sin(\pi t), \pi\cos(\pi t)\rangle\, dt$$

$$= \int_0^{1/2} -\pi\sin^2(\pi t) - \pi\cos^2(\pi t)\, dt = -\pi\int_0^{1/2} dt = -\tfrac{\pi}{2}.$$

21. Parametrization of the line segment: $\mathbf{r}(t) = \langle 1-t, t\rangle$.
$\mathbf{r}'(t) = \langle -1, 1\rangle;\qquad \|\mathbf{r}'(t)\| = \sqrt{2}$.

$\oint_{C'} f\, ds = \sqrt{2}\int_0^1 (1-t)^2 t^3\, dt = \tfrac{\sqrt{2}}{60}$. Yes, it is different.

24. Whether we move *against* or *with* a force will affect the work we perform.

Section 16.4

1. $\mathbf{r}(u,v) = \langle u, v, u^3 + v^2\rangle$;
D = circle with radius 1 and center at $(0,0)$.

4. $\mathbf{r}(u,v) = \langle u, v, -u^3 - v^3\rangle$;
D = circle with radius 2 and center at $(0,0)$.

7. $\mathbf{T}_u = \langle -v\sin u, v\cos u, 0\rangle;\qquad \mathbf{T}_u(\pi, -\pi) = \langle 0, \pi, 0\rangle;$
$\mathbf{T}_v = \langle \cos u, \sin u, 1\rangle;\qquad \mathbf{T}_v(\pi, -\pi) = \langle -1, 0, 1\rangle;$
$\mathbf{T}_u \times \mathbf{T}_v = \langle v\cos u, v\sin u, -v\rangle;\qquad \mathbf{T}_u \times \mathbf{T}_v(\pi, -\pi) = \langle \pi, 0, \pi\rangle.$
The equation of tangent plane is of the form
$$\pi(x - \pi) + \pi(z + \pi) = 0.$$

10. $\mathbf{T}_u = \langle 0, 0, 1\rangle;\quad \mathbf{T}_v = \langle -\sin v, \cos v, 0\rangle;\quad \mathbf{T}_v(1, \tfrac{\pi}{4}) = \langle -\tfrac{1}{\sqrt{2}}, \tfrac{1}{\sqrt{2}}, 0\rangle;$
$\mathbf{T}_u \times \mathbf{T}_v = \langle -\cos v, -\sin v, 0\rangle;\quad \mathbf{T}_u \times \mathbf{T}_v(1, \tfrac{\pi}{4}) = \langle -\tfrac{1}{\sqrt{2}}, -\tfrac{1}{\sqrt{2}}, 0\rangle.$
The equation of the tangent plane is of the form
$$-\tfrac{1}{\sqrt{2}}(x - \tfrac{1}{\sqrt{2}}) - \tfrac{1}{\sqrt{2}}(y - \tfrac{1}{\sqrt{2}}) = 0.$$

13. $\mathbf{T}_u = \langle 2u, 0, 2u\rangle;\quad \mathbf{T}_u(1,1) = \langle 2, 0, 2\rangle;$
$\mathbf{T}_v = \langle 0, 2v, 2v\rangle;\quad \mathbf{T}_v(1,1) = \langle 0, 2, 2\rangle;$
$\mathbf{T}_u \times \mathbf{T}_v = \langle -4uv, -4uv, 4uv\rangle;\quad \mathbf{T}_u \times \mathbf{T}_v(1,1) = \langle -4, -4, 4\rangle.$
The equation of the tangent plane is of the form $-x - y + z = 0$.

16. $\mathbf{T}_u = \langle 1, \frac{1}{2}(u+v)^{-1/2}, 0 \rangle;\quad \mathbf{T}_u(\frac{1}{8}, \frac{1}{8}) = \langle 1, 1, 0 \rangle;$

$\mathbf{T}_v = \langle 0, \frac{1}{2}(u+v)^{-1/2}, 1 \rangle;\quad \mathbf{T}_v(\frac{1}{8}, \frac{1}{8}) = \langle 0, 1, 1 \rangle;$

$\mathbf{T}_u \times \mathbf{T}_v(\frac{1}{8}, \frac{1}{8}) = \langle 1, -1, 1 \rangle.$

The unit vector is $\mathbf{u} = \frac{\langle 1, -1, 1 \rangle}{\sqrt{3}}$.

19. $\mathbf{T}_u = \langle -4u, 1, 0 \rangle;\quad \mathbf{T}_u(1/2, 0) = \langle -2, 1, 0 \rangle;$

$\mathbf{T}_v = \langle 2v, 0, 1 \rangle;\quad \mathbf{T}_v(\frac{1}{2}, 0) = \langle 0, 0, 1 \rangle;$

$\mathbf{T}_u \times \mathbf{T}_v(\frac{1}{2}, 0) = \langle 1, 2, 0 \rangle.$

An equation of the tangent plane is $(x - \frac{5}{2}) + 2(y - \frac{1}{2}) = 0.$

Section 16.5

2. $\mathbf{T}_u = \langle \cos v, \sin v, 1 \rangle;\quad \mathbf{T}_v = \langle -u \sin v, u \cos v, 0 \rangle;$

$\mathbf{T}_u \times \mathbf{T}_v = \langle -u \cos v, -u \sin v, u \rangle;\quad \|\mathbf{T}_u + \mathbf{T}_v\| = \sqrt{2}u.$

Surface area:

$$\int_0^{\pi/2} \int_0^1 \sqrt{2}u \, du \, dv = \sqrt{2} \int_0^{\pi/2} \left(\frac{u^2}{2}\right)\Big]_0^1 dv = \sqrt{2} \int_0^{\pi/2} \frac{1}{2} dv = \frac{\sqrt{2}}{2}v \Big]_0^{\pi/2} = \frac{\pi\sqrt{2}}{4}.$$

$$\iint_S f \, dS = \int_0^{\pi/2} \int_0^1 \sqrt{2}u^3 \, du \, dv = \sqrt{2} \int_0^{\pi/2} \left(\frac{u^4}{4}\right)\Big]_0^1 dv$$

$$= \sqrt{2} \int_0^{\pi/2} \frac{1}{4} dv = \frac{\sqrt{2}}{4} v \Big]_0^{\pi/2} = \frac{\pi\sqrt{2}}{8}.$$

6. $\mathbf{T}_u = \langle 1, -1, 1 \rangle;\quad \mathbf{T}_v = \langle -1, 1, 1 \rangle;$

$\mathbf{T}_u \times \mathbf{T}_v = \langle -2, -2, 0 \rangle;\quad \|\mathbf{T}_u \times \mathbf{T}_v\| = \sqrt{8}.$

Surface area: $\int_1^2 \int_0^1 \sqrt{8} \, du \, dv = \sqrt{8}.$

$$\iint_S f \, dS = \int_1^2 \int_0^1 \sqrt{8}((v-u)^2 + (u+v)^2) \, du \, dv$$

$$= \sqrt{8} \int_1^2 \int_0^1 (2u^2 + 2v^2) \, du \, dv = \sqrt{8} \int_1^2 \frac{2u^3}{3} + 2v^2 u \Big]_0^1 dv$$

$$= \sqrt{8} \int_1^2 \frac{2}{3} + 2v^2 \, dv = \sqrt{8}\left(\frac{2v + 2v^3}{3}\right)\Big]_1^2 = \frac{16\sqrt{8}}{3}.$$

9. Using the parametrization in exercise 8,

$\mathbf{T}_u = \langle -\sin u \sin v,\ 0,\ \cos u \sin v \rangle; \quad \mathbf{T}_v = \langle \cos u \cos v,\ -\sin v,\ \sin u \cos v \rangle;$

$\mathbf{T}_u \times \mathbf{T}_v = \langle \cos u \sin^2 v,\ \sin v \cos v,\ \sin u \sin^2 v \rangle; \quad \|\mathbf{T}_u \times \mathbf{T}_v\| = |\sin v| = \sin v.$

(We can remove the absolute value signs since $\sin v \geq 0$ for $\frac{\pi}{2} \leq v \leq \pi$.)

The surface area of the hemisphere is

$$\int_{\pi/2}^{\pi} \int_0^{2\pi} \sin v \, du \, dv = \int_{\pi/2}^{\pi} u \sin v \Big]_{u=0}^{u=2\pi} dv$$

$$= \int_{\pi/2}^{\pi} 2\pi \sin v \, dv = -2\pi \cos v \Big]_{v=\pi/2}^{v=\pi} = 2\pi.$$

ANSWERS TO SELECTED EXERCISES FROM CHAPTER 17

Section 17.1

1. True; a potential is $\varphi(x,y) = e^{xy} + \ln(y)$.

4. True; F is conservative, so $\oint_C P\,dx + Q\,dy = \oint_C \mathbf{F} \cdot \mathbf{T}\,ds = 0$.

7. True; div(curl **F**) = div(**0**) = 0.

10. True; **F** is conservative, so curl **F** = **0**.

13. True; **F** is conservative.

16. Let $P(x,y) = e^x - \cos y$ and $Q(x,y) = e^y - x\sin y$. Then
$$\frac{\partial P}{\partial y} = \sin(y) \neq \frac{\partial Q}{\partial x} = -\sin(y)$$
and **H** is not conservative.

19. Since **F** is conservative, we should have the same answer as in exercise 18: $\frac{16}{3}$.

Section 17.2

1. $\mathbf{T}_u = \langle 1, 0, 2u \rangle$; $\mathbf{T}_v = \langle 0, 1, 2v \rangle$;
$$\mathbf{T}_u \times \mathbf{T}_v = \begin{vmatrix} \mathbf{i} & \mathbf{j} & \mathbf{k} \\ 1 & 0 & 2u \\ 0 & 1 & 2v \end{vmatrix} = \langle -2u, -2v, 1 \rangle; \quad \|\mathbf{T}_u \times \mathbf{T}_v\| = \sqrt{4u^2 + 4v^2 + 1}.$$

Using polar coordinates, we have
$$\int_{-1}^{1} \int_{-\sqrt{1-v^2}}^{\sqrt{1-v^2}} \sqrt{4u^2 + 4v^2 + 1}\,du\,dv = \int_0^{2\pi} \int_0^1 \sqrt{4r^2 + 1}\,r\,dr\,d\theta$$
$$= \tfrac{1}{8} \int_0^{2\pi} \tfrac{2}{3}(4r^2 + 1)^{3/2} \Big]_0^1 d\theta = \tfrac{1}{12}(5^{3/2} - 1)\theta \Big]_0^{2\pi} = \tfrac{\pi}{6}(5^{3/2} - 1).$$

4. $\nabla \times \mathbf{F} = \begin{vmatrix} \mathbf{i} & \mathbf{j} & \mathbf{k} \\ \frac{\partial}{\partial x} & \frac{\partial}{\partial y} & \frac{\partial}{\partial z} \\ y & -x & z - x^2 - y^2 \end{vmatrix} = \langle -2y, 2x, -2 \rangle$.

7. S is parametrized by $\mathbf{r}(u,v) = \langle u\cos v, u\sin v, u\rangle$ for $0 \leq u \leq 4$, $0 \leq v \leq 2\pi$.

$\mathbf{T}_u = \langle \cos v, \sin v, 1\rangle$; $\mathbf{T}_v = \langle -u\sin v, u\cos v, 0\rangle$;

$\mathbf{T}_u \times \mathbf{T}_v = \langle -u\cos v, -u\sin v, u\rangle$; $\nabla \times \mathbf{F} = \langle -2y, 1, 2\rangle$.

$$\iint_S (\nabla \times F) \cdot \mathbf{n}\, dS = \int_0^{2\pi} \int_0^4 \langle -2u\sin v, 1, 2\rangle \cdot \langle -u\cos v, -u\sin v, u\rangle\, du\, dv$$

$$= \int_0^{2\pi} \int_0^4 2u^2 \sin v \cos v - u\sin v + 2u\, du\, dv$$

$$= \int_0^{2\pi} \tfrac{128}{3}\sin v \cos v - 8\sin v + 16\, dv$$

$$= \tfrac{128}{6}\sin^2 v + 8\cos v + 16v \Big]_0^{2\pi} = 32\pi.$$

The curve of intersection of the cone and plane is the circle of radius 16 at height $z = 4$, parametrized by $\mathbf{r}(t) = \langle 4\cos t, 4\sin t, 4\rangle$;

$\mathbf{r}'(t) = \langle -4\sin t, 4\cos t, 0\rangle$.

$$\oint_C \mathbf{F} \cdot \mathbf{T}\, ds = \int_0^{2\pi} \langle 4, 8\cos t, -16\sin^2 t\rangle \cdot \langle -4\sin t, 4\cos t, 0\rangle\, dt$$

$$= \int_0^{2\pi} -16\sin t + 32\cos^2 t\, dt$$

$$= 16\cos t + 32\left(\tfrac{1}{2}t + \tfrac{1}{4}\sin 2t\right)\Big]_0^{2\pi} = 32\pi.$$

10. The portion of the plane bounded by the curve C is parametrized by $\mathbf{r}(u,v) = \langle u\cos v, u\sin v, u\cos v + 2\rangle$ for $0 \leq u \leq 2$ and $0 \leq v \leq 2\pi$.

$\mathbf{T}_u = \langle \cos v, \sin v, \cos v\rangle$; $\mathbf{T}_v = \langle -u\sin v, u\cos v, -u\sin v\rangle$;

$\mathbf{T}_u \times \mathbf{T}_v = \langle -u, 0, u\rangle$; $\nabla \times \mathbf{F} = \langle 1 - 3x, 0, 3z\rangle$.

By Stoke's theorem:

$$\oint_C \mathbf{F} \cdot \mathbf{T}\, ds = \iint_S (\nabla \times \mathbf{F}) \cdot \mathbf{n}\, dS$$

$$= \int_0^{2\pi} \int_0^2 \langle 1 - 3u\cos v, 0, 3u\cos v + 6\rangle \cdot \langle -u, 0, u\rangle\, du\, dv$$

$$= \int_0^{2\pi} \int_0^2 -u + 3u^2\cos v + 3u^2\cos v + 6u\, du\, dv$$

$$= \int_0^{2\pi} \int_0^2 5u + 6u^2\cos v\, du\, dv = 20\pi.$$

13. The portion of the sphere bounded by the curve C is parametrized by
$\mathbf{r}(u,v) = \langle u\cos v, u\sin v, \sqrt{1-u^2}\rangle$ for $0 \le u \le 1$ and $0 \le v \le \frac{\pi}{2}$.

$\mathbf{T}_u = \langle \cos v, \sin v, -\frac{u}{\sqrt{1-u^2}}\rangle$; $\quad \mathbf{T}_v = \langle -u\sin v, u\cos v, 0\rangle$;

$\mathbf{T}_u \times \mathbf{T}_v = \langle \frac{u^2\cos v}{\sqrt{1-u^2}}, \frac{u^2\sin v}{\sqrt{1-u^2}}, u\rangle$; $\quad \nabla \times \mathbf{F} = \langle -z, 0, 0\rangle$;

$\nabla \times \mathbf{F}(\mathbf{r}(u,v)) = \langle -\sqrt{1-u^2}, 0, 0\rangle$.

$$\oint_C \mathbf{F}\cdot\mathbf{T}\,ds = \iint_S (\nabla\times\mathbf{F})\cdot\mathbf{n}\,dS = \int_0^{\pi/2}\int_0^1 -u^2\cos v\,du\,dv$$

$$= \int_0^{\pi/2} -\tfrac{1}{3}\cos v\,dv = -\tfrac{1}{3}\sin v\Big]_0^{\pi/2} = -\tfrac{1}{3}.$$

Section 17.3

1. $\iint_D 6 - 4\,dx\,dy = \iint_D 2\,dx\,dy = \int_0^{2\pi}\int_0^4 2r\,dr\,d\theta = 32\pi.$

4. The path C consists of the straight line segment from $(0,0)$ to $(1,0)$, the straight line segment from $(1,0)$ to $(1,1)$, and the straight line segment from $(1,1)$ to $(0,0)$; let $P = 0$ and $Q = xy$;

Parametrize C_1: $\mathbf{r}_1(t) = \langle t, 0\rangle$ for $0 \le t \le 1$, so $\mathbf{r}_1'(t) = \langle 1, 0\rangle$, and

$$\oint_{C_1} P\,dx + Q\,dy = \int_0^1 \langle 0, 0\rangle \cdot \langle 1, 0\rangle\,dt = 0;$$

Parametrize C_2: $\mathbf{r}_2(t) = \langle 1, t\rangle$ for $0 \le t \le 1$, so $\mathbf{r}_2'(t) = \langle 0, 1\rangle$ and

$$\oint_{C_2} P\,dx + Q\,dy = \int_0^1 \langle 0, t\rangle \cdot \langle 0, 1\rangle\,dt = \int_0^1 t\,dt = \tfrac{t^2}{2}\Big]_0^1 = \tfrac{1}{2};$$

Parametrize C_3: $\mathbf{r}_3(t) = \langle 1-t, 1-t\rangle$ for $0 \le t \le 1$, so $\mathbf{r}_3'(t) = \langle -1, -1\rangle$ and

$$\oint_{C_3} P\,dx + Q\,dy = \int_0^1 \langle 0, (1-t)^2\rangle \cdot \langle -1, -1\rangle\,dt$$

$$= -\int_0^1 1 - 2t + t^2\,dt = -t + t^2 - \tfrac{t^3}{3}\Big]_0^1 = -\tfrac{1}{3}.$$

Hence, $\oint_C P\,dx + Q\,dy = 0 + \tfrac{1}{2} - \tfrac{1}{3} = \tfrac{1}{6}.$

8. Parametrize $C_1 : \mathbf{r}(t) = \langle t, t^2\rangle$ for $0 \le t \le 1$.

$$\oint_{C_1} \mathbf{F}\cdot\mathbf{T}\,ds = \int_0^1 \langle t+t^2, t^3\rangle\cdot\langle 1, 2t\rangle\,dt = \int_0^1 t+t^2+2t^4\,dt = \tfrac{1}{2}+\tfrac{1}{3}+\tfrac{2}{5} = \tfrac{37}{30}.$$

Parametrize $C_2 : \mathbf{r}(t) = \langle t^2, t\rangle$ for $0 \le t \le 1$. (Orientation is reversed, wo we'll need to change signs.)

$$\oint_{C_2} \mathbf{F}\cdot\mathbf{T}\,ds = -\int_0^1 \langle t+t^2, t^3\rangle\cdot\langle 2t, 1\rangle\,dt = -\int_0^1 2t^2+3t^3\,dt = -\left(\tfrac{2}{3}+\tfrac{3}{4}\right) = -\tfrac{17}{12}.$$

(continued on the next page)

8. (continued from previous page)

So, $\oint_C P\,dx + Q\,dy = \frac{37}{30} - \frac{17}{12} = -\frac{11}{60}$.

On the other hand,

$$\iint_D \frac{\partial Q}{\partial x} - \frac{\partial P}{\partial y}\,dA = \int_0^1 \int_{x^2}^{\sqrt{x}} y - 1\,dy\,dx = \int_0^1 \frac{1}{2}y^2 - y\Big]_{x^2}^{\sqrt{x}}\,dx$$

$$= \int_0^1 \frac{1}{2}x - \sqrt{x} - \frac{1}{2}x^4 + x^2\,dx = \frac{1}{4} - \frac{2}{3} - \frac{1}{10} + \frac{1}{3} = -\frac{11}{60}.$$

11. $\dfrac{\partial Q}{\partial x} - \dfrac{\partial P}{\partial y} = 2x - \sin y$.

$$\iint_D \frac{\partial Q}{\partial x} - \frac{\partial P}{\partial y}\,dA = \int_0^1 \int_x^{2-x^2} 2x - \sin y\,dy\,dx \approx -0.10243 \text{ using a machine}$$

to compute.

14. $C_1 : \frac{1}{2}\oint_{C_1} x\,dy - y\,dx = \frac{1}{2}\int_0^1 2t^2 - t^2\,dt = \frac{1}{2}\int_0^1 t^2\,dt = \frac{1}{6}$.

$C_2 : \frac{1}{2}\oint_{C_2} x\,dy - y\,dx = \frac{1}{2}\int_0^1 t^2 - 2t^2\,dt = -\frac{1}{6}$

but since the orientation is reversed for C_2, the area is $\frac{1}{6} + \frac{1}{6} = \frac{1}{3}$.

17. $\oint_C x\,dy = \int_0^{2\pi} \langle 0, a\cos t\rangle \cdot \langle -a\sin t, b\cos t\rangle\,dt = \int_0^{2\pi} ab\cos^2 t\,dt$

$$= \int_0^{2\pi} ab\frac{(1+\cos 2t)}{2}\,dt = \frac{ab}{2}\left(t + \frac{\sin 2t}{2}\right)\Big]_0^{2\pi} = \frac{ab}{2}(2\pi) = \pi ab.$$

$-\oint_C y\,dx = -\int_0^{2\pi} \langle b\sin t, 0\rangle \cdot \langle -a\sin t, b\cos t\rangle\,dt = \int_0^{2\pi} ab\sin^2 t\,dt$

$$= ab\int_0^{2\pi} \frac{1-\cos 2t}{2}\,dt = \frac{ab}{2}\left(t - \frac{\sin 2t}{2}\right)\Big]_0^{2\pi} = \frac{ab}{2}(2\pi) = \pi ab.$$

Section 17.4

3. $\iiint_R 2 + 3 + 4\,dV = \iiint_R 9\,dV = 9(3\pi) = 27\pi$.

6. $\nabla \cdot \mathbf{F} = 1$ so $\iiint_R \nabla \cdot \mathbf{F} \, dV$ = volume of cone = $\frac{8}{3}\pi$.

On the other hand, S_1, the plane portion of S, is parametrized by $\mathbf{r}(u,v) = \langle u\cos v, u\sin v, 2\rangle$ for $0 \leq u \leq 2$ and $0 \leq v \leq 2\pi$.

$\mathbf{T}_u = \langle \cos v, \sin v, 0\rangle$; $\mathbf{T}_v = \langle -u\sin v, u\cos v, 0\rangle$; $\mathbf{T}_u \times \mathbf{T}_v = \langle 0, 0, u\rangle$.

$$\iint_{S_1} \mathbf{F} \cdot \mathbf{n} \, dS = \int_0^{2\pi} \int_0^2 u^2 \sin v \, du \, dv = 0.$$

S_2, the cone portion of S, is parametrized by $\mathbf{r}(u,v) = \langle u\cos v, u\sin v, u\rangle$ for $0 \leq u \leq 2$ and $0 \leq v \leq 2\pi$.

$\mathbf{T}_u = \langle \cos v, \sin v, 1\rangle$; $\mathbf{T}_v = \langle -u\sin v, u\cos v, 0\rangle$;

$\mathbf{T}_u \times \mathbf{T}_v = \langle -u\cos v, -u\sin v, u\rangle$.

Change sign for outward normal:

$$-\iint_{S_2} \mathbf{F} \cdot \mathbf{n} \, dS$$

$$= -\int_0^{2\pi} \int_0^2 \langle u\cos v, 3u^2\cos v, u\sin v\rangle \cdot \langle -u\cos v, -u\sin v, u\rangle \, du \, dv$$

$$= \int_0^{2\pi} \int_0^2 u^2 \cos^2 v + 3u^3 \cos v \sin v - u^2 \sin v \, du \, dv$$

$$= \int_0^{2\pi} \tfrac{8}{3} \cos^2 v \, dv = \tfrac{8}{3}\left(\tfrac{1}{2}v + \tfrac{1}{4}\sin 2v\right)\Big]_0^{2\pi} = \tfrac{8}{3}\pi.$$

9. Sphere: $z = \sqrt{4-r^2} + 2$, paraboloid: $z = r^2$.

$$\iiint_R \nabla \cdot \mathbf{F} \, dV = \iiint_{R_1} \nabla \cdot \mathbf{F} \, dV - \iiint_{R_2} \nabla \cdot \mathbf{F} \, dV$$

where R_1 is the entire region above the paraboloid, inside the sphere, and R_2 is the portion that lies below the plane $z = 1$.

$$\text{flux} = \iint_S \mathbf{F} \cdot \mathbf{n} \, dS$$

$$= \iiint_R \nabla \cdot \mathbf{F} \, dV \text{ (by divergence theorem)}$$

$$= \int_0^{2\pi} \int_0^{\sqrt{3}} \int_{r^2}^{\sqrt{4-r^2}+2} r \, dz \, dr \, d\theta - \int_0^{2\pi} \int_0^1 \int_{r^2}^1 r \, dz \, dr \, d\theta$$

$$= \int_0^{2\pi} \int_0^{\sqrt{3}} (\sqrt{4-r^2} + 2 - r^2) r \, dr \, d\theta - \int_0^{2\pi} \int_0^1 r - r^3 \, dr \, d\theta$$

$$= \int_0^{2\pi} \int_0^{\sqrt{3}} r\sqrt{4-r^2} + 2r - r^3 \, dr \, d\theta - \int_0^{2\pi} \tfrac{1}{2}r^2 - \tfrac{1}{4}r^4\Big]_0^1 d\theta$$

$$= 2\pi \left(-\tfrac{1}{3}(4-r^2)^{3/2} + r^2 - \tfrac{1}{4}r^4\right)\Big]_0^{\sqrt{3}} - \tfrac{\pi}{2}$$

$$= 2\pi \left(-\tfrac{1}{3} + 3 - \tfrac{9}{4} + \tfrac{8}{3}\right) - \tfrac{\pi}{2} = \tfrac{17\pi}{3}.$$

12. By the divergence theorem, $\iint_S \mathbf{F} \cdot \mathbf{n} \, dS = \iiint_R \nabla \cdot \mathbf{F} \, dV$.

We have $\nabla \cdot \mathbf{F} = 1$.

The two spheres intersect when $z = 1$ in a circle of radius $\sqrt{3}$.

$$\iiint_R dV = \int_0^{2\pi} \int_0^{\sqrt{3}} \int_{2-\sqrt{4-r^2}}^{\sqrt{4-r^2}} r \, dz \, dr \, d\theta$$

$$= \int_0^{2\pi} \int_0^{\sqrt{3}} (\sqrt{4-r^2} - 2 + \sqrt{4-r^2}) r \, dr \, d\theta$$

$$= \int_0^{2\pi} \int_0^{\sqrt{3}} 2r\sqrt{4-r^2} - 2r \, dr \, d\theta$$

$$= \int_0^{2\pi} \left[-\tfrac{2}{3}(4-r^2)^{3/2} - r^2 \right]_0^{\sqrt{3}} d\theta = \int_0^{2\pi} \tfrac{5}{3} d\theta = \tfrac{10\pi}{3}.$$

ANSWERS TO SELECTED EXERCISES FROM THE APPENDICES

Appendix 1. TRIANGLE TRIGONOMETRY

1. $\sin(\theta) = \frac{5}{13}$.

4. $\csc(\theta) = \frac{13}{5}$.

7. $\frac{\pi}{5} = 36°$.
$\sin(\frac{\pi}{5}) \approx 0.5878$,
$\cos(\frac{\pi}{5}) \approx 0.8090$,
$\tan(\frac{\pi}{5}) \approx 0.7265$,
$\sec(\frac{\pi}{5}) \approx 1.2361$,
$\csc(\frac{\pi}{5}) \approx 1.7013$,
$\cot(\frac{\pi}{5}) \approx 1.3764$.

8. $\frac{7\pi}{2} = 630°$.
$\sin(\frac{7\pi}{2}) = -1$,
$\cos(\frac{7\pi}{2}) = 0$,
$\tan(\frac{7\pi}{2})$ is undef.,
$\sec(\frac{7\pi}{2})$ is undef.,
$\csc(\frac{7\pi}{2}) = -1$,
$\cot(\frac{7\pi}{2}) = 0$.

9. $-\frac{5\pi}{4} = -225°$.
$\sin(-\frac{5\pi}{4}) \approx 0.7071$,
$\cos(-\frac{5\pi}{4}) \approx -0.7071$,
$\tan(-\frac{5\pi}{4}) = -1$,
$\sec(-\frac{5\pi}{4}) \approx -1.4142$,
$\csc(-\frac{5\pi}{4}) \approx 1.4142$,
$\cot(-\frac{5\pi}{4}) = -1$.

13. $100° = \frac{5\pi}{9}$.
$\sin(100°) \approx 0.9848$,
$\cos(100°) \approx -0.1736$,
$\tan(100°) \approx -5.6713$,
$\sec(100°) \approx -5.7588$,
$\csc(100°) \approx 1.0154$,
$\cot(100°) \approx -0.1763$.

14. $-45° = -\frac{\pi}{4}$.
$\sin(-45°) = -\frac{1}{\sqrt{2}}$,
$\cos(-45°) = \frac{1}{\sqrt{2}}$,
$\tan(-45°) = -1$,
$\sec(-45°) = \sqrt{2}$,
$\csc(-45°) = -\sqrt{2}$,
$\cot -45°) = -1$.

15. $270° = \frac{3\pi}{2}$.
$\sin(270°) = -1$,
$\cos(270°) = 0$,
$\tan(270°)$ is undef.,
$\sec(270°)$ is undef.,
$\csc(270°) = -1$,
$\cot(270°) = 0$.

19. angle $C = 100°$, side $a \approx 4.077$ cm, side $b \approx 5.026$ cm.

22. angle $A \approx 44°$, angle $B \approx 52.6°$, angle $C \approx 83.4°$.

25. ≈ 76.84 miles.

28. ≈ 48 miles.

Appendix 2. TECHNIQUES OF INTEGRATION

1. $\sin(x) - \frac{1}{3}\sin^3(x) + C$.

4. $\frac{1}{8}(\frac{5}{2}x - 2\sin(2x) + \frac{3}{8}\sin(4x) + \frac{1}{6}\sin^3(2x)) + C$

7. $\frac{1}{5}\tan^5 x - \frac{1}{3}\tan^3 x + \tan x - x + C$.

10. $\frac{\cos^3(x)}{3} - \cos(x) + C$.

13. $-\frac{1}{5}\cot^5(x) - \frac{1}{7}\cot^7(x) + C$.

16. $\sin(x) - \frac{2\sin^3(x)}{3} + \frac{\sin^5(x)}{5} + C$.

19. $\frac{\sqrt{x^2-25}}{25x} + C$.

22. $\frac{1}{432}(\arctan(\frac{x}{6}) + \frac{6x}{36+x^2}) + C$.

25. $\frac{1}{243}(9x^2 + 49)^{3/2} - \frac{49}{81}(9x^2 + 49)^{1/2} + C$.

28. $\arcsin(\frac{x-2}{2}) + C$.

31. $\frac{1}{2}(\arctan(x+2) + \frac{x+2}{x^2+4x+5}) + C$.   **34.** $\frac{1}{2}x^2 + 8\ln|x| - 8x^{-2} + C$.

Appendix 3. METHOD OF PARTIAL FRACTIONS

1. $3\ln|x| + 2\ln|x-4| + C$, or $\ln|x|^3(x-4)^2 + C$.

4. $3\ln|x-2| - 2\ln|x+4| + C$, or $\ln\frac{|x-2|^3}{(x+4)^2} + C$.

7. $5\ln|x| - \frac{2}{x} + \frac{3}{2x^2} - \frac{1}{3x^3} + 4\ln|x+3| + C$.

11. $\ln(x^2+1) - \frac{4}{x^2+1} + C$.

14. $2\ln|x+4| + 6(x+4)^{-1} - 5(x-3)^{-1} + C$.

Appendix 4. POLAR COORDINATES

For 1-16, coordinates are approximate.

1. $(1.5, 2.5981)$.   **4.** $(1.75, -3.0311)$.

7. $(0,0)$.   **10.** $(7.0711, -135°)$; $(7.0711, -2.3562)$.

13. $(4.4429, 45°)$; $(4.4429, \frac{\pi}{4})$.   **16.** $(1000, 0°)$; $(1000, 0)$.

19. $(2a, 0)$ is the furthest right-hand edge of the cardioid. $(0, \pm a)$ are the y-intercepts.

22. $x = 1$ is an asymptote. $(1 \pm a, 0)$ are x-axis intercepts. For $a \geq 1$, $(0,0)$ is also a point on the graph.

25. $(a, 0)$ is the x-intercept. $x = a$ is an asymptote.

28. A point at $(0,0)$.

31. The outer tip of each leaf is a distance of a from the origin.

34. $r = 3\cos(5\theta)$: The outer tip of each leaf is a distance of 3 from the origin. The number of leaves is 5. The first leaf is on the positive x-axis. $r = 1.5\sin(6\theta)$: The outer tip of each leaf is a distance of 1.5 from the origin. The number of leaves is 12. The center of the first leaf is in the direction of $\frac{\pi}{12}$.

37. $\sin(\pi - \theta) = \sin(\theta)$, so for any equation with θ only in terms of $\sin(\theta)$ or $\csc(\theta)$, the new and old graphs are identical. $\cos(\pi - \theta) = -\cos\theta$, so for any equation with θ only in terms of $\cos(\theta)$ or $\sec(\theta)$, the new graph is a reflection of the old graph across the origin.

40. For any equation with θ only in terms of $\sin(\theta)$ or $\csc(\theta)$, the new graph is a reflection of the old graph across the origin. For any equation with θ only in terms of $\cos(\theta)$ or $\sec(\theta)$, the new graph is the same as the old graph.

43. $(x^2 + y^2)^2 - 3x^2 y = 0;$ $\quad x(t) = 3\sin(t)\cos^3(t),$
$\quad y(t) = 3\sin^2(t)\cos^2(t).$

46. $(4-x)^2(x^2+y^2) - x^2 = 0;$ $\quad x(t) = 4 + \cos(t),$
$\quad y(t) = 4\tan(t) + \sin(t).$

49. $(x^2 + y^2 - x)^2 - 9(x^2+y^2) = 0;$ $\quad x(t) = (-3+\cos(t))\cos(t),$
$\quad y(t) = (-3+\cos(t))\sin(t).$

52. $\frac{1}{2}\int_{\pi/6}^{\pi/4} 9\sin^2(\theta)\cos^4(\theta)\, d\theta \approx 0.1674.$

55. $\frac{1}{2}\int_{\pi/6}^{\pi/4} (4\sec(\theta) + 1)^2\, d\theta \approx 4.8404.$

58. $\frac{1}{2}\int_{\pi/6}^{\pi/4} (\cos(\theta) - 3)^2\, d\theta \approx 0.6390.$

61. $\frac{1}{2}\int_{0}^{2\pi} (1 - \sin(\theta))^2\, d\theta \approx 4.7124.$

64. $\frac{1}{2}\int_{0}^{2\pi} (\cos(\theta) - 1)^2\, d\theta \approx 4.7124.$

67. $\frac{1}{2}\int_{0}^{2\pi} (1 - \sin(2\theta))^2\, d\theta \approx 4.7124.$

70. $\frac{1}{2}\int_{7\pi/6}^{11\pi/6} (1 + 2\sin(\theta))^2\, d\theta \approx 0.5435.$

73. $\frac{1}{2}\int_{\pi/4}^{3\pi/4} \cos^2(2\theta)\, d\theta \approx 0.3927 \quad (= \frac{\pi}{8}).$

76. $\frac{1}{2}\int_{0}^{\pi/3} \sin^2(3\theta)\, d\theta \approx 0.2618 \quad (= \frac{\pi}{12}).$

79. $\frac{1}{2}\int_{\pi/2k}^{3\pi/2k} a^2 \cos^2(k\theta)\, d\theta = \frac{a^2}{2}\left(\frac{\theta}{2} + \frac{\sin(2k\theta)}{4k}\right)\Big]_{\pi/(2k)}^{3\pi/(2k)} = \frac{a^2}{2}\left(\frac{3\pi}{4k} - \frac{\pi}{4k}\right) = \frac{a^2\pi}{4k}$.

Appendix 5. COMPLEX NUMBERS

1. $z \approx 8.06 \text{ cis } 5.23$.

4. $-8 - 11i$.

7. With $z_1 = 6 - 5i \approx 7.81 \text{ cis } 5.59$, and $z_2 = -4 - 2i \approx 4.47 \text{ cis } 3.61$, $z_1 z_2 \approx 34.91 \text{ cis } 9.20$. With $z_1 = 6 - 5i$ as above and $z_2 = 2 - 4i \approx 4.47 \text{ cis } 5.18$, $\frac{z_1}{z_2} \approx 1.75 \text{ cis } 0.41$.

10. $3^{15} \text{ cis}(-60°)$.

13. 2.

16. $z_1 \approx 1.38$, $z_2 \approx 1.12 + .81i$, $z_3 \approx 0.43 + 1.31$, $z_4 \approx -0.43 + 1.31i$, $z_5 \approx -1.12 + 0.81i$, $z_6 \approx -1.38$, $z_7 \approx -1.12 - 0.81i$, $z_8 \approx -0.43 - 1.31i$, $z_9 \approx 0.43 - 1.31i$, $z_{10} \approx 1.12 - 0.81i$.

19. $z_1 \approx 1.08 + 1.87i$, $z_2 \approx -2.15$, $z_3 \approx 1.08 - 1.87i$.

20. $e^{\pi i} = \text{cis}(\pi) = \cos(\pi) + i\sin(\pi) = -1$ so $e^{\pi i} + 1 = 0$.

Appendix 6. TAYLOR'S FORMULA

1. $x \approx -0.89361$.

4. $x \approx 0.23026$.

7. $x \approx -0.77015$.

10. $x \approx 4.0953$.

12. The Taylor series for $y = \sqrt{x}$ at $a = 1$ is
$$p(x) = 1 + \tfrac{1}{2}(x-1) - \tfrac{1}{4}\frac{(x-1)^2}{2!} + \tfrac{3}{8}\frac{(x-1)^3}{3!} - \tfrac{15}{16}\frac{(x-1)^4}{4!} + \cdots$$
Since this is an alternating series, if $|c_{N+1}| < 0.001$, then $P_N(x)$ is accurate to within 0.001.
$\frac{3}{8}\frac{(1.2-1)^3}{3!} = 0.000\overline{49} < 0.001$. Hence,
$P_2(1.2) = 1 + \tfrac{1}{2}(.2) - \tfrac{1}{4}\frac{(0.2)^2}{2} = 1.095$ approximates $\sqrt{1.2}$ within 0.001.

13. $R_N(1) = \frac{f^{(N+1)}(c)}{(N+1)!}(1-0)$ is the error term.
$f^{(N+1)}(c) = e^c$, where $0 < c < 1$, so $|R_N(1)| = \left|\frac{e^c}{(N+1)!}\right| < \frac{3}{(N+1)!}$.
When $N = 6$, $\frac{3}{7!} \approx .0005952.001$, and $P_6(x) = 1 + x + \frac{x^2}{2!} + \frac{x^3}{3!} + \frac{x^4}{4!} + \frac{x^5}{5!} + \frac{x^6}{6!}$.
$P_6(1) = 1 + 1 + \tfrac{1}{2} + \tfrac{1}{6} + \tfrac{1}{24} + \tfrac{1}{120} + \tfrac{1}{720} = 2.7180\overline{5}$ approximates e within 0.001.

DIFFERENTIATION PRACTICE

1. $\dfrac{ds}{dt} = 8t + 4.$

4. $\dfrac{dy}{dx} = 15x^{1/2} - \dfrac{3}{2}x^{-3/2}.$

7. $\dfrac{dN}{dt} = \dfrac{2}{3}t^{-1/3} + \dfrac{1}{3}t^{-4/3}.$

10. $\dfrac{dx}{dw} = 6(2w+1)^2.$

13. $\dfrac{dy}{dt} = \dfrac{3}{\sqrt{6t+5}}.$

16. $\dfrac{dy}{dv} = \dfrac{-3(4v^3+14v)}{2(v^4+7v^2)^{5/2}}.$

19. $\dfrac{dy}{dx} = \dfrac{2}{3}t^{-1/3} + \dfrac{3}{2}t^{-5/2}.$

22. $\dfrac{dy}{dx} = 5(8x^3 - 2x^2 + x - 7)^4(24x^2 - 4x + 1).$

25. $\dfrac{dy}{dx} = 4x(3x^2 - 1)^3.$

28. $\dfrac{dP}{dv} = -3(v^{-1} - 2v^{-2})^{-4}(-v^{-2} + 4v^{-3}).$

31. $\dfrac{dy}{dx} = 4(7x + (x^2+6)^{1/2})^3(7 + x(x^2+6)^{-1/2}).$

34. $\dfrac{ds}{dt} = \dfrac{1}{2}(t^2 + t + 1)^{-1/2}(2t+1)(4t-9)^{1/3} + (t^2+t+1)^{1/2}(\dfrac{1}{3})(4t-9)^{-2/3}(4).$

37. $\dfrac{dy}{dx} = 7(x^2+1)(5x^2+1).\ .$

40. $\dfrac{dA}{ds} = \dfrac{s}{2}(s^2+9)^{-3/4}.$

43. $\dfrac{dy}{dx} = 7 + x(x^2+6)^{-1/2}.$

46. $\dfrac{dy}{dx} = ((x-6)^{-1} + 1)^{-2}(x-6)^{-2}.$

49. $\dfrac{dy}{dx} = \dfrac{5x^{3/2}}{2(x+1)^{7/2}}.$

52. $\dfrac{dy}{ds} = \dfrac{4(8s^2-4)^3(72s^4-108s^2+16s)}{(1-9s^3)^5}$.

55. $\dfrac{dx}{dy} = \tfrac{3}{2}(y^2-2)^{-1/4}(25y^2+2y-20)$.

58. $\dfrac{dy}{dx} = -24(3x-8)^{-3}(7x^2+4)^{-4}(7x^2-14x+1)$.

61. $\dfrac{dA}{dr} = \tfrac{1}{2}r^{-1/2}(r+1)^{-1/2}(r+2)^{-1/2}(3r^2+6r+2)$.

64. $\dfrac{dy}{dx} = \tfrac{1}{4}(3x^3-x+1)^{-3/4}(9x^2-1)(x^{-2}+2x^{1/3})^3$
$\qquad +(3x^3-x+1)^{1/4}3(x^{-2}+2x^{1/3})^2(-2x^{-3}+\tfrac{2}{3}x^{-2/3})$.

67. $\dfrac{dy}{dx} = \tfrac{12}{5}(3x+5)^{-1/5}$.

70. $\dfrac{dy}{dx} = \dfrac{-3x^4+53x^2+2x-6}{2(3x^3-x+1)^{3/2}(x^2-6)^{1/2}}$.

73. $\dfrac{dy}{dx} = \dfrac{27z^2-60z+5}{(6z+1)^4}$.

76. $\dfrac{dA}{ds} = \dfrac{-10(-5+14s)}{(8-5s+7s^2)^{11}}$.

79. $\dfrac{dy}{dx} = \dfrac{7(x^2-1)(3x^3+50x^2+9x-10)}{(3x+10)^5}$.

82. $f'(x) = \left(\tfrac{4}{3}\right)\sin^{-2/3}(4x)\cos(4x)$.

85. $f'(x) = 24x\sec^3(4x^2-8)\tan(4x^2-8)$.

88. $f'(x) = \dfrac{\sec(\sqrt[3]{x})\tan(\sqrt[3]{x})}{3\sqrt[3]{x^2}}$.

91. $f'(x) = 3e^x\tan^2(e^x)\sec^2(e^x)$.

94. $f'(x) = \cos(e^{2x}+e^{-2x})(2e^{2x}-2e^{-2x})$.

97. $f'(x) = \cos(\tfrac{1}{x}) + \tfrac{1}{x}\sin(\tfrac{1}{x})$.

100. $f'(x) = \sec^2\left(\dfrac{x^2-1}{x^3+1}\right)\left(\dfrac{-x^4+3x^2+2x}{(x^3+1)^2}\right)$.

INTEGRATION PRACTICE

1. $\frac{x^2}{2} \arcsin x - \frac{1}{4} \arcsin x + \frac{x}{4}\sqrt{1-x^2} + C.$

4. $(\frac{1}{5}) \sec^5(x) + C.$

7. $2\ln|x-1| - \ln|x| - \frac{x}{(x-1)^2} + C.$

10. $3(x+8)^{1/3} + \ln[(x+8)^{1/3} - 2]^2 - \ln[(x+8)^{2/3} + 2(x+8)^{1/3} + 4]$
$- \frac{6}{\sqrt{3}} \arctan \frac{(x+8)^{1/3}+1}{\sqrt{3}} + C.$

13. $-\sqrt{4-x^2} + C.$

16. $\ln|\sec e^x + \tan e^x| + C.$

19. $\frac{2}{3}(1+e^x)^{3/2} + C.$

22. $-x\csc(x) + \ln|\csc(x) - \cot(x)| + C.$

25. $\frac{1}{2}e^{2x} - e^x + \ln(1+e^x) + C.$

28. $(\frac{11}{2})\ln|x+5| - (\frac{15}{2})\ln|x+7| + C.$

31. $\frac{1}{\sqrt{5}} \ln\left|\sqrt{5}x + \sqrt{7+5x^2}\right| + C.$

34. $\frac{x^3}{3} - \frac{\tanh(4x)}{4} + C.$

37. $-\frac{1}{7}\cos(7x) + C.$

40. $(-\frac{1}{x})\sqrt{9-4x^2} - 2\arcsin(\frac{2x}{3}) + C.$

43. $-2\sqrt{1+\cos(x)} + C.$

46. $\ln(x^2+4) - (\frac{3}{2})\arctan(\frac{x}{2}) + (\frac{7}{\sqrt{5}})\arctan(\frac{x}{\sqrt{5}}) + C.$

49. $\frac{3}{64}(2x+3)^{8/3} - \frac{9}{20}(2x+3)^{5/3} + \frac{27}{16}(2x+3)^{2/3} + C.$

Guide for the Study of

DIFFERENTIAL AND INTEGRAL CALCULUS

to accompany

CALCULUS

or

CALCULUS OF A SINGLE VARIABLE

by Dick & Patton

Prepared by

John McGraw

Oregon State University

Preface to the Study Guide

This guide was written by John McGraw and adapted from student materials used in the introductory differential and integral calculus courses at Oregon State University. It is intended to complement your reading of textbooks, your lectures, and your recitation and/or laboratory activity in learning calculus. The study guide is closely tied to Chapters 1-8 (and Section 9.5) of *Calculus* and *Calculus of a Single Variable* by Dick & Patton. It is organized into "lessons" where, generally, each lesson corresponds to a section of the textbook.

Each lesson starts with a list of important vocabulary and notation introduced in that section. This is followed by several "Questions for Understanding." These questions are meant to probe your understanding of the key concepts in the section and to highlight important points (page references are included). The list of suggested exercises given is not intended to be a minimal or model homework assignment. Rather, these exercises were chosen to give you some indication of what level of understanding you are expected to reach. Certainly, your instructor may assign different or additional problems for homework. The "Notes on Exercises" are often actual exercises worked out as additional examples. Generally, these are problems with solutions and/or approaches not readily apparent from the discussion in the textbook. Those suggested exercises that are similar to examples already discussed in the textbook are rarely treated in these notes.

Since the textbook takes advantage of the use of technology, some sections depend heavily on the use of a graphing calculator (or computer algebra system if you have access to one). Other sections do not depend on the use of a graphing calculator at all. We have included "Graphing Calculator Hints" where appropriate. These include particular instructions or strategies for using TI or HP graphing calculators (specifically, the TI-82, TI-85, HP-48G, and HP-48S). Your course may involve what are commonly called "gateway exams." These are usually tests of basic paper-and-pencil skills of differentiation and integration where no calculator or computer is allowed. Included in this study guide are sample gateway differentiation and integration exams similar to those used at Oregon State University.

The textbook was written with the intent that you actually *read* it, and not just use it as a source of examples and exercises. The *Preface to the Student* gives some advice on how best to use the textbook and includes a discussion of a powerful method for solving mathematical problems outlined by the master teacher, George Polya. Ultimately, what you get from your calculus course depends on you. Read your text, attend lecture, and participate fully in recitation or labs. Use this study guide to check and extend your understanding of the material. And don't hesitate to ask questions, both of yourself and your instructor!

Study Guide for Sections 0.1, 0.3
Real Numbers and Intervals

§0.1 Real Numbers
Key Terms & Notation
Real numbers   Rational numbers   Irrational numbers

Questions for Understanding
- Are integers rational numbers? Can $\sqrt{2}$ be represented *exactly* in decimal form? What about π or $\pi/2$?   (page 3)

- Are fractions *always* rational numbers?   (page 3)

§0.3 Absolute Value and Interval Notation
Key Terms & Notation

Absolute Value	$\mathbb{R}$	Interval notation
Set notation $\{\}$	Empty set $\emptyset$	Endpoints
Element	Union $\cup$	Bounded/Unbounded
Roster notation	Intersection $\cap$	Open/Closed
Rule notation	Subset	Half-open/closed

Questions for Understanding
- Look at EXAMPLE 6, page 14. How does the implied definition for the *absolute value of a function* compare to the one given in Definition 1 for a *real number*?   (page 13)

- Identify the meaning of each of the following notations (pages 16-18)

$$\{\ \mathbb{R},\quad \emptyset,\quad A\cup B,\quad A\cap B,\quad [a,b],\quad (-\infty,b],\quad [a,\infty),\quad (a,b)\ \}$$

- Each of the following is a way to express some interval. What do they all have in common? (Hint: Sketch on a number line)   (pages 16-19)

$$(-3, 17] \qquad \{x : x > -3, x \leq 17\} \qquad \{x : -3 < x \leq 17\}$$

$$\{x : |x| \leq 17\} \cap \{x : -3 < x\} \qquad \{x : |x| < 3\} \cup \{x : 0 \leq x \leq 17\}$$

Suggested Exercises

§0.1 (pages 6-7):   8, 9, 10
§0.3 (pages 20-21):   6, 9, 12, 13, 16, 24, 30, 36

Notes on Exercises

- §0.3 EXAMPLE 10, an alternate method

  Observe that $|x+2| \leq 5$ is satisfied if

  $$-5 \leq (x+2) \leq 0 \quad \text{or} \quad 0 \leq (x+2) \leq 5$$

  which can be rewritten as $\quad -5 \leq x+2 \leq 5$

  which is equivalent to $\quad -5-2 \leq (x+2)-2 \leq 5-2$

  and which simplifies to $\quad\quad\quad -7 \leq x \leq 3$

- §0.3 Problem 24

  Remember, a and δ represent numbers that remain fixed. They are treated in the same way as 2 and 5 in EXAMPLE 10, page 19.

Graphing Calculator Hints

- Somtimes graphing both sides of an inequality as separate expressions on the same screen can help build your intuition about inequalities. As an example, consider problem 13, page 20.

  HP-48x
      Enter {'X^2-4' '5'} as the expression to plot.

  Ti-8x
      Enter x^2-4 for y1= (or Y1=) and 5 for y2= (or Y2=).

Study Guide for Section 0.4
Functions

Key Terms & Notation

Function
Domain
$f : D \to \mathbb{R}$
First/second coordinate
Numerical representation
Graphical representation
Symbolic representation

Origin $(0,0)$
Graph
Ordered pair

Independent variable
Dependent variable
Argument
Constant
Parameter
Discriminant

Questions for Understanding

- Suppose $y = f(x) = x^2$. How are $(3, f(3))$ and $(3, 9)$ related? How does the $f(x)$-axis relate to the y-axis? (pages 24-26)

- What can be said about C if it is a *constant*? Does the value of a *parameter* depend on the particular x-value (*input* value) under consideration? (pages 30-31)

- How would you relate the following three functions to the idea of a "family of functions?" How are they the same? How are they different? (page 32)

$$f(x) = (x-2)^3 + 7 \qquad g(x) = (x-2)^3 - 6 \qquad h(x) = (x-2)^3 - b$$

- How does one tell whether the symbol (a, b) means a point on a graph or an open interval on the real number line? (page 31)

Suggested Exercises

§0.4 (pages 33-35): 1, 2, 10, 16, 19, 25, 26

Notes on Exercises

- Problem 3
  Want:
  $$S: D \to \mathbb{R} \text{ such that } d \mapsto S(d) \quad \text{(See page 23.)}$$
  Need:
  S in terms of $d = S(d) \implies$ need *some* kind of expression for S as the surface area of a sphere.
  Know:
  $$S(r) = S \text{ in terms of radius } r = 4\pi r^2 \text{ } r \in [2, 7]$$
  ("*r is an element of* the interval $[2, 7]$")
  $$d = 2r \implies \begin{cases} d \in [4, 14] \implies D = [4, 14] \\ 4\pi r^2 = \pi(2r)^2 = \pi d^2 \end{cases}$$
  Solution
  $$S: [4, 14] \to \mathbb{R} \qquad S(d) = \pi d^2$$

Graphing Calculator Hints

- Numerical investigation of equations and varying parameters using the expression $(x + a)^2 + b$ from EXAMPLE 20, page 32.

  ### HP-48G
  Enter the expression as the calculator's EQ variable. (Using the PLOT feature is one way.) Get into the SOLVR menu (which is not listed on the keyboard), by pressing ⬒, SOLVE, [ROOT], [SOLVR]. Enter various values for X, A, or B and obtain the result by pressing EXPR=. See SOLVR environment in your *User's Guide* index. *Note*: the **HP-48S** is similar and actually easier. (Further exploration: Get into PLOT and graph the function)

  ### TI-85
  Enter the expression under GRAPH, $y(x) =$. Press 2nd, SOLVER, choose the appropriate function name at the bottom of the view screen. Press ENTER for the next screen. Move the cursor to each entry you want to change and enter the number. Move the cursor to exp= and press [SOLVE] (the F5 key) See SOLVER in your *Guidebook* index. (Further exploration: Press the GRAPH (F1) key at the bottom of the view screen.)

TI-82

Choose initial values for a, b and store them as A, B by using the following sequence from the HOME screen.

>*number for a*, STO ▷, ALPHA, A
>*number for b*, STO ▷, ALPHA, B

Enter the expression under Y= and plot using ,GRAPH. Bring up the CALC menu screen with 2nd, CALC. Choose **value**, press ENTER. Enter the desired x-value, press ENTER, and read the results from the listed coordinates. See Variables, CALC menu in your *Guidebook* index.

Study Guide for Section 1.1
Polynomial, Rational, and Power Functions

Key Terms & Notation

Polynomial function Identity function Power function
Coefficient   Linear function   Even/odd functions
Term   Quadratic function Symmetry
Degree   Cubic function   Rational function
Constant function   Line of reflection

Questions for Understanding

- Given the symbolic form of a *polynomial function*

$$a_n x^n + a_{n-1} x^{n-1} + \ldots +_1 x + a_0,$$

can you see that the exponents must be integers? (Compare them with the subscripts for the coeffiecients.) (page 48)

- Which of the following are polynomial functions? (page 48)

$$f(x) = x^{12} + x \qquad g(x) = x^3 - x^{3/2} + 1 \qquad h(x) = x^{-2} - x - 7$$

- If $x^4 - x = a_5 x^5 + a_4 x^4 + a_3 x^3 + a_2 + x^2 + a_1 x + a_0$ what must each a_i be? (page 48)

- How are each of the following power functions related? (page 54)

$$\sqrt[3]{x^2} \qquad (x^2)^{1/3} \qquad x^{2/3} \qquad (x^{1/3})^2 \qquad (\sqrt[3]{x})^2$$

- If $f(2) = f(-2)$, could f be *symmetric* about the $y-axis$? Could f be an *even* function, an *odd* function, or neither? (pages 55-57)

Study Guide for Section 1.1 193

Suggested Exercises

§1.1 (pages 57-59): 3*, 5*, 6, 7, 11, 19, 20, 21, 31, 34
*Explain your answers in complete sentences

Notes on Exercises

- Problem 35
  Consider the symmetry of the parabola. We know the maximum height must occur at the t-value representing the midpoint of the time interval in the air. This interval is between the two points when $h(t) = 0$. Hence, $0 = 1000t - 16t^2$

  $\implies 0 = 16t^2 - 1000t = t(t - 1000)$

  $\implies t = 0$ or $t = \dfrac{125}{2}$

  $\implies$ maximum height occurs when

  $$t = \frac{1}{2}\left(\frac{125}{2}\right) = \frac{125}{4}$$

  Therefore, maximum height is

  $$h\left(\frac{125}{4}\right) = 1000\left(\frac{125}{4}\right) - 16\left(\frac{125}{4}\right)^2 = 250(125) - 125^2$$
  $$= 2 \cdot 125^2 - 125^2 = 125^2(2 - 1) = 125^2$$

Graphing Calculator Hints

- Graphing functions with odd roots
  These types of functions require special handling. As examples, consider $x^{1/3}$ and $x^{2/3}$. If you enter simply X^(1/3) or X^(2/3), then the calculator may not graph anything for x-values less than zero.
  HP-48x
  For any odd roots, you must use the XROOT() feature. It is located above the $\sqrt{x}$ key and is denoted by $\sqrt[x]{y}$. Assuming you are ready to enter $x^{1/3}$ for graphing, press ⌈r⌉, $\sqrt[x]{y}$, then enter 3,X between the parentheses. When you are done, it should look like

XROOT(3,X) To graph $x^{2/3}$, enter XROOT(3,X^2) The general format for $x^{p/q}$ is XROOT(q,X^p)

TI-8x

You can enter X^(1/3) for $x^{1/3}$, but the same pattern doesn't work for $x^{2/3}$. That must be entered as (X^2)^(1/3), which is the equivalent of $(x^2)^{1/3}$ in typical mathematical notation. The recommended, general form for $x^{p/q}$ is (X^p)^(1/q).

Study Guide for Section 1.2
Exponential and Logarithmic Functions

Key Terms & Notation

Exponential function	Base	Half-life
Logarithmic function	$\log_a x$	
Common logarithmic function	$\log x$	
Natural exponential function	e^x	$\exp(x)$
Natural logarithmic function	$\ln x$	

Questions for Understanding

- If $A_0 =$ the initial amount of radioactive material present when $t = 0$, what is $\frac{1}{2}A_0$? What is the term for the time it takes A_0 to decay to $\frac{1}{2}A_0$? (page 61)

- Is e a variable like x? (page 62)

- How are a *logarithm* and an *exponent* related? (page 63)

- How are the following two statements related? (page 63)

    $\log_a x$ "the exponent that I raise a to that gives me x."

Suggested Exercises

§1.2 (pages 65-67): 8, 11, 12, 14

Notes on Exercises

- Problem 3
$$\begin{aligned}
\log_a 5 &= \log_a\left(\frac{10}{2}\right) = \log_a 10 - \log_a 2 \\
&= 2.0115 - .6055 = 1.406
\end{aligned}$$

- Problem 13

  Want:
  Time for waste to decay to 1% of original (or *initial*) amount.

  Need:
  Expression for waste in terms of time
  $\implies$ Let $q(t) = A_0 \cdot 2^{-t/k}$, where $A_0 =$ initial amount, $k =$ half-life.

  Know:
  half-life $= k = 20 =$ time for A_0 to decay to $\frac{1}{2} A_0$
  $\implies q(t) = A_0 \, 2^{-t/20}$
  $A_0 =$ initial amount $\implies$ 1% of initial amount $= .01 A_0$.

  Want:
  Specific time t so $q(t) = .01\, A_0$ is satisfied.
  $\implies$ Find t such that $.01\, A_0 = q(t) = A_0\, 2^{-t/20}$
  $\implies .01 = 2^{-t/20}$

  Using the natural logarithm function
  $$\ln .01 = \ln 2^{-t/20} = -\frac{t}{20} \ln 2 = t\left(-\frac{\ln 2}{20}\right)$$
  $$\implies t = \frac{\ln .01}{\left(-\frac{\ln 2}{20}\right)} = -\frac{20 \ln .01}{\ln 2} \approx 132.87712 \text{ years.}$$

Graphing Calculator Hints

- Your calculator cannot directly evaluate a logarithm with an arbitrary base. You must convert to either the base 10 logarithm, $\log_{10}$, or the natural logarithm, ln. This requires using the properties of logarithms on page 64.

Study Guide for Section 1.3
Trigonometric Functions

Key Terms & Notation

$\sin\theta$ $\csc\theta$ Frequency
$\cos\theta$ $\sec\theta$ Amplitude
$\tan\theta$ $\cot\theta$ Phase shift
Unit circle   Periodic   Fundamental period
Trigonometric identities   Common trigonometric values

Questions for Understanding

- Is 5π in the domain of $\sec\theta$?   (page 69)

- Which of the following equations are true?   (page 71)

$$\frac{\sin\theta}{\theta} = \sin \qquad \sin^2\theta = (\sin\theta)^2 \qquad (\sin\theta)^2 = (\sin(\theta))^2$$

$$\sin^2\theta = \sin\theta^2 \qquad \sin\theta^2 = \sin(\theta^2)$$

- Find each of the following   (page 72)

$$\sin 2\theta \text{ by evaluating } \sin(\theta+\theta) \text{ with a } \textit{sum formula}$$

$$\cos^2\frac{\theta}{2} \text{ by evaluating } \cos\left(\frac{\theta}{2}+\frac{\theta}{2}\right) \text{ with a } \textit{sum formula}$$

- What do you get when you divide both sides of the trigonometric identity $\sin^2\theta + \cos^2\theta = 1$ by $\sin^2\theta$? by $\cos^2\theta$?   (page 72)

- The "triangle" form of the definition of $\sin\theta$ is $\sin\theta = \dfrac{opp}{hyp}$. What are the respective forms for $\cos\theta$ and $\tan\theta$?

Suggested Exercises

§1.3 (pages 72-74): 5, 6, 7, 8

Note: For all graphing problems, include enough labeling to make them readily identifiable.

Notes on Exercises

- Problems 5-8, 13

    Recall that $-1 \leq \sin\theta \leq 1$ *always*

    $s(-1) = -3 \cdot (-1) - 1 = 2$ and $s(1) = -3 \cdot (1) - 1 = -4$
    $\implies -4 \leq s(x) \leq 2$

    $\implies$ amplitude of $s = \dfrac{|2 - (-4)|}{2} = 3$

Figure 1: $s(x) = -3\sin 2(x - \pi/3) - 1$

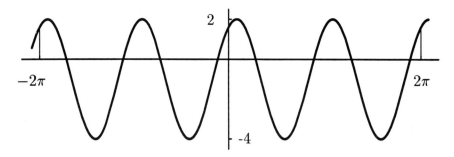

Note: For all graphing problems, include enough labeling to make them readily identifiable.

Study Guide for Section 1.3

Graphing Calculator Hints

- To numerically evaluate $\sec\theta, \csc\theta, \cot\theta$, remember the definitions

$$\sec\theta = \frac{1}{\cos\theta} \implies \begin{cases} \textbf{HP-48x} & \text{Enter } \theta, \text{ press COS key,} \\ & \text{then 1/x key} \\ \textbf{TI-8x} & \text{Press COS, enter } \theta, \\ & \text{press ENTER,} \\ & \text{Press 2nd, } x^{-1}, \text{then ENTER} \end{cases}$$

$$\csc\theta = \frac{1}{\sin\theta} \implies \begin{cases} \textbf{HP-48x} & \text{Enter } \theta, \text{ press SIN key,} \\ & \text{then 1/x key} \\ \textbf{TI-8x} & \text{Press SIN, enter } \theta, \\ & \text{press ENTER,} \\ & \text{Press 2nd, } x^{-1}, \text{then ENTER} \end{cases}$$

$$\cot\theta = \frac{1}{\tan\theta} \implies \begin{cases} \textbf{HP-48x} & \text{Enter } \theta, \text{ press TAN key,} \\ & \text{then 1/x key} \\ \textbf{TI-8x} & \text{Press TAN, enter } \theta, \\ & \text{press ENTER,} \\ & \text{Press 2nd, } x^{-1}, \text{then ENTER} \end{cases}$$

Caution: The COS^{-1} key is used for the *inverse cosine* (or *arccos*) function, *not* the reciprocal.

Study Guide for Section 1.4
New Functions From Old

Key Terms & Notation

Composition $f+g$ fg
f/g $g \circ f$ $g(f(x))$
Chain of composition

Questions for Understanding

- If $f(x) = \dfrac{1}{2}x$, $g(x) = \dfrac{1}{3}x$, what is $(f+g)(\pi)$? What is $(f \circ g)(\pi)$? What about $(f \circ g)(\text{apple } \pi)$?   (pages 75-77)

- In EXAMPLE 11, how does the numerator affect the domain of f? What about the denominator?   (page 80)

- In EXAMPLE 12, suppose $g(x) = \sqrt{\cot^2 2x}$. Does this simplify to $\cot \sqrt{2x}$? Diagram the chain of composition for $g(x) = \sqrt{\cot^2 2x}$.   (page 80)

Suggested Exercises

§1.4 (pages 81-84):   1, 2, 8, 11, 14, 18, 28, 38

Notes on Exercises

- Problem 7

$$x \;\mapsto\; \underbrace{\frac{1}{x-4}}_{j(x)}^{u} \;\mapsto\; \underbrace{\sin\left(\frac{1}{x-4}\right)}_{h(u)}^{v} \;\mapsto\; \underbrace{\sin^2\left(\frac{1}{x-4}\right)}_{f(v)}^{w}$$

$$\mapsto\; \underbrace{3\sin^2\left(\frac{1}{x-4}\right)}_{g(w)}$$

Study Guide for Section 1.4

$$p(x) = g(w) = g(f(v)) = g(f(h(u))) = g(f(h(j(x))))$$
$$= (g \circ f \circ h \circ j)(x) \implies p = (g \circ f \circ h \circ j)$$

- Problem 13
$$(f+g)(4) = f(4) + g(4) = (-1) + (2) = 1$$

- Problem 19
$$(g \circ f)(4) = g(f(4)) = g(-1) = 3$$

Graphing Calculator Hints

- To help you see why the order of function composition is important, let
$$f(x) = x^2 \quad \text{and} \quad g(x) = e^x.$$
Use your calculator to graph
$$f \circ g \quad \text{and} \quad g \circ f$$
for comparison.

Study Guide for Section 1.5
Inverse Functions

Key Terms & Notation

Inverse	arcsin	arccsc
Inverse function	arccos	arcsec
f^{-1}	arctan	arccot
$\sin^{-1} x$	$\cos^{-1} x$	$\tan^{-1} x$

Questions for Understanding

- In what ways are the inverse trig *functions* restricted and why? (pages 22, 87-89)

- In general, does $f^{-1}(x) = \dfrac{1}{f(x)}$? (page 90)

- What is the important exception to $\sin^n \theta = (\sin \theta)^n$? (pages 90-91)

- Consider the graphical representation of an inverse. Do any of the following have inverses? Why or Why not? (page 86)

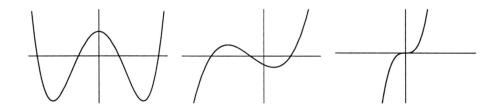

Figure 2: Interpreting inverses graphically

Suggested Exercises

§1.5 (pages 91-94): 1, 6, 18, 22, 27 (Explain your answer.)

Study Guide for Section 1.5

Notes on Exercises

- **Problem 25**

$$\arctan(\sqrt{3}) = \text{angle } \theta \text{ such that } \tan \theta = \sqrt{3} = \frac{\text{opp}}{\text{adj}} = \frac{\sin \theta}{\cos \theta}$$

Angles associated with $\sqrt{3}$, in the proper domain for θ, are

$$\frac{\pi}{6} \text{ with } \begin{cases} \sin(\pi/6) = 1/2 \\ \cos(\pi/6) = \sqrt{3}/2 \end{cases} \text{ and}$$

$$\frac{\pi}{3} \text{ with } \begin{cases} \sin(\pi/3) = \sqrt{3}/2 \\ \cos(\pi/3) = 1/2 \end{cases}$$

Observe that

$$\tan \pi/3 = \frac{\sin(\pi/3)}{\cos(\pi/3)} = \frac{\sqrt{3}/2}{1/2} = \frac{\sqrt{3}}{2} \frac{2}{1} = \sqrt{3}$$

$$\Longrightarrow \quad \arctan(\sqrt{3}) = \theta = \frac{\pi}{3}$$

Graphing Calculator Hints

- Do problems 32, 34, 35, 36, on pages 92-93. How would you compare the domains of

$$\text{arcsin vs. arccsc} \qquad \text{arccos vs. arcsec}$$
$$\text{arctan vs. arccot}$$

Study Guide for Section 2.1
The Language of Limits

Key Terms & Notation

Limit
$\lim_{x \to a} f(x)$
Continuous
Continuous from the right

Left-hand limit
$\lim_{x \to a^-} f(x)$
Continuous from the left

Right-hand limit
$\lim_{x \to a^+} f(x)$

Questions for Understanding

- Read the **Note** ... on page 104. Why do you think the authors emphasized the word **THE**? How would this apply to Requirement 1 on page 107? (pages 104-107)

- Would you say **Definition 1** and Requirements 1-3 are equivalent sets of conditions? (page 107)

- If $\lim_{x \to a} f(x) = f(a)$, what conclusion can you make about the points $(x, f(x))$ and $(a, f(a))$ as x approaches a? (page 107)

- Can you identify all of the features of Figure 2.3 that illustrate the limit and continuity concepts of this section? (page 107)

Suggested Exercises

§2.1 (pages 109-111): 1, 2, 13, 14, 15, 23, 27, 34, 36

Notes on Exercises

- Problems 27, 28

  Make sure that you zoom in far enough to determine the actual behavior of $\lim\limits_{x \to 0^-} f(x)$ and $\lim\limits_{x \to 0^+} f(x)$.

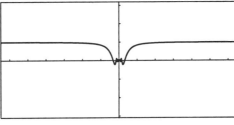

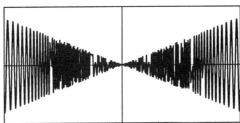

Figure 3: $f(x) = x \sin(1/x)$

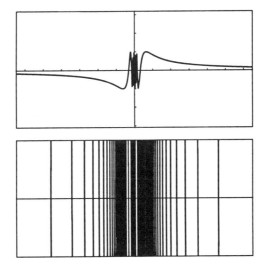

Figure 4: $f(x) = \sin(1/x)$

Graphing Calculator Hints

- Problem 35

    Split functions are entered into your calculator as a sum of the individual functions, each multiplied by their respective domains.

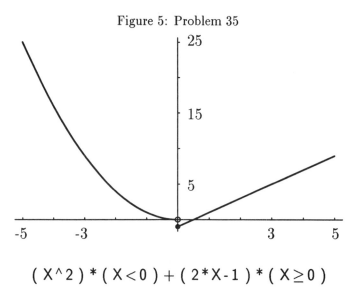

Figure 5: Problem 35

$$(X^\wedge 2)*(X<0)+(2*X-1)*(X\geq 0)$$

HP-48x

The inequality signs are found under PRG, TEST.

TI-8x

The inequality signs are found under TEST.

Study Guide for Section 2.2
Exploring Limits Numerically

Key Terms & Notation

Precision

Questions for Understanding

- In each of the examples in this section, why do you think the authors chose x-values on both sides of the proposed limiting x-values?

- How does the caution in bold type on page 113 relate to problem 27 from §1.5, page 111?

Suggested Exercises

§2.2 (pages 115-116): 2, 8, 10

Notes on Exercises

- Problem 7

    Note that $\lim_{x \to 0^-} f(x) = -\infty$ and $\lim_{x \to 0^+} f(x) = +\infty$.

    Consequently, the function has no limit as $x \to 0$.

x	$f(x)$	x	$f(x)$
-0.1	-102.4	0.1	102.4
-0.01	-1.267865 x 10^{28}	0.01	1.267865 x 10^{28}
-0.001	-1.071509 x 10^{298}	0.001	1.071509 x 10^{298}
0	undefined	0	undefined

Graphing Calculator Hints

- Using the SOLVR, SOLVER or CALC feature makes generating tables fairly easy.

HP-48G

1. Enter the expression as the EQ variable.
2. Move to the SOLVR menu by pressing [↑], SOLVE, [ROOT], [SOLVR].

TI-85

1. Enter the expression under GRAPH, $y(x) =$.
2. Move to the SOLVER by pressing 2nd, SOLVER.
3. Choose the appropriate function at the bottom of the view screen.
4. Press ENTER to move to the next screen.
5. Move cursor to x=, enter x-value.
6. Move cursor to exp=, press SOLVE.

TI-82

1. Enter the expression under Y= and plot with GRAPH.
2. Press 2nd, CALC. Select value, press ENTER.
3. Enter desired x-value, press ENTER.

Study Guide for Section 2.3
Analyzing Local Behavior

Key Terms & Notation

Discontinuity   Essential discontinuity
Vertical asymptote   Removable discontinuity
                         Jump discontinuity

Questions for Understanding

- Intuitively, what do the words *removable* and *essential* suggest? Does this make sense in their mathematical context, as well?
  (page 117)

- Consider each discontinuity in Figure 2.5. Of what type is each discontinuity that occurs? Which of the requirements on page 107 fails in each case?
  (page 117)

- Does a vertical asymptote represent a discontinuity? (page 117)

- Can you sketch a picture illustrating each of the four conditions at the top of page 121?

- Does it seem reasonable that the plot range settings on a graphing calculator could be adjusted so that the discontinuities of the function in EXAMPLE 18 will appear on the calculator plot of the function? What about a function like $g(x) = \dfrac{(x-\pi)^2}{x-\pi}$? (pages 122-123)

Suggested Exercises

§2.3 (pages 124-126):   5, 6, 8, 14, 19, 22, 26, 28

Notes on Exercises

- Problem 2

  $f(0) = \dfrac{\sin(2 \cdot 0)}{5 \cdot 0}$ is undefined, so fails Requirement 1, page 107, for continuity. A numerical table and the graph of $f(x)$ look like

x	$f(x)$	x	$f(x)$
-0.1	0.39734	0.1	0.39734
-0.01	0.39997	0.01	0.39997
-0.001	0.39999	0.001	0.39999
-0.0001	0.40000	0.0001	0.40000
0	0.4	0	0.4

  Table 1: $f(x) = \dfrac{\sin 2x}{5x}$

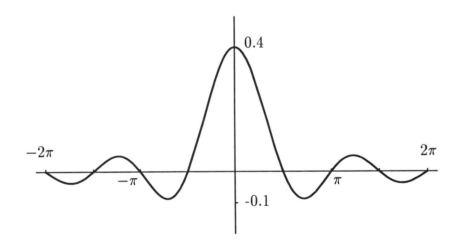

Figure 6: $f(x) = \dfrac{\sin 2x}{5x}$

Study Guide for Section 2.3

- **Problem 33**

    The function will be discontinuous where the denominator is zero. Using the Quadratic Formula to solve $16x^2 - 48x - 32 = 0$, we have

    $$x = \frac{48 \pm \sqrt{48^2 - 4(16)(-32)}}{2(16)} = \ldots = \frac{3 \pm \sqrt{7}}{2}.$$

    At these two x-values, the function is discontinuous. Using graphical or numeric methods, we see each represents a vertical asymptote.

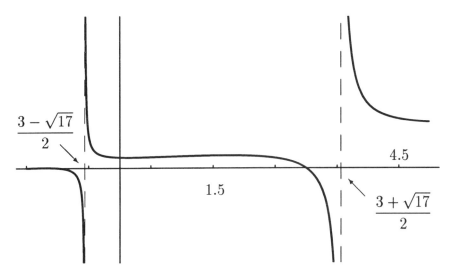

Figure 7: $f(x) = \dfrac{2x^3 - x^2 - 12x - 9}{16x^2 - 48x - 32}$

x	$f(x)$	x	$f(x)$
-0.563	-30.5524	3.5	-5.6250
-0.562	-99.2019	3.56	-252.3675
-0.5616	-941.3695	3.561	-710.2724
-0.56156	-6181.3387	3.5615	-7441.9805
-0.5615528...	$\to -\infty$	3.5615528...	$\to -\infty$

x	$f(x)$	x	$f(x)$
-0.560	28.7581	3.6	10.9960
-0.561	80.5133	3.57	47.3024
-0.5615	841.3732	3.562	879.7572
-0.56154	3467.5771	3.5616	8330.8358
-0.5615528...	$\to +\infty$	3.5615528...	$\to +\infty$

Table 2: $f(x) = \dfrac{2x^3 - x^2 - 12x - 9}{16x^2 - 48x - 32}$

Graphing Calculator Hints

- If you encounter a situation like Figure 2.11, page 124, changing your plot mode out of "connected" *may* make a clearer picture.
  HP-48G
  $\boxed{r}$, PLOT, OPTS, move to CONNECT, CHK to deselect
  TI-85
  GRAPH, MORE, FORMT, choose DrawDot
  TI-82
  MODE, choose Dot

Study Guide for Section 2.4
Continuity and its Consequences

Key Terms & Notation

Continuous on an open/closed interval
Intermediate Value Theorem
Intermediate Zero Theorem
Bisection method
Extreme Value Theorem

Questions for Understanding

- Why are the Intermediate Value Theorem and the Extreme Value Theorem *consequences* of continuity? (page 129)

- Both the Intermediate Value Theorem and the Extreme Value Theorem require continuity on a *closed* interval. Which of the examples in Table 2.8 satisfy the conclusions of the Intermediate and Extreme Value Theorems on the half-closed interval $(0,1]$? (pages 126, 129)

- As you read through Steps 1-5 of the *bisection method*, identify where each part of the Intermediate Value Theorem applies. (pages 131-132)

Suggested Exercises

§2.4 (pages 136-138): 1, 2, 8, 14, 19

Notes on Exercises

- Problem 5
  $Ax+2$ is continuous for every x. $1-x$ is continuous for every x. So $k(x)$ will be continuous for every x if $k(x)$ is continuous at $x=3$. From the Requirements on page 107, we need

  $$\lim_{x \to 3^-} k(x) = \lim_{x \to 3^+} k(x) = k(3)$$

  (and all three *must* exist)

Observe that

$$\lim_{x \to 3^-} k(x) = \lim_{x \to 3^-} (Ax + 2) = 3A + 2 \quad \text{and}$$

$$\lim_{x \to 3^+} k(x) = \lim_{x \to 3^+} (1 - x) = -2 = k(3)$$

Suppose the two limits are equal, then

$$3A + 2 = -2 \quad \Longrightarrow \quad 3A = -4 \quad \Longrightarrow \quad A = -\frac{4}{3}$$

By back substitution, we verify that

$$\lim_{x \to 3^-} k(x) = \lim_{x \to 3^-} \left(-\frac{4}{3}x + 2\right) = -\frac{4}{3} \cdot 3 + 2$$

$$= -2 = \lim_{x \to 3^+} (1 - x) = \lim_{x \to 3^+} k(x) = k(3)$$

Graphing Calculator Hints

- Remember that "split" functions must be entered as the sum of two functions, each multiplied by its domain. For example, consider

$$f(x) = \begin{cases} x^3 - x, & \text{if } x < 0 \\ |x|, & \text{if } x \geq 0 \end{cases}$$

which is entered as   (X^3-X)*(X < 0)+ABS(X)*(X ≥ 0)

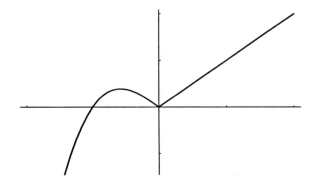

Figure 8: The "split" function $f(x)$

Study Guide for Section 2.5
Global Behavior of Functions

Key Terms & Notation

Global behavior   Horizontal asymptote   Oblique asymptote

Questions for Understanding

- What kind of limit does a horizontal asymptote indicate?

- In the solution for EXAMPLE 25, are the following really equal? (page 140)

$$f(x) = \frac{-2x^2 - 3x + 5}{4x^3 + 6x^2 - 7} = \frac{x^2}{x^2} \cdot \frac{-2 - \frac{3}{x} + \frac{5}{x^2}}{4x + 6 - \frac{7}{x^2}} = \frac{-2 - \frac{3}{x} + \frac{5}{x^2}}{4x + 6 - \frac{7}{x^2}}$$

- Consider the caution at the top of page 141 and Figure 2.16. Although $f(x)$ does cross a horizontal asymptote, is that crossing more related to global behavior or local behavior?   (pages 140-141)

- Suppose $f(x) = \dfrac{p(x)}{q(x)}$ and $\lim\limits_{x \to -\infty} f(x) = C = \lim\limits_{x \to +\infty} f(x)$ where C is some constant. What type of global behavior do you expect to see? How do the highest degree terms of $p(x)$ and $q(x)$ compare in this situation?   (page 143)

- Let $f(x) = \left(1 + \dfrac{1}{x}\right)^x$, use your calculator to find $f(10^{10})$ and $f(10^{14})$. Do these answers make sense?   (page 145)

Suggested Exercises

§2.5 (pages 146-147):   2, 7, 8, 9, 15, 16, 18, 24, 31

Notes on Exercises

- Problem 22

$$f(x) = \frac{4x^3 + 2x - 5}{8 - 9x^3} = \frac{4x^3 + 2x - 5}{-9x^3 + 8} = \frac{x^3}{x^3} \frac{4 + \frac{2}{x^2} - \frac{5}{x^3}}{-9 + \frac{8}{x^3}}$$

$\Longrightarrow$ for $|x|$ large, we have

$$f(x) \approx -\frac{4}{9} \Longrightarrow y = C = -\frac{4}{9}$$

Viewing window ranges such that the graph of $y = f(x)$ is indistinquishable from $y = -\frac{4}{9}$ are x-range $= [-300, 300]$ and y-range $= [-2, 1]$

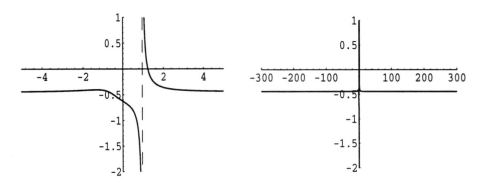

Local view   Global view

Figure 9: $f(x) = \dfrac{4x^3 + 2x - 5}{8 - 9x^3}$

Graphing Calculator Hints

- Each time you investigate a function's global limits, reset your calculator to its standard x-range and y-range values.

Study Guide for Section 2.6
Formal Definition of Limit

Key Terms & Notation

Limit
$|f(x) - L| < \epsilon$
Squeezing principle

epsilon ϵ
$|x - a| < \delta$

delta δ
Continuous

Questions for Understanding

- How are the following expressions related? (page 148)

$$|f(x) - L| < \epsilon \qquad -\epsilon < f(x) - L < \epsilon$$
$$L - \epsilon < f(x) < L + \epsilon$$

- Consider Figure 2.20 and suppose your calculator view window is set to $[a - \delta, a + \delta]$ by $[L - \epsilon, L + \epsilon]$. What portion of the printed figure would appear on your calculator screen? (page 150)

- Let $f(x) = \dfrac{x^2 + 3x}{x}$, $g(x) = \dfrac{\sin 2x}{2x}$. What are the following limits? (page 152)

$$\lim_{x \to 0} f(x) \qquad \lim_{x \to 0} g(x) \qquad \lim_{x \to 0} \frac{f(x)}{g(x)}$$

- Restate the *Squeezing Principle* (also called the *Squeeze Law*) in your own words. (page 155)

- How could you use the Squeezing Principle and the function $\dfrac{\sin^2 x}{x}$ to show that $\lim\limits_{x \to 0} \dfrac{1 - \cos x}{x} = 0$? (pages 155-156, 158)

Suggested Exercises

§2.6 (pages 156-158):   2, 10, 13, 14, 17, 20,

Notes on Exercises

- Problem 3
  (a) By substitution, we know

$$|f(x) - L| = \left|\frac{4x^3 + 4x^2 + x}{2x^2 + x} - 0\right| < .01$$

and we want some δ that guarantees this whenever

$$\left|x - \left(-\frac{1}{2}\right)\right| = \left|x + \frac{1}{2}\right| < \delta$$

Using algebra, we solve for $\left|x + \frac{1}{2}\right|$.

$$\left|\frac{4x^3 + 4x^2 + x}{2x^2 + x} - 0\right| = \left|\frac{x}{x} \cdot \frac{4x^2 + 4x + 1}{2x + 1}\right| < .01$$

$$\implies \left|\frac{(2x+1)^2}{2x+1}\right| = |2x + 1| < .01 \implies -.01 < 2x + 1 < .01$$

$$\implies -\frac{.01}{2} < x + \frac{1}{2} < \frac{.01}{2} = \frac{1}{2}\frac{1}{100} = \frac{1}{200} = .005$$

$$\implies \left|x + \frac{1}{2}\right| < .005$$

$$\implies \text{any } \delta \text{ such that } 0 < \delta \leq .005 \text{ will work}$$

Study Guide for Section 2.6 219

Viewing rectangle values are obtained from

$$[L - .01, L + .01] = [0 - .01, 0 + .01] = [-.01, .01]$$

$$[a - \delta, a + \delta] = [-\frac{1}{2} - .005, -\frac{1}{2} + .005]$$
$$= [-.505, -.495]$$

(Do you see that this is the largest viewing rectangle, if $\epsilon = .01$?)

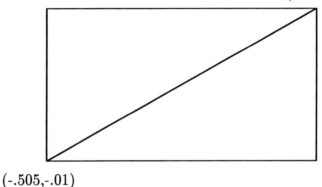

(-.495, .01)

(-.505, -.01)

Figure 10: $f(x) = \dfrac{4x^3 + 4x^2 + x}{2x^2 + x}$

(b) In $|f(x) - L| < .01$, if you replace .01 with the generic ϵ, then

$$|f(x) - L| = \left|\dfrac{4x^3 + 4x^2 + x}{2x^2 + x} - 0\right| = \ldots = |2x + 1| < \epsilon$$

$$\Rightarrow \quad -\epsilon < 2x + 1 < \epsilon \quad \Rightarrow \quad -\dfrac{\epsilon}{2} < x + \dfrac{1}{2} < \dfrac{\epsilon}{2}$$

$$\Rightarrow \quad \left|x + \dfrac{1}{2}\right| < \dfrac{\epsilon}{2}$$

Therefore, let $\delta = \dfrac{\epsilon}{2}$ and state the proof formally:

Proof:
Choose any $\epsilon > 0$. Let $\delta = \dfrac{\epsilon}{2}$, then

$$\left|x + \dfrac{1}{2}\right| < \delta = \dfrac{\epsilon}{2} \quad \Rightarrow \quad -\dfrac{\epsilon}{2} < x + \dfrac{1}{2} < \dfrac{\epsilon}{2}$$

$$\Rightarrow \quad -\epsilon < 2x+1 < \epsilon \quad \Rightarrow \quad |2x+1| < \epsilon$$

$$\Rightarrow \quad \left|(2x+1)\left(\frac{2x+1}{2x+1}\right)\right| = \left|\frac{4x^2+4x+1}{2x+1}\right| = \left|\frac{x}{x}\cdot\frac{4x^2+4x+1}{2x+1}\right|$$

$$= \left|\frac{4x^3+4x^2+x}{2x^2+x} - 0\right| < \epsilon$$

$$\Rightarrow \quad \lim_{x \to -\frac{1}{2}} \frac{4x^3+4x^2+x}{2x^2+x} = 0$$

- **Problem 19**
  Finding the limit L

  Let $g(x) = \left(1 + \frac{1}{x}\right)^x$. From EXAMPLE 28, page 145, we know $\lim_{x \to \infty} g(x) = e$. Further investigation shows that $\lim_{x \to -\infty} g(x) = e$. From Problem 31, page 147, we know

  $$\lim_{x \to \infty} g(x) = \lim_{x \to 0^+} g\left(\frac{1}{x}\right) \quad \text{and} \quad \lim_{x \to -\infty} g(x) = \lim_{x \to 0^-} g\left(\frac{1}{x}\right)$$

  $$\text{But } g\left(\frac{1}{x}\right) = \left(1 + \frac{1}{1/x}\right)^{1/x} = (1+x)^{1/x} = f(x)$$

  $$\Rightarrow \quad e = \lim_{x \to \infty}\left(1 + \frac{1}{x}\right)^x = \lim_{x \to 0}(1+x)^{1/x} = \lim_{x \to 0} f(x)$$

  So the limit L exists and $L = e$.

  Finding $[-\delta, \delta]$

  Setting the y-range to $[L - .01, L + .01] = [e - .01, e + .01]$, we find one δ we can use is

  $$\delta = 0.0065 \quad \Rightarrow \quad [-\delta, \delta] = [-0.0065, 0.0065]$$

  (*Question*: Is this the *largest* δ that could be used, given the $\epsilon = .01$ condition?)

Study Guide for Section 2.6   221

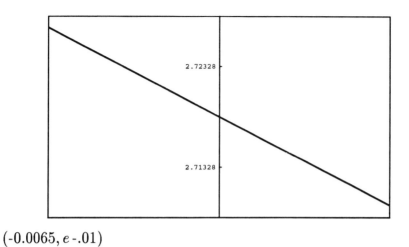

$(0.0065, e + .01)$

2.72328

2.71328

$(-0.0065, e - .01)$

Figure 11: $f(x) = (1+x)^{1/x}$

Graphing Calculator Hints

- To find $[a - \delta, a + \delta]$ viewing rectangles on your calculator

    1. Use the centering feature to center the viewing rectangle about the point (a, L)

    2. Set the y-range to $[L - \epsilon, L + \epsilon]$

    3. Each calculator has the capability to zoom in and out on only the x-range. Use this feature to obtain a viewing rectangle where the function enters the screen somewhere on the left edge and exits somewhere on the right edge. The resulting x-range will be $[a - \delta, a + \delta]$ for some appropriate δ.

Study Guide for Section 3.1
What is a Derivative?

Key Terms & Notation

Constant rate $d(t)$ $(t, d(t))$

Instantaneous rate $(t_0, d(t_0))$

Average rate $\dfrac{d(t_1) - d(t_0)}{t_1 - t_0}$

Questions for Understanding

- On a graph of distance travelled vs. elapsed time, what feature (or characteristic) of the graph represents the *derivative*? (page 160)

- How are *average speed* and *instantaneous speed* different? (page 162)

- What happens to the quantity $t_1 - t_0$ as average speed approaches instantaneous speed? (page 162)

- Consider the traveling automobile example of this section. Fill in the blanks

$$\frac{\text{the derivative}}{\text{(i.e., speed)}} = \frac{\text{instantaneous change in \underline{\hspace{2cm}}}}{\text{instantaneous change in \underline{\hspace{2cm}}}}$$

- What do you think the following expression might represent? (pages 162-163)

$$\lim_{t \to t_0} \frac{d(t) - d(t_0)}{t - t_0}$$

Suggested Exercises

§3.1 (pages 164-166): 2, 3, 4, 7, 10 (Why?), 14, 15, 16, 18, 21, 23

Notes on Exercises

- Problem 22

    Plug out, water off $\implies$ water level dropping fairly steeply $\implies$ steep negative slope.

    Water turned on $\implies$ two possibilities:

    (1) water level dropping at slower rate

    (2) water level increasing.

    Since rate of water in is greater than rate of water out, we have water level increasing $\implies$ positive slope. But plug still out $\implies$ water level increasing at slower rate than with plug in $\implies$ less steep positive slope. Therefore, need time when steeply negative slope changes to moderate positive slope $\implies$ at 14 minute mark.

Graphing Calculator Hints

- Your calculator will be useful later when we begin applying the concepts of this section to more specific situations.

Study Guide for Section 3.2
Locally Linear Functions

Key Terms & Notation

Two-point formula   Point-slope form   Locally linear
$(x_1, y_1), (x_2, y_2)$ Taylor form   Slope at the input
$Ax + By + C = 0$ Angle of inclination   Step function
Slope-intercept form   Angle of incidence   Floor function
Piece-wise linear   Angle of reflection   Ceiling function
Parallel wrt* slope   Greatest integer function
Perpendicular wrt* slope

* (wrt = with respect to)

Questions for Understanding

- Given the *point-slope* form of a line, $(y - y_0) = m(x - x_0)$. What does the b equal when y is put into *slope-intercept* form? (pages 168-169)

- How are parallel lines related? How are perpendicular lines related? (pages 168-169)

- Can you obtain the *Taylor* form of a linear function from the point-slope form? (pages 169-170)

- Given the table to the right, what would indicate that f may be a linear function? (pages 168-169)

x	$f(x)$
-2	15
-1	11
0	7
1.75	0

- Plot $f(x) = x^2$ on your calculator and consider the point $(2, f(2))$. Can you zoom in $(2, f(2))$ so $f(x)$ looks *locally linear* about the point? Approximate the resulting slope. (pages 174)

- What do the symbols $\lfloor \ \rfloor$ and $\lceil \ \rceil$ mean? What processes do they represent? (pages 179-180)

Study Guide for Section 3.2

Suggested Exercises

§3.2 (pages 176-180): 1, 2, 3, 4, 6, 15, 19, 21, 22, 26, 29, 41, 48

Notes on Exercises

- Problem 17
  (a) Graph

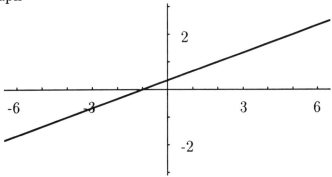

Figure 12: $y = \dfrac{x^3 + x^2 + x + 1}{3x^2 + 3}$

(Question: Is this a continuous function? Why or why not?)

(b) $\quad x = 0 \implies y = \dfrac{1}{3} \quad$ and $\quad x = 1 \implies y = \dfrac{2}{3}$

$$\implies m = \dfrac{\frac{2}{3} - \frac{1}{3}}{1 - 0} = \dfrac{1}{3} \implies$$

$$y - \dfrac{2}{3} = \dfrac{1}{3}(x - 1) = \dfrac{1}{3}x - \dfrac{1}{3} \implies y = \dfrac{1}{3}x + \dfrac{1}{3}$$

(c) $\quad y - \dfrac{2}{3} = \dfrac{1}{3}(x - 1) \implies x_0 = 1, f(x_0) = \dfrac{2}{3}$

$$\implies f(x) - \dfrac{2}{3} = \dfrac{1}{3}(x - 1) \implies f(x) = \dfrac{2}{3} + \dfrac{1}{3}(x - 1)$$

is the Taylor form

(d)

x_n	y_n	m
.9	.63333	.33333
.99	.66333	.33333
.999	.66633	.33333
.9999	.66663	.33333
.99999	.66666	.33333

(e) Comparing with $m = \frac{1}{3}$ shows all x-values give the most accurate calculator slope value of .33333 (Why are they all the same?)

Graphing Calculator Hints

- To graph *ceiling* $\lceil \ \rceil$ and *floor* $\lfloor \ \rfloor$ functions, FIRST, make sure you are out of connected graphing mode.

HP-48G

MTH, REAL, $\begin{cases} \text{CEIL}(\sin x) = \lceil \sin x \rceil \\ \text{FLOOR}(\sin x) = \lfloor \sin x \rfloor \end{cases}$

HP-48S

MTH, PARTS, $\begin{cases} \text{CEIL}(\sin x) = \lceil \sin x \rceil \\ \text{FLOOR}(\sin x) = \lfloor \sin x \rfloor \end{cases}$

TI-85

2nd, MATH, NUM, $\begin{cases} \text{int}(\sin x + .99999) = \lceil \sin x \rceil \\ \text{int}(\sin x) = \lfloor \sin x \rfloor \end{cases}$

TI-82

MATH, arrow to NUM,

arrow to int, ENTER, $\begin{cases} \text{int}(\sin x + .99999) = \lceil \sin x \rceil \\ \text{int}(\sin x) = \lfloor \sin x \rfloor \end{cases}$

Study Guide for Sections 3.3
Defining and Computing the Derivative

Key Terms & Notation

Difference quotient	Increments	Derivative
Δx	Secant line	Left-hand derivative
Δy	Differentialble at x_0	Right-hand derivative

$$f'(x_0) = \lim_{x \to x_0} \frac{f(x) - f(x_0)}{x - x_0} \qquad f'(x_0) = \lim_{h \to 0} \frac{f(x_0 + h) - f(x_0)}{h}$$

$$y'(x) = \lim_{\Delta x \to 0} \frac{\Delta y}{\Delta x}$$

Questions for Understanding

- Does the term *difference quotient* seem a logical choice for the expressions on the right in Definition 2? (page 182)

- How do Figure 3.13 and the concept *approximately locally linear* tie together? (pages 181, 185)

- As you study the "Alternate forms of the derivative," how are the denominators $x - x_0$, Δx, and h related? (pages 186-187)

- Compare Definition 4 with the case for limits in §2.1. How does this left/right idea apply to EXAMPLE 16? (pages 104, 189, 191)

Suggested Exercises

§3.3 (pages 192-193):   6, 9, 12, 21, 22, 29, 30, 37

Notes on Exercises

- **Problem 10**

  $f'(0) \approx$ slope of graph once the graph approaches a straight line. Use the TRACE feature on your calculator to obtain two points near the extreme ends of the graph. Use these points to calculate the slope.

  $$f'(0) \approx \frac{.0414 - (-.0353)}{1.1000 - .9219} \approx .43067$$

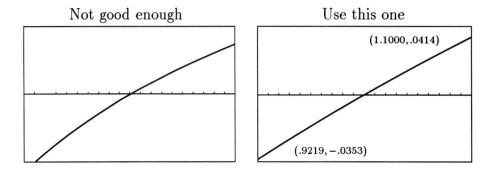

Figure 13: $y = \log_{10} x$

(See Graphing Calculator Hints on how to obtain the graph)

- **Problem 24**

  For $h = -0.001$,

  $$\frac{f(x + (-.001)) - f(x)}{-.001} = \frac{f(x - .001) - f(x)}{-.001}$$

  $$= \frac{\sqrt{2(x - .001) + 3} - \sqrt{2x + 3}}{-.001} = \frac{\sqrt{2x + 2.998} - \sqrt{2x + 3}}{-.001}$$

  For $x = 3$, this gives

  $$\frac{\sqrt{2(3) + 2.998} - \sqrt{2(3) + 3}}{-.001} = \frac{\sqrt{8.998} - \sqrt{9}}{-.001} \approx .33335$$

For $h = 0.001$,

$$\frac{f(x+.001)-f(x)}{.001} = \frac{\sqrt{2(x+.001)+3}-\sqrt{2x+3}}{.001}$$

For $x = 3$, this gives

$$\frac{\sqrt{2(3.001)+3}-\sqrt{2(3)+3}}{.001} = \frac{\sqrt{9.002}-\sqrt{9}}{.001} \approx .33331$$

$$\text{Average} = \frac{.33335+.33331}{2} = .33333 \implies f'(3) \approx .33333$$

- Problem 32

$$\lim_{h \to 0} \frac{f(x+h)-f(x)}{h} = \lim_{h \to 0} \frac{\sqrt{2(x+h)+3}-\sqrt{2x+3}}{h}$$

$$= \lim_{h \to 0} \left(\frac{\sqrt{2(x+h)+3}-\sqrt{2x+3}}{h} \cdot \frac{\sqrt{2(x+h)+3}+\sqrt{2x+3}}{\sqrt{2(x+h)+3}+\sqrt{2x+3}} \right)$$

$$= \lim_{h \to 0} \frac{2(x+h)+3-(2x+3)}{h\left(\sqrt{2(x+h)+3}+\sqrt{2x+3}\right)}$$

$$= \lim_{h \to 0} \frac{2x+2h+3-2x-3}{h\left(\sqrt{2(x+h)+3}+\sqrt{2x+3}\right)}$$

$$= \lim_{h \to 0} \frac{2h}{h\left(\sqrt{2(x+h)+3}+\sqrt{2x+3}\right)}$$

$$= \lim_{h \to 0} \frac{2}{\sqrt{2(x+h)+3}+\sqrt{2x+3}}$$

$$= \frac{2}{2\sqrt{2x+3}} = \frac{1}{\sqrt{2x+3}}$$

$$x = 3 \implies f'(3) = \frac{1}{\sqrt{2(3)+3}} = \frac{1}{9} = \frac{1}{3} = .333\overline{33}$$

which is exactly the same as from Problem 24, up to machine precision.

Graphing Calculator Hints

- For *exercises 1-10*, you need to zoom in far enough on the point $(1, f(1))$ until $y = f(x)$ looks exactly like a straight line. To accomplish this

    1. Use your calculator's *centering* feature to center the point $(1, f(1))$ on the view screen.

    2. Set both the x-zoom and y-zoom factors to the same values
       HP-48G
           With view window showing, [ZOOM], [ZFACT]
       HP-48S
           With view window showing, [ZOOM], [XY]
       TI-85
           With view window showing, [ZOOM], [ZFACT]
       TI-82
           ZOOM, go to MEMORY, go to 4:SetFactors..., ENTER

    3. Zoom in on $(1, f(1))$ until the graph truly appears straight.

    For *exercises 10-20*, zoom in on the point $(0, f(0))$.

Study Guide for Sections 3.4
The Derivative Function

Key Terms & Notation

Derivative function   Derivative of sine   Derivative of a^x
Monomial function   Derivative of cosine   Derivative of e^x
nx^{n-1}

Questions for Understanding

- As we move from the *value* $f'(x_0)$ to the *function* f', does this mean f' may also be investigated for function properties of its own (e.g., continuity, existence of a derivative)?

- After reading the material on pages 194-195, what can you say is the relationship between the derivative of $f(x)$ and the slopes of lines tangent to $f(x)$?

- As a *function*, can f' be graphed? Can it be written as an expression in x?   (pages 194-196)

- Using your knowledge of exponents, can you express the constant 2 as $2x^p$ for some p? What is the exponent of x in $5x$? How could this be used when applied to the box on page 198?

- Do you think the two boxes on page 200 might be important to memorize?

- If EXAMPLE 27 contains a "remarkable" result, would it be worth remembering?   (page 201)

Suggested Exercises

§3.4 (pages 202-205):   2, 11, 12, 21, 22, 23, 24, 29, 32, 37, 38

Notes on Exercises

- Problems 21-28

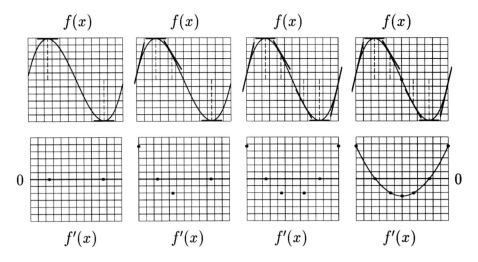

Figure 14: Graphing $f'(x)$ from $f(x)$

Graphing Calculator Hints

- Your calculator's *solver* feature works nicely for numerical investigations like problems 35 and 36.

HP-48G

Enter the expression as the calculator's EQ variable. (Using the PLOT feature is one way.) Get into the SOLVR menu (which is not listed on the keyboard), by pressing [↑], SOLVE, [ROOT], [SOLVR]. Enter various values for A and H and obtain the result by pressing EXPR=.
Note: the **HP-48S** is similar and actually easier.

TI-85

Enter the expression under GRAPH, $y(x) =$ using x in place of a. Press 2nd, SOLVER, choose the appropriate function name at the bottom of the view screen. Press ENTER for the next screen. Move the cursor to each entry you want to change and enter the number. Move the cursor to **exp=** and press [SOLVE] (the F5 key).

TI-82

Choose an initial value for h and store it as H by using the following sequence from the HOME screen.

number for h, STO ▷, ALPHA, H

Enter the expression under Y=, using x for a and plot using ,GRAPH. Bring up the CALC menu screen with **2nd**, **CALC**. Choose **value**, press ENTER. Enter the desired x-value, press ENTER, and read the results from the listed coordinates.

Study Guide for Section 3.5
Interpreting Derivatives

Key Terms & Notation

$\dfrac{dy}{dx}$ $\qquad\qquad dy = y'(x) \cdot dx \qquad\qquad Df,\ Dy,\ D_x y$

$\lim\limits_{\Delta x \to 0} \dfrac{\Delta y}{\Delta x}$ $\qquad\qquad$ Differential $\qquad\qquad A'(r),\ \dfrac{dA}{dr}$

$\dfrac{d}{dx}(y),\ y'(x)$ $\qquad\qquad \left.\dfrac{dy}{dx}\right|_{x=x_0} \qquad\qquad$ Velocity

Questions for Understanding

- What are the advantages of the *Leibniz notation*? (pages 206-207)

- Does the *differential* seem to be an outgrowth of Leibniz notation? (pages 206-207)

- How are $f'(x_0)$ and $\left.\dfrac{dy}{dx}\right|_{x=x_0}$ related? (pages 207)

- Consider the area A of a circle. How is $\dfrac{dA}{dr}$ related to the circumference of a circle? (pages 208)

- Does the explanation beginning at the bottom of page 210 seem "reasonable," as the authors suggest?

- Suppose you were tutoring a fellow student. How would you explain the following sequence of symbols? (pages 213-214)

$$\text{average velocity} = \dfrac{\Delta s(t)}{\Delta t} \to \dfrac{d\,s(t)}{dt} = \text{instantaneous velocity}$$
$$\text{as } \Delta t \to 0$$

What about

$$v(t) = \lim_{\Delta t \to 0} \dfrac{\Delta s(t)}{\Delta t} = \dfrac{d\,s(t)}{dt} = s'(t)\ ?$$

Study Guide for Section 3.5   235

- The volume V and surface area S_A of a sphere in terms of its radius r are

$$V(r) = \frac{4}{3}\pi r^3 \qquad S_A(r) = 4\pi r^2$$

Can you express V as a function of S_A? In other words, what would the expression $V(S_A)$ look like? Could you find $\dfrac{dV}{dS_A}$?

Suggested Exercises

§3.5 (pages 217-218):   4, 5, 6, 10, 12, 13, 14, 18, 20, 23, 24, 27, 29

Notes on Exercises

- **Problem 8**

  Want:

  (a) $\left.\dfrac{dA}{dx}\right|_{x=3.6cm}$,   (b) $\left.\dfrac{dP}{dx}\right|_{x=3.6cm}$

  Need:
  Expressions for area A and perimeter P of the computer chip.

  Know:
  $x =$ length of chip side, chip is square
  $\Longrightarrow A = x^2, \quad P = 4x$

  $\Longrightarrow$ (a) $\dfrac{dA}{dx} = 2x \Longrightarrow \left.\dfrac{dA}{dx}\right|_{x=3.6cm}$
  $= 2(3.6) = 7.2 \text{ cm}^2/\text{cm}$

  $\Longrightarrow$ (b) $\dfrac{dP}{dx} = 4 \Longrightarrow \left.\dfrac{dP}{dx}\right|_{x=3.6cm} = 4 \text{ cm/cm}$

- Problem 25

  Want:
  Velocity $v(t)$ when object hits ground
  $\implies v(t)$ when $s(t) = 0$

  Need:
  Expression for velocity and time t when $s(t) = 0$

  Know:
  $s(t) = 15 - 6.729t^2 \implies v(t) = -13.458t$
  (from Problem 24)

  $s(t) = 0 \implies 15 - 6.729t^2 = 0 \implies t^2 = \dfrac{15}{6.729}$

  $\implies t \approx 1.493 \implies v(t)$ when $s(t) = 0$ is approximately

  $v(1.493) \approx -20.093$ meters/sec.

  (Why is this slightly different from the answer in the back of the text?)

Graphing Calculator Hints

- There are just some things you can't get a machine to do.

Study Guide for Section 3.6
New Deriviatives From Old

Key Terms & Notation

Linearity rule

$(af + bg)'$

$\dfrac{d}{dx}(au + bv)$

Product rule

$(fg)'$

$\dfrac{d(uv)}{dx}$

Quotient Rule

$\left(\dfrac{f}{g}\right)'$

$\dfrac{\dfrac{du}{dx}v - u\dfrac{dv}{dx}}{v^2}$

Derivative of tangent

$\dfrac{d}{dx}(\tan x)$

Derivative of secant

$\dfrac{d}{dx}(\sec x)$

Derivative of cosecant

$\dfrac{d}{dx}(\csc x)$

Derivative of x^{-n}

Questions for Understanding

- Why do you think $(fg)'$ and $\left(\dfrac{f}{g}\right)'$ are not included in the box on page 219? (pages 219, 223, 225)

- Give an example of why $(fg)' \neq f'g'$. (pages 221-222)

- In the *product rule*, commutativity of multiplication lets us write it in several ways:
$$(fg)' \;=\; f'g + fg' \;=\; f'g + g'f \;=\; fg' + gf'$$
Can we do the same with the *quotient rule*? (page 225)

- Do you think the three boxes on page 227 might be worth memorizing?

- As you look at the *exercises*, does EXAMPLE 48 take on more significance? (page 228)

- Using the general derivative forms and rules, along with the following information, compute the derivatives at the indicated points.

$$\begin{array}{llll} f(1) = 7 & f'(1) = 4 & g(1) = 3 & g'(1) = \sqrt{2} \\ f(2) = 23 & f'(2) = 1/2 & g(2) = 2 & g'(2) = \pi \\ f(3) = 1 & f'(3) = 6 & g(3) = 1 & g'(3) = e \end{array}$$

1. $F'(3)$ where $F(x) = f(x)g(x)$
2. $F'(1)$ where $F(x) = \dfrac{f(x)}{g(x)}$
3. $G'(2)$ where $G'(x) = g(x)f(x)$

Suggested Exercises

§3.6 (pages 228-230):   2, 8, 12, 19, 21, 22, 31, 36

Notes on Exercises

- Problem 1

    Recognize that $y = (x^3 + 3x^2 + 1)(2x^2 + 8x - 5)$ is a product of two functions. We can label

    $$u = x^3 + 3x^2 + 1 \qquad v = 2x^2 + 8x - 5$$

    and we have

    $$y = uv \implies \frac{dy}{dx} = \frac{d(uv)}{dx} = \frac{du}{dx}v + u\frac{dv}{dx}$$

    $$\frac{du}{dx} = 3x^2 + 6x \qquad \frac{dv}{dx} = 4x + 8 \implies$$

    $$\frac{dy}{dx} = (3x^2 + 6x)(2x^2 + 8x - 5) + (x^3 + 3x^2 + 1)(4x + 8)$$

Study Guide for Section 3.6

- Problem 8 (Here is an example *like* problem 8)

    Suppose $y = (5x^2 - 3)(2x^3 + 7x)(x - 9)$

    We let $\quad u = (5x^2 - 3) \quad\quad v = (2x^3 + 7x) \quad\quad w = (x - 9)$

    Then $\quad y = uvw = (uv)w$

    $$\implies y' = [uvw]' = [(uv)w]' = [(uv)]'w + (uv)w'$$

    Next we find $[uv]'$ using the product rule again to get

    $$[uv]' = u'v + uv'$$

By substitution, we have

$$y' = [uv]'w + (uv)w' = [u'v + uv']w + (uv)w'$$

$$= \left[\underbrace{[5x^2 - 3]'}_{[u]'} \overbrace{(2x^3 + 7x)}^{v} + \overbrace{(5x^2 - 3)}^{u} \underbrace{[2x^3 + 7x]'}_{[v]'} \right] \overbrace{(x - 9)}^{w}$$

$$+ \overbrace{(5x^2 - 3)(2x^3 + 7x)}^{uv} \underbrace{[x - 9]'}_{[w]'}$$

$$= \left[\underbrace{(10x)}_{u'}(2x^3 + 7x) + (5x^2 - 3) \underbrace{(6x^2 + 7)}_{v'} \right](x - 9)$$

$$+ (5x^2 - 3)(2x^3 + 7x) \underbrace{1}_{w'}$$

$$= (20x^4 + 70x^2 + 30x^4 + 35x^2 + 18x^2 + 21)(x - 9)$$

$$+ (10x^5 + 35x^3 - 6x^3 - 21x)$$

$$= 10x^5 + 50x^4 + 29x^3 + 123x^2 - 21x + 21$$

- Problems 31-40

$$f(x) = \tan x \sin x \implies f'(x) = [\tan x]' \sin x + \tan x [\sin x]'$$

$$[\tan x]' = \sec^2 x \qquad [\sin x]' = \cos x$$

$$\implies f'(x) = \sec^2 x \sin x + \tan x \cos x$$

$$= \frac{1}{\cos^2 x} \sin x + \frac{\sin x}{\cos x} \cos x$$

$$= \tan x \sec x + \sin x$$

Graphing Calculator Hints

- Same as Study Guide for Section 3.5.

Study Guide for Section 3.7
The Chain Rule

Key Terms & Notation

Chain rule $\quad (f \circ g)'(x_0) \quad\quad \dfrac{dy}{du}\dfrac{du}{dx}$

Power rule $\quad \dfrac{d(u^r)}{dx} \quad\quad ru^{r-1} \cdot \dfrac{du}{dx}$

Questions for Understanding

- Suppose $y = \sqrt{(x^4+1)(x-2)}$. The "outside" function in the composition is the square root function. The value of the square root function depends on the value of the variable under the square root sign. Call this variable u. Next, the "inside" function u is as a product of two variables. The value of u depends on the values of the two variables describing it. Call these v and w. Last, both v and w are functions in x.

    Find $\dfrac{dy}{dx}$ by letting $y = \sqrt{u}$, $u = vw$, $v = (x^4+1)$, $w = (x-2)$, and

    1. Finding (a) $\dfrac{dy}{du}$ (b) $\dfrac{du}{dx}$ in terms of $\dfrac{dv}{dx}$ and $\dfrac{dw}{dx}$
    2. Using the chain rule to evaluate $\dfrac{dy}{dx} = \dfrac{dy}{du}\dfrac{du}{dx}$
    3. Substituting $\dfrac{dv}{dx}$ and $\dfrac{dv}{dx}$, where appropriate.

- Using the exact notation in the box for the *Power rule for differentiation*, substitute x for u. How does this reconcile with the box on the previous page? (pages 232-233)

- Recall that x_0 generally implies a specific value. If we rewrite the five rules on page 234 by replacing x_0 with x, would we have a set of rules for general differentiable *functions* as well?

- Would there be an advantage to memorizing the more general "chain rule versions" of the basic derivative formulas? What would u equal to obtain a basic formula from a chain rule version? (pages 234-235)

- Do you see how an "outside-in" approach to the chain rule applies in EXAMPLE 54? (page 235)

- Using the laws of exponents, we can rewrite

$$\left(\frac{f}{g}\right)' = \left(\frac{u}{v}\right)' = (uv^{-1})'.$$

Can you use the chain rule, the product rule, the general power rule, and some algebra to obtain the same result as the quotient rule? (pages 225, 233)

- How does the *Derivative Formula*

$$\frac{dy}{dx} = a^x(\ln a) \quad \text{for} \quad y = a^x \ (a > 0, \ a \neq 1)$$

apply to e and the "remarkable" result of §3.4? (pages 201, 234)

Suggested Exercises

§3.7 (pages 236-238): 4, 5, 10, 22, 28, 34, 46

Notes on Exercises

- Problem 13

$$F'(x) = [f((g(x))^2)]' = f'(\overbrace{(g(x))^2}) (\underbrace{}_{[(g(x))^2]'})$$

$$= f'(\overbrace{(g(x))^2}) (\underbrace{}_{2g(x)}) g'(x) = f'((g(x))^2) \, 2g(x) \, g'(x)$$

$$F'(2) = [f((g(2))^2)]' = f'(\overbrace{4}^{(g(2))^2}) (\underbrace{2(2)}_{2g(2)}) g'(2)$$

$$= (-4) \, 2 \, (2) \, (-0.3) = 4.8$$

- Problem 29
  Using the chain rule, we have

$$F'(x) = [f(g(x))]' = f'(g(x)) \, g'(x)$$
$$\implies F'(-3) = f'(g(-3)) \, g'(-3)$$

Remember that the graphical interpretation of the derivative is the slope of the tangent line at each point

$$\implies f'(g(-3)) = \text{slope of } y = f(x) \text{ at } x = g(-3) = 1$$

$$x = g(-3) = 1 \implies (x, f(x)) = (g(-3),$$
$$f(g(-3))) = (1, f(1)) = (1, 3)$$

The slope of $y = f(x)$ at the point $(1, 3)$ is $0 \implies f'(g(-3)) = 0$.

$$g'(-3) = (\text{ slope of } g(x) \text{ at } x = -3)$$
$$= (\text{ the slope at the point } (-3, 3)) \implies g'(-3) = 1$$

Therefore, $F'(-3) = f'(g(-3)) g'(-3) = 0 \cdot 1 = 0$

- **Problem 47 (modified)**
  Suppose $f(x) = (x^3+4)\cos(x^2-3x+4)$, then

$$f'(x) = [\underbrace{(x^3+4)}_{u}]' \underbrace{\overbrace{\cos(x^2-3x+4)}^{w}}_{v}$$

$$+ \underbrace{(x^3+4)}_{u} [\underbrace{\overbrace{\cos(x^2-3x+4)}^{w}}_{v}]'$$

$$= \underbrace{(3x^2)}_{u'} \underbrace{\overbrace{\cos(x^2-3x+4)}^{w}}_{v}$$

$$+ \underbrace{(x^3+4)}_{u} [\underbrace{-\sin(\overbrace{x^2-3x+4}^{w})\overbrace{(2x-3)}^{w'}}_{v'}]$$

$$= 3x^2\cos(x^2-3x+4) - (x^3+4)(2x-3)\sin(x^2-3x+4)$$

Graphing Calculator Hints

- Same as Study Guide for Section 3.5

Study Guide for Section 4.1
Linear Approximations

Key Terms & Notation

Best linear approximation Tangent line Normal line
Local approximation

Questions for Understanding

- How does the equation in Definition 1 relate to the *point-slope* form and the *Taylor* form of a line equation? (pages 169-170, 241)

- What point do the tangent line, normal line and the curve $y = f(x)$ have in common when $x = x_0$? (pages 243-244)

- How are $f'(x_0)$, the tangent line at $(x_0, f(x_0))$, and the normal line at $(x_0, f(x_0))$ related? (pages 243-244)

- Can you give an example of a function that has a tangent line that is tangent to infinitely many points on the functions's graph? (page 245)

Suggested Exercises

§4.1 (pages 246-248): 2, 3, 18, 20, 24, 27

Notes on Exercises

- Problem 1
  The most convenient way to find the tangent line is to use the point-slope form of the line equation along with the information we can glean from the problem. We want to satisfy the equation

  $$y - f(x_0) = f'(x_0)(x - x_0)$$

Because this tangent line is also parallel to the line $2y + 8x - 5 = 0$, we solve this equation for y and get

$$y = \frac{1}{2}(-8x + 5) = -4x + \frac{5}{2}$$

which indicates that the slope of the parallel line is -4. This is also the slope of the tangent line, so $f'(x_0) = -4$, as well. From the original equation

$$y = x^3 + 2x^2 - 4x + 5$$

we obtain the derivative *function*

$$f'(x) = 3x^2 + 4x - 4$$

We can now obtain x_0 by substitution

$$-4 = f'(x_0) = 3x_0^2 + 4x_0 - 4 \implies 3x_0^2 + 4x_0 = x_0(3x_0 + 4) = 0$$

$$\implies x_0 = 0 \quad \text{or} \quad x_0 = -\frac{4}{3}$$

Again, by substitution, we can find the two associated tangent lines.

1. $y - f(0) = f'(0)(x - 0) \implies y = 5 - 4x$

2. $y - f\left(-\frac{4}{3}\right) = f'\left(-\frac{4}{3}\right)\left(x - \left(-\frac{4}{3}\right)\right)$

   $\implies y = \frac{311}{27} - 4\left(x + \frac{4}{3}\right)$

- Problems 21-30
    See EXAMPLES 2-4

Graphing Calculator Hints

- Plotting two expressions together using $f(x) = x^2$, $f'(x) = 2x$
    HP-48x
    Enter the two equations in the form {'X^2' '2*X'}

    TI-8x
    Enter x^2 for **y1** $=$ (or **Y1** $=$) and $2x$ for **y2** $=$ (or **Y2** $=$). Make sure both are selected and plot.

Study Guide for Section 4.2
Newton's Method

Key Terms & Notation

Newton's method   Seed   Iteration
Golden ratio

Questions for Understanding

- What do n, $n+1$ represent in the subscripting notation (pages 249)
$$x_i, \quad i = 0, 1, 2, \ldots, n, n+1, \ldots$$

- What does each x_i represent in *Newton's method* and the *iterating* function?
(pages 249)

- In §4.1, we found the *best linear approximation* to a differentiable function f to be the line
$$y = f(x_0) + m(x - x_0)$$
To find a root, we want $y = 0$. Can you derive the general iterating function for Newton's method by using algebra, letting $x_0 = x_n$, $x = x_{n+1}$, and making the appropriate substitution for m?
(pages 241, 249-250)

- Finding roots means finding x-values that satisfy the equation $f(x) = 0$. Suppose
$$x^2 + 6 = 5x$$
How can you solve this equation by forming a function $f(x)$ and finding its roots? (page 250)

- How do you know when to stop with Newton's method? (page 252)

- Suppose the initial *seed* value for EXAMPLE 13 is 0.25, then the third iteration of the iterating function ≈ -0.6777188. Without finding x_0, x_1, x_2, can you find x_4? (pages 249, 254)

- Could you illustrate finding a root graphically (similar to Figure 4.4) starting with the function plotted at right and the indicated seed value? (page 250)

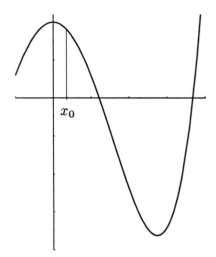

Figure 15: Newton's method

- What are the reasons that Newton's method can fail in EXAMPLES 12-14? Can Newton's method *diverge* away from a root? (pages 254-255)

Suggested Exercises

§4.2 (pages 256-258):   2, 6, 9, 11, 12, 13, 14, 20 (graph it)

Notes on Exercises

- Find a zero of the following function within the indicated interval, to at least four decimal places of accuracy.

$$f(x) = x^5 - 2x^3 - 1 \qquad [1.1,\ 1.6]$$

Since the function is continuous, and $\left.\begin{array}{l}f(1.1) = -2.05149 \\ f(1.6) = 1.29376\end{array}\right\}$ a zero exists in the interval.

Study Guide for Section 4.2

n	x_n	x_{n+1}
0	1.35	1.60328
1	1.60328	1.52657
2	1.52657	1.51325
3	1.51325	1.51288
4	1.51288	1.51288

$$x_0 = \frac{1.1 + 1.6}{2} = 1.35$$

$$f'(x) = 5x^4 - 6x^2$$

$$x_{n+1} = x_n - \frac{x_n^5 - 2x_n^3 - 1}{5x_n^4 - 6x_n^2}$$

$$x \approx 1.51288$$

Graphing Calculator Hints

- Is $(x^2 + 3)/2x$ the same as $(x^2 + 3)/(2x)$ in your calculator?

- Suppose $f(x) = 5x^2 - 4x$, then $f'(x) = 10x - 4 \implies$ the iterating formula

$$x_{n+1} = x_n - \frac{f(x_n)}{f'(x_n)} = x_n - \frac{5x_n^2 - 4x_n}{10x_n - 4}$$

which gives us an iterating function of

$$y = x - \frac{5x^2 - 4x}{10x - 4}$$

which can be entered in your calculator as

$$\text{X-(5*X}\wedge\text{2-4*X)/(10*X-4)}$$

- Your calculator's SOLVER is perfect for the iterating function of Newton's method

HP-48G
Enter the expression as the calculator's EQ variable. Get into SOLVR menu by pressing [↰], 7, [ROOT], [SOLVR]. (See Study Guide for §0.1 for a more in-depth discussion.) Enter your initial seed value, press [X], then press [EXPR=]

(See next page)

1. Write down the EXPR= value.
2. Press [X] to enter the new value.
3. Press [EXPR=] to obtain the next iteration value.

Continue 1-3 until EXPR= remains the same to the required degree of accuracy.

TI-85

To make entering the next x_n easier each time, you first need to create a link between the variable **exp** and another single-letter variable that is easy to keep entering. In the following explanation, the variable used will be **A**.

On the HOME screen, enter A= with [ALPHA], LOG, and [ALPHA], STO. Get into VARS menu with [2nd], 3. Choose [REAL] with F2 key. Cursor down to **exp** and press ENTER. Press ENTER again to complete the variable assignment. You have just linked the contents of the variable **A** to the contents of the variable **exp** in the second section of the SOLVER menu.

To use the link with the solver, Enter expression under GRAPH, [y(x) =]. Get into SOLVER with [2nd], GRAPH, enter appropriate function name at bottom of view screen. Press ENTER for next screen. Arrow down to **x =** and enter your initial seed value. Arrow up to **exp =** and press [SOLVE].

1. Write down the **exp =** value.
2. Place the cursor on the **x =** line.
3. Press [ALPHA], LOG and A will appear next to **x =** in the SOLVER.
4. Arrow up to **exp =** and press [SOLVE] for the next iteration.

Continue 1-4 until **exp =** remains the same to the required degree of accuracy.

TI-82

Since there is no SOLVER-type capability, you must do a little finger gymnastics. Under Y =, enter the original expression on the Y_1 = line. On the Y_2 = line, enter X-Y_1/nDeriv(Y_1,X,X). You will find the correct Y_1 under [2nd], VARS, 1:Function. Press ENTER to bring up the list containing Y₁. Press ENTER to insert Y₁ where you originally started from. The nDeriv( ) function is item 8 under MATH. Once the Y₂ = expression is entered, exit from Y = to the HOME screen. Store your initial seed value in the X variable by entering the value on the screen and pressing STO>, X, ENTER.

1. Execute [2nd], VARS, 1:Function, ENTER, Y₂, ENTER.
2. Press ENTER to evaluate Y₂, then write down the result.
3. Execute STO>, X, ENTER to store the new value.

Continue 1-3 until Y₂ remains the same to the required degree of accuracy.

Study Guide for Section 4.3
Analyzing Function Behavior

Key Terms & Notation

"Smooth"   Decreasing   Local maximum
Increasing   Strictly decreasing   Local minimum
Strictly increasing   Monotonic   Local extremum
Critical value   Strictly monotonic

Questions for Understanding

- Do you agree with the opening paragraph of this section?

  (page 258)

- How are "smooth," *differentiable*, and the box on page 259 related?

  (pages 258-259)

- As you read Definition 2, picture a graph. As you go from left to right on the x-axis, what happens to $f(x)$? (page 260)

- Does *monotonic* mean flat? (page 260)

- Is a *critical value* in the domain of $f(x)$, the range of $f(x)$, or neither? (page 260)

- How are Figure 4.7, EXAMPLE 15, and the caution on page 261 related?

  (pages 260-262)

- How do *local extrema* and critical values influence each other?

  (pages 262-263)

Suggested Exercises

§4.3 (pages 265-269): 6, 8, 21, 25, 30

Notes on Exercises

- §4.3 Problems 1-10 with $f(x) = x^4 - 20x^2 + 64$

  (a) Graph $f(x)$

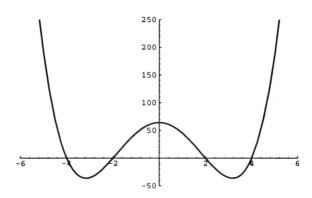

Figure 16: $f(x) = x^4 - 20x^2 + 64$

(b) $f'(x) = 4x^3 - 40x$, and $f'(x) = 0$
$$\implies 4x(x^2 - 10) = 4x(x + \sqrt{10})(x - \sqrt{10}) = 0$$
$$\implies x = -\sqrt{10}, 0, \sqrt{10}$$

Choose any convenient $x < -\sqrt{10} \approx -3.16228$
$$x = -10 \implies f'(x) = f'(-10) = 4(-1000) - 40(-10) < 0$$
$$\implies f(x) \text{ decreasing on } (-\infty, -\sqrt{10})$$

Choose any convenient x such that $-\sqrt{10} < x < 0$
$$x = -1 \implies f'(x) = f'(-1) = 4(-1) - 40(-1) > 0$$
$$\implies f(x) \text{ increasing on } (-\sqrt{10}, 0)$$

Choose any convenient x such that $0 < x < \sqrt{10}$
$$x = 1 \implies f'(x) = f'(1) = 4(1) - 40(1) < 0$$
$$\implies f(x) \text{ decreasing on } (0, \sqrt{10})$$

Choose any convenient $x > \sqrt{10} \approx 3.16228$
$$x = 10 \implies f'(x) = f'(10) = 4(1000) - 40(10) > 0$$
$$\implies f(x) \text{ increasing on } (\sqrt{10}, \infty)$$

(c) Graph $f'(x)$

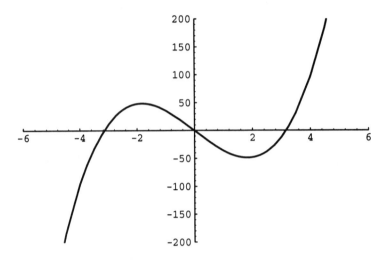

Figure 17: $f'(x) = 4x^3 - 40x$

(d) From (b), we have
$$f'(x) > 0 \text{ on } (-\sqrt{10}, 0) \cup (\sqrt{10}, \infty)$$
$$f'(x) < 0 \text{ on } (-\infty, -\sqrt{10}) \cup (0, \sqrt{10})$$

(e) From (b), we have critical values of $x = -\sqrt{10}, 0, \sqrt{10}$

(f) $f(-\sqrt{10}) = -36 \implies x = -\sqrt{10}$ represents a local min
$f(0) = 64 \implies x = 0$ represents a local max
$f(\sqrt{10}) = -36 \implies x = \sqrt{10}$ represents a local min

Graphing Calculator Hints

- Practice plotting $f(x)$ and $f'(x)$ together and recognizing the features on each that indicate the extrema of $f(x)$.

Study Guide for Section 4.4
The Derivative Test

Key Terms & Notation

Derivative test   Absolute maximum   Absolute minimum

Questions for Understanding

- Can you sketch a picture of each condition in "The derivative test for local extrema?" (page 264)

- Can an *absolute* extremum also be a local extremum? (page 270)

- In the "Strategy for Finding Absolute Extrema of f over $[a, b]$," when is $f(x)$ used and when is $f'(x)$ used? (pages 270-271)

- What do you visualize graphically when you see the symbolic representation $f'(x) = 0$?

Suggested Exercises

§4.4 (pages 272-275):   6, 10, 11, 21, 22, 23, 25

Notes on Exercises

- §4.4 Problem 26

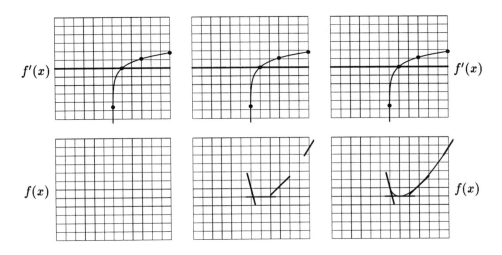

Figure 18: Local min at $(1, f(1))$

Graphing Calculator Hints

- Same as Study Guide for Section 4.3.

Study Guide for Section 4.5
Finding Extrema Under Constraints

Key Terms & Notation

Constraint

Questions for Understanding

- How are functions, causes, and effects related in this section's opening paragraph? (page 276)

- What does the word *constraint* mean in a general sense? How is that related to the mathematical model used here? (page 276)

- This section and the *exercises* are split into two related types of problems. What are they?

- When the worker presents you with each of the two problems, how much informataion does she provide? How much are you expected to supply? Does this type of interaction strike you as a "real world" situation?

- Why do we want the radius r in terms of h? (pages 276-277)

- How is the subsection "Finding an extemum" related to the four step strategy of §4.4? (pages 270, 279-280)

- Do you see why the maximization problem required answering the previous problem of finding volume in terms of height? (page 279)

Suggested Exercises

§4.5 (pages 280-281):   1, 6, 9, 10, 11, 16, 19, 20

Notes on Exercises

- Don't hesitate to write out your thinking processes. The written work done to solve these types of problems could easily look like a brief outline of what is printed in this section of your text.

- Problem 8

Want: Perimeter P as function of b
$$\implies P = P(b)$$
Know: $P = 2a + b$
Need: a in terms of b

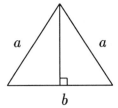

Know: $y^2 + \left(\dfrac{b}{2}\right)^2 = r^2 = 4^2 = 16 \implies$

$y = \sqrt{16 - \left(\frac{b}{2}\right)^2} = \frac{1}{2}\sqrt{64 - b^2}$

$a^2 = (y+4)^2 + \left(\frac{b}{2}\right)^2 \implies$

$a = \sqrt{(y+4)^2 + \left(\frac{b}{2}\right)^2}$
$ = \frac{1}{2}\sqrt{4(y+4)^2 + b^2}$

$ = \frac{1}{2}\sqrt{4\left(\frac{1}{2}\sqrt{64-b^2}+4\right)^2 + b^2}$

$ = \ldots = 2\sqrt{8 + \sqrt{64-b^2}} \implies$

$P(b) = 2\left(2\sqrt{8 + \sqrt{64-b^2}}\right) + b$

$ = 4\sqrt{8 + \sqrt{64-b^2}} + b$

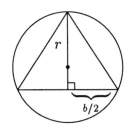

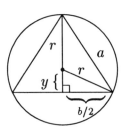

Figure 19: Problem 8

- Problem 18

$$P(b) = 4\sqrt{8 + \sqrt{64 - b^2}} + b \implies$$

$$P'(b) = 1 + 4\left(\tfrac{1}{2}\right)(8 + (64-b^2)^{1/2})^{-1/2}\left(\tfrac{1}{2}\right)(64-b^2)^{-1/2}(-2b)$$

$$= 1 - \frac{2b}{(8 + (64-b^2)^{1/2})^{1/2}(64-b^2)^{1/2}}$$

$$P'(b) = 0 \implies 2b = (8 + (64-b^2)^{1/2})^{1/2}(64-b^2)^{1/2} = 0$$

$$\implies \quad \ldots \text{(algebra)} \ldots$$
$$\implies \quad b^4(b^2 - 48) = 0$$
$$\implies \quad b = 0 \text{ or } b = \pm\sqrt{48} = \pm 4\sqrt{3} \approx \pm 6.9820323$$

Since domain of P is the *open* interval $(0, 8)$, the maximum perimeter occurs at the critical value $b = +4\sqrt{3} \implies$ maximum perimeter $= P(4\sqrt{3}) = 12\sqrt{3}$cm

Graphing Calculator Hints

- Once the modeling function and its domain are determined, plotting the function may help you see if the max or min you obtained computationally makes sense. You can obtain similar information by plotting the derivative.

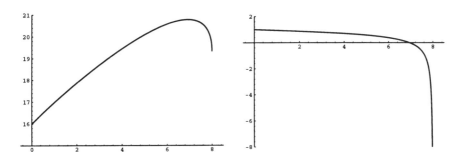

Figure 20: $P(b)$ Figure 21: $P'(b)$

Study Guide for Section 4.6
The Mean Value Theorem

Key Terms & Notation

Mean   Rolle's theorem   Mean value theorem

Questions for Understanding

- How is the idea of a continuous function applied in the explanation of the traveling car? (pages 281-282)

- Can you find any examples among *exercises 21-28* that fail to satisfy either one or both of the hypotheses of the *Mean value theorem*? (pages 283, 287-289)

- Although this section mainly discusses the Mean value theorem, what other existence theorems/properties from previous sections also apply simply because of the continuity hypothesis? (pages 129-130, 285)

- Part of the "Reasoning" for *Theorem 4.4* states, "From our discussion above, we can conclude that $f(x) - g(x)$ is constant." How does the "discussion above" guarantee this? Where does the Mean value theorem fit in? (pages 285-286)

- Both $f(x) = x^2$ and $g(x) = x^2 + 2$ have the same derivative of $2x$. They differ by the constant 2. How many other functions with derivative of $2x$ lie between $f(x)$ and $g(x)$? (pages 286)

Suggested Exercises

§4.6 (pages 286-291):   3, 4, 8, 12, 18, 22, 24, 28

Notes on Exercises

- Problem 2

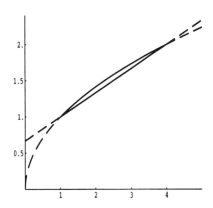

Figure 22: Graphs for (a) & (b)

(c) $(1, f(1)) = (1, 1)$, $(4, f(4)) = (4, 2)$

$\implies \dfrac{f(4) - f(1)}{4 - 1} = \dfrac{2 - 1}{4 - 1} = \dfrac{1}{3}$

$\implies$ want $1 < x_0 < 4$ such that the tangent line through $(x_0, f(x_0))$ is parallel to the line described by $y - 1 = \dfrac{1}{3}(x - 1)$

i.e., has slope $= \dfrac{1}{3} \implies f'(x_0) = \dfrac{1}{3}$

$f'(x) = [\sqrt{x}]' = [x^{1/2}]' = \dfrac{1}{2}x^{1/2-1} = \dfrac{1}{2x^{1/2}} = \dfrac{1}{2\sqrt{x}}$

$\implies f'(x_0) = f'(x)|_{x=x_0} = \dfrac{1}{2\sqrt{x_0}}$

If $f'(x_0) = \dfrac{1}{2\sqrt{x_0}} = \dfrac{1}{3}$, then $3 = 2\sqrt{x_0}$

$\implies \dfrac{3}{2} = \sqrt{x_0} \implies x_0 = \dfrac{9}{4} = 2\dfrac{1}{4}$

Since $1 < \dfrac{9}{4} < 4$, we have that $x_0 = \dfrac{9}{4}$ satisfies the Mean value theorem.

Graphing Calculator Hints

- Remember that entering $f(x)$ into your calculator for plotting also makes it available for the *solver* feature of your calculator. You can use it to check $f(a)$ and $f(b)$ values. (See Study Guide for §4.2).

Study Guide for Section 5.1
Optimization – Part 1

Key Terms & Notation

Optimization   Strategy for solving optimization problems

Questions for Understanding

- Why do you think the authors detailed their six step "Strategy for Solving Optimization Problems?" (pages 294-295)

- In EXAMPLE 1
    1. What does the problem *want* you to do? How is this related to the *dependent* variable?
    2. What do you *know* about the situation? Are the *constraints* part of this knowledge? Is everything you know about this problem given to you explicitly in the problem's description?
    3. What else might you *need* so you can analyze and/or continue solving the problem? Would this include expressing the dependent variable in terms of an appropriate *single independent* variable?

    Where would a picture help? How might the max value for $A(w)$ be related to w_0 if $A'(w_0) = 0$? (pages 296-297)

- What is significantly different between the problems posed in EXAMPLES 1 and 2? What processes remain the same?
(pages 296-300)

- Do you agree that changing the written words into mathematical representations is basically translating one language into another?

Suggested Exercises

§5.1 (pages 308-318):   1, 11, 17, 20, 21

Notes on Exercises

- Here is a synopsis of the box on page 308.

> **Step 1.** Identify the dependent variable.
>
> **Step 2.** Identify the constraints.
>
> **Step 3.** Express the dependent variable as a function of a single independent variable and identify its domain.
>
> **Step 4.** Graph the function over its domain if you can.
>
> **Step 5.** Identify the locations and values of the absolute maximum and/or minimum. Examine endpoints (if any) and critical values of the domain.
>
> **Step 6.** Interpret the results and answer the question posed.

Graphing Calculator Hints

- Be careful to enter your function correctly. For convenience, you may want to use x as the independent variable in your calculator. It may also help to first write out the calculator version of the function by hand.

$$A = 30w - w^2 \quad \xrightarrow{becomes} \quad y = \underbrace{30 * X - X \,\char`\^\, 2}_{calculator}$$

- Since *optimization problems* depend on their domain, make sure your calculator's x-range values are set for the domain of permissible values.

Study Guide for Section 5.1
Optimization – Part 2

Key Terms & Notation

Optimization   Strategy for solving optimization problems

Questions for Understanding

- Word problems (or "story" problems) are generally difficult. Would you expect "real world" problems to be easier?

- In EXAMPLE 4
    1. Is $C = 25\sin\dfrac{\theta}{2}\cos\dfrac{\theta}{2}$ still a function of a single independent variable? Is θ or $\dfrac{\theta}{2}$ that variable?
    2. What are the *triangle-based* definitions of sine and cosine?
    3. Why can't the domain of values θ for C be $\mathbb{R} = (-\infty, \infty)$?
    4. The "second derivative test" mentioned at the end of **Step 5** is discussed in the next section. Does this suggest that there may be more than one approach you could use?   (pages 302-305)

- Does EXAMPLE 5 give you some insight into the value of calculus to economics and business?   (pages 305-307)

Suggested Exercises

§5.1 (pages 308-318):   2, 3, 10, 19, 31

Notes on Exercises

- **EXAMPLE 5, page 305**

  Minimizing cost is one of the more common applications of calculus. The major hurdle for most students is one of translation. In this example, the cost for the *material* for the bottom and top is $0.06 *per in^2*. Hence the

  (total cost for bottom and top)
  = (cost per in^2)(number of in^2 in the bottom and top)

  = (cost per in^2)×
  [(number of in^2 on bottom) + (number of in^2 on top)]

  = $0.06 [(surface area of bottom) + (surface area of top)]

  Similarly, the *materials* cost for the rest of the can is $0.03 *per square inch*, which gives us

  (total cost for the curved surface)
  = (cost per square inch) ×
  (number of square inches in the curved surface)
  = $0.03 (surface area of the curved surface)

  But the surface area is

  $$\underbrace{\pi r^2}_{\text{top}} + \underbrace{\pi r^2}_{\text{bottom}} + \underbrace{2\pi r}_{\text{curved surface}} \qquad \text{where } d = 2r$$

  So the cost C of the surface area is

  $$C = \underbrace{(\$0.06)\pi r^2}_{\text{top}} + \underbrace{(\$0.06)\pi r^2}_{\text{bottom}} + \underbrace{(\$0.03)2\pi r}_{\text{curved surface}}$$

  $$= (.06)2\pi r^2 + (.03)2\pi r$$

  (The $ can be dropped until obtaining the final answer, because everything is now expressed in terms of U. S. currency.) b

---

Graphing Calculator Hints

- Same as Study Guide for Section 5.1 – Part 1

Study Guide for Section 5.2
Higher Order Derivatives

Key Terms & Notation

Second derivative   Concavity f'', f''', $f^{(n)}$

Point of inflection   Concave up/down $\dfrac{d^2 f}{dx^2}$, $\dfrac{d}{dx}\left(\dfrac{df}{dx}\right)$, $\dfrac{d^n f}{dx^n}$

Questions for Understanding

- How does the box on page 320 compare with the bold-faced statements in the middle of page 260?

- Figure 23 below represents the *1st derivative* of a function f. Use it to provide the requested information in items 1-6.

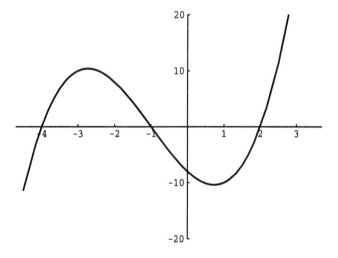

Figure 23: $y = f'(x)$

1. Find all critical points of the **original function** f.

2. Estimate the intervals over which the **original function** f is increasing and decreasing.

3. Estimate the intervals over which the graph of the **original function** f is concave up and concave down.

4. Estimate the x-coordinates of all local maximum and minimum points of the **original function** f.

5. Estimate the x-coordinates of all inflection points of the **original function** f.

6. Sketch a graph of the **original function** f.

- Still using Figure 23 above, consider f' as a function in its own right. Then the **first derivative of** f' is also the **second derivative** f'' of the *original function* f. Use this information for the following.

    1. Sketch a graph of $y = f''(x)$.

    2. Complete each of items 1-5 above, replacing "the **original function** f" with "the **1st derivative function** f'" and label the appropriate areas on your sketched graph of the second derivative f''.

Suggested Exercises

§5.2 (pages 325-329):   1, 4, 7, 21, 24, 35, 44, 49

Notes on Exercises

- Problems 21-28

    1. Can you also estimate the intervals over which the **original function** f is increasing and/or decreasing?

    2. Remember, derivative values represent slope values for the original function. Recall the sketching strategy illustrated toward the end of the Study Guide for Section 4.3.

- Problem 35

    What are the physical interpretations of the 1st and 2nd derivatives?

- Problems 43-50

    (a) Look at $f'(x)$ and $f''(x)$.
    (b) From problems 39-42, we find that for each real number c s.t. $f'(c) = 0$, if $f''(c)$ exists, then

    (i) $f''(c) > 0 \implies$ graph is concave up (ccup)
    $\implies f(c)$ is a local min.

    (ii) $f''(c) < 0 \implies$ graph is concave down (ccdown)
    $\implies f(c)$ is a local max.

    (iii) $f''(c) = 0 \implies$ ambiguous case
    $\implies$ use 1st derivative test.

Graphing Calculator Hints

- Sometimes the *solver* feature of your calulator can be used to make it easier to determine where $f''(x) < 0$ or $f''(x) > 0$.

Study Guide for Section 5.3
Implicit Differentiation

Key Terms & Notation

Locus   Implicitly   Implicit differentiation

Questions for Understanding

- Consider EXAMPLE 12. The chain rule gives $\dfrac{d(y^2)}{dx} = 2y\dfrac{dy}{dx}$. Applying this same reasoning to x^2, we get

$$\frac{d(x^2)}{dx} = 2x\frac{dx}{dx} \implies 2x + 2y\frac{dy}{dx} = 0 = 2x\frac{dx}{dx} + 2y\frac{dy}{dx}$$

  How can you reconcile this?   (page 333)

- Suppose $x^2y^2 = \dfrac{y^3}{x} + 66$. How does this equality extend to (page 335)

$$\frac{d}{dx}(x^2y^2) \quad \text{and} \quad \frac{d}{dx}\left(\frac{y^3}{x} + 66\right) \ ?$$

- How is the *power rule* for derivatives used in EXAMPLE 15?
                                                                 (page 336)

- In finding the **Derivative of an inverse function**, we start with $x = f^{-1}(y)$. Why don't we use the *power rule* to obtain
                                                             (page 333-337)

$$1 = (-1)f^{-1-1}(y)\frac{dy}{dx} = -\frac{dy}{dx}\frac{1}{f^2(y)}$$

- Since $\sin x$, $\cos x$, $\tan x$, are the most commonly used trig functions, which *arctrig* derivatives should you be sure to know?
                                                                 (page 339)

Study Guide for Section 5.3

Suggested Exercises

§5.3 (pages 339-340):   1, 6, 12, 16, 18 (follow EXAMPLE 18)

Notes on Exercises

- Problem 15

$$xy + 16 = 0 \implies \frac{d}{dx}(xy) + \frac{d}{dx}(16) = \frac{d}{dx}(0)$$

$$\implies \frac{d}{dx}(xy) + 0 = 0$$

$$\frac{d}{dx}(xy) = \left(\frac{d}{dx}x\right)y + x\left(\frac{d}{dx}y\right) = \left(\frac{dx}{dx}\right)y + x\left(\frac{dy}{dx}\right) = y + x\frac{dy}{dx}$$

$$\implies -y = x\frac{dy}{dx} \quad \text{and} \quad \frac{dy}{dx} = -\frac{y}{x}$$

Find $\frac{d^2y}{dx^2}$ implicitly by

$$\frac{d}{dx}(-y) = \frac{d}{dx}\left(x\frac{dy}{dx}\right) \implies -\frac{dy}{dx} = \frac{dx}{dx}\frac{dy}{dx} + x\frac{d}{dx}\left(\frac{dy}{dx}\right)$$

$$= \frac{dy}{dx} + x\frac{d^2y}{dx^2} \implies \frac{d^2y}{dx^2} = -\frac{2}{x}\frac{dy}{dx}$$

$$\implies \left.\frac{d^2y}{dx^2}\right|_{(-2,8)} = \left[-\frac{2}{x}\frac{dy}{dx}\right]_{(-2,8)} = \left[-\frac{2}{x}\left(\frac{-y}{x}\right)\right]_{(-2,8)}$$

$$= \left.\frac{2y}{x^2}\right|_{(-2,8)} = \frac{16}{4} = 4$$

- Problem 25

$$y = (x-5)^{1/3} \implies \frac{d}{dy}(y) = \frac{d}{dy}(x-5)^{1/3}$$

$$\frac{d}{dy}(y) = \frac{dy}{dy} = 1$$

$$\frac{d}{dy}(x-5)^{1/3} = \frac{d}{d(x-5)}(x-5)^{1/3} \frac{d(x-5)}{dx} \frac{dx}{dy} = \frac{1}{3}(x-5)^{-2/3}(1)\frac{dy}{dx}$$

$$\implies 1 = \frac{d}{dy}(y) = \frac{d}{dy}(x-5)^{1/3} = \frac{1}{3}(x-5)^{-2/3}\frac{dy}{dx}$$

$$\implies \frac{dy}{dx} = 3(x-5)^{2/3}$$

If $\frac{dx}{dy} = 0$, then $3(x-5) = 0 \implies x = 5 \implies y = (5-5)^{1/3} \implies$

$(5, 0)$ is the only point where the tangent line is vertical

---

Graphing Calculator Hints

- Some equations, such as those representing conics, can be plotted on your calculator. If you are interested, check out your manual.

Study Guide for Section 5.4
Related Rates

Key Terms & Notation

Related rates

Questions for Understanding

- In this section, we differentiate with respect to time t, rather than x. What does the term *related rate* indicate about t and any other variables involved? (pages 341, 342)

- Do you see how the chain rule and implicit differentiation work together in related rates problems?

- In EXAMPLE 23, Step 2, t does not appear anywhere in the equation. It is known to be involved *implicitly*. How does this relate to the differential equation of Step 3? (page 346)

- What difference does it make if one ignores the warning on page 347?

- One of the descriptions of derivative includes the word rate. How does that apply to this section?

Suggested Exercises

§5.4 (pages 347-349):   1, 2, 4, 13, 16, 23, 24

Notes on Exercises

- Try to generalize the approaches and concepts in the EXAMPLES.

- Here is a synopsis of the **Strategy for Solving Related Rates Problems**.

> **Step 1.** Identify the variable and constraint quantities involved in the process, along with any rates of change.
>
> **Step 2.** Find a relationship between the variables.
>
> **Step 3.** Differentiate with respect to time to find a relationship between the variables and their rates of change.
>
> **Step 4.** Substitute all known variable and rate values for the specific instant of time in question, and determine the unknown rates of change.
>
> **Step 5.** Interpret the results and use them to answer the particular questions posed by the problem.

Graphing Calculator Hints

- Your calculator's *solver* feature can help you check your numerical results once the appropriate differential equation is obtained.

Study Guide for Section 5.5
Parametric Equations and Particle Motion

Key Terms & Notation

Parametric equations Parameterized curve Parameter
Intersection point   "Collision" point

Questions for Understanding

- Is a parameter in a parametric equation the same as a parameter for a family of curves? (pages 31, 350)

- What is significant about EXAMPLES 24 and 25? (pages 350-351)

- Can we find the slope $\dfrac{dy}{dx}$ of the curve $x^2+y^2 = 9$ by using parametric equations and finding $\dfrac{dx}{dt}$, $\dfrac{dy}{dt}$? (pages 352-353)

- What are the general parametric equations for an ellipse? What is the general rectangular equation for an ellipse? (page 354)

- What is the difference between an intersection point and a "collision" point? (page 355)

- **NOTE:** In the middle of page 356, the description of a flight path contains an error. The description for $y(t)$ should be

   $y(t) = $ *Vertical* position of the projectile at time t

- What are the general parametric equations for projectile motion? What substitutions would you make so the following equations represent the same situation as projectile motion? (e.g., substitute θ for α.) (page 356)

$$\cos\alpha = \frac{x}{r} \qquad \sin\alpha = \frac{y}{r}$$

Suggested Exercises

§5.5 (pages 358-360): 2, 8, 11, 20, 26

Notes on Exercises

- Problem 19

    Since arccsc is the inverse function of csc, we have

    $$\text{arccsc}\,(\csc\,\theta) = \theta.$$

    If we let $x = \csc t$, then $y = \text{arccsc}\,(\csc t) = t \implies$

    $$x = \frac{1}{\sin t}, \quad y = t$$

    is the desired pair of parametric equations.

Graphing Calculator Hints

- **HP-48G**

    In the PLOT screen, arrow to Parametric, select [OK], arrow to EQ:. Enter the two expressions for $x(t)$ and $y(t)$ in the form

    $$'(x(t), y(t))'$$

    For example, $x(t) = 3\sin(3t)$ and $y(t) = 2\sin(4t)$ are entered as

    $$'(3*SIN(3*T), 2*SIN(4*T))'$$

    Arrow to INDEP: and change to T. H-VIEW and V-VIEW still represent xy-range viewing rectangle. To set upper and lower bounds for T and step-size for choices of T, select [OPTS] and enter the appropriate values. Dflt stands for the calculator default values. (See Plot Types in the *User's Guide*)

- **HP-48S**

    Parametric functions must be entered in complex number form; e.g., $x(t) = 3\sin(3t)$ and $y(t) = 2\sin(4t)$ are entered on the stack as

    $$'3*SIN(3*T) + i*(2*SIN(4*T))'$$

    (The complex number **i** is entered with ALPHA, ⌐, CST.) Store the expression as the EQ variable. Bring up the PLOT menu and set the plot type to parametric by pressing [PTYPE], [PARA]. Press [PLOTR], set the independent variable to T, and specify the upper and lower bounds by entering {T lower upper} (e.g., {T -3 3} is entered by ⌐, {}, T, SPC, 3, +/-, SPC, 3, ENTER, [INDEP]). Pressing [AUTO] then produces a graph with autoscaling. Pressing [DRAW] produces a graph at the current settings. (See More About Plotting and Graphics Objects in the *Owner's Manual*)

- **TI-85**

    Select [MODE] with 2nd, MORE; arrow to PARAM and press ENTER, then EXIT. Press GRAPH, [E(t)]; enter $x(t)$ as **xt1=** and $y(t)$ as **yt1=**. Choose the t-variable with the appropriate function key at the bottom of the screen. Press GRAPH to plot the functions. Upper and lower t-values, xMin/Max, yMin/Max are entered under the [RANGE] menu. (See Parametric Graphing in the *Guidebook*.)

- **TI-82**

    Select MODE; arrow to Par and press ENTER. Press Y=, then enter $x(t)$ as **X1T=** and $y(t)$ as **Y1T=**. The T-variable automatically appears when you press the X,T,θ key while in Par mode. Press GRAPH to plot the functions. Press WINDOW to change Tmin/mas, Tstep, Xmin/max, Ymin/max settings. (See Parametric Graphing in the *Guidebook*.)

Study Guide for Section 6.1
What is a Definite Integral?

Key Terms & Notation

Partition

Δt

Definite integral

Inscribed

Circumscribed

Summation notation

$\sum_{n=1}^{k} f(n)$

Index variable

Method of exhaustion

Questions for Understanding

- Suppose $v(t)$ is a function representing the velocity of a boat. What type of quantity would the area under the graph $y = v(t)$ for $t =$ 8:30 AM to $t =$ 10:05 AM represent? (pages 362-363)

- In Figure 6.3, how would you best approximate the distance traveled by the car from $t = 0.5$ to $t = 1.5$ hours: by using an average of the two endpoints or by using an average over several points? (page 363)

- Graph the function $y = x^2$ over the interval $[-2, 2]$ and consider Figure 6.10. Which of the rectangles are inscribed and which are circumscribed? (page 369)

- Given $y = x^2$, which of the following best describes the summation? (pages 368, 370-371)

$$\sum_{n=1}^{8} \left(1 + n \cdot \frac{1}{8}\right) \left(\frac{1}{8}\right)$$

  1. 8 inscribed rectangles on the interval $[1, 8]$.
  2. 8 inscribed rectangles on the interval $[1, 2]$.
  3. 8 circumscribed rectangles on the interval $[0, 1]$.
  4. 8 circumscribed rectangles on the interval $[1, 2]$.
  5. 8 circumscribed rectangles on the interval $[1, 8]$.

Study Guide for Section 6.1

- How does the subsection on *methods of exhaustion* relate to the material in the main part of this section? (pages 375-379)

Suggested Exercises

§6.1 (pages 372-375): 3, 8, 22, 26, 36, 46

Notes on Exercises

- Problem 7

  The distance traveled d is the area under the curve between $t = 1.75$ and $t = 2.0$. This can be approximated by the area of the rectangle A_R with

  $$\text{base } b = \text{elapsed time} = 2.0 - 1.75 = .25$$
  $$\text{height } h = \text{average speed} \approx \frac{55 + 30}{2} = 42.5$$
  $$\implies A_R = bh = (.25)(42.5) = 10.625 \text{ miles}$$

- Problems 21-30

  Find the area bounded by $y = \sqrt{4 - (x-1)^2}$, $y = x - 3$, the x-axis, $x = 1$, and $x = 5$.

  (a) Graph of problem

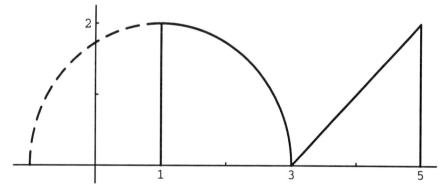

Figure 24: Graph of problem situation

(b) $\text{Area} = \frac{1}{4}\pi r^2 + \frac{1}{2}bh$

$= \frac{1}{4}\pi(2)^2 + \frac{1}{2}(5-3)(5-3)$

$= \pi + 2$

≈ 5.14159

Graphing Calculator Hints

- Figure 24, above, is a partial graph of the split function

$$f(x) = \begin{cases} \sqrt{4-(x-1)^2}, & x \leq 3 \\ x-3, & 3 < x \end{cases}$$

which can be entered on your calculator as

$(\sqrt{(}4\text{-}(X\text{-}1)^\wedge 2))*(X{\leq}3) + (X\text{-}3)*(X{>}3)$

HP-48x
    The inequality signs are found under PRG, [TEST].
TI-8x
    The inequality signs are found under TEST.

Study Guide for Section 6.2
Computing Definite Integrals – Part 1

Key Terms & Notation

Riemann sum   Lower Riemann sum   Upper Riemann sum

Regular partition $\sum_{i=1}^{n} f(x_i)\Delta x$ $\int_a^b f(x)\,dx$

Definite integral   Riemann integrable   Integrand

Limits of integration   Variable of integration

Additivity property   Linearity property

Constant multiple property   Additive function property

Comparison property

Questions for Understanding

- Why does the *extreme value theorem* guarantee a max and min for a continuous function on a closed interval?   (pages 129-130, 382)

- Let $f(x) = x^2$ and consider the graph of $y = f(x)$ over the interval $[-3, 3]$. Using a *regular partition* of size 12, and Figure 25 as a reference, find   (pages 379-383)

  1. The set of left endpoints
  2. The set of right endpoints
  3. The set of lower Riemann sum endpoints
  4. The set of upper Riemann sum endpoints

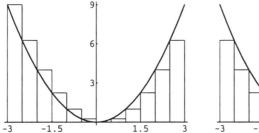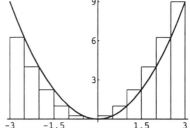

Figure 25: $f(x) = x^2$

- Referring to the four sets of endpoints above
    1. Which of them would give an overestimate of the actual area?
    2. Which of them would give an underestimate of the actual area?
    3. If the number of partitions was increased, how would that affect the value of the overestimate? The underestimate? How would the relationship between the overestimate and the underestimate change? (pages 382-383)

Suggested Exercises

§6.2 (pages 391-395): 1, 4, 5, 12, 14, 16, 24 (Is this the same as 23?), 29, 34

Notes on Exercises

- Here is a synopsis of the steps for producing a Riemann sum for a function f on an interval $[a, b]$ with partition size n.

> **Step 1.** Form a regular partition of $[a, b]$ of size n with subinterval lengths $\Delta x = \dfrac{b - a}{n}$.
>
> **Step 2.** Choose one point x_i ($i = 1, 2, \ldots, n$) from each of the n subintervals.
>
> **Step 3.** Calculate $\displaystyle\sum_{i=1}^{n} f(x_i)\Delta x$
>
> $= f(x_1)\Delta x + f(x_2)\Delta x + \ldots + f(x_n)\Delta x = \Delta x \displaystyle\sum_{i=1}^{n} f(x_i)$

Study Guide for Section 6.2– Part 1

- Problem 8 (See Graphing Calculator Hints, also)

  $f(x)$ is a monotonically increasing function on the interval $[1,4]$, so we have the

  left endpoints give the lower Riemann sum L_5
  right endpoints give the upper Riemann sum U_5

  $$n = 5, \ [a,b] = [1,4] \implies \Delta x = \frac{b-a}{n} = \frac{4-1}{5} = \frac{3}{5}$$

  (a) Lower Riemann sum L_5
  Left end points are
  $$x_1 = 1 + 0 \cdot \Delta x = 1 + 0\left(\frac{3}{5}\right) = 1$$
  $$x_2 = 1 + 1 \cdot \Delta x = 1 + 1\left(\frac{3}{5}\right) = 1.6$$
  $$x_3 = 1 + 2 \cdot \Delta x = 1 + 2\left(\frac{3}{5}\right) = 2.2$$
  $$x_4 = 1 + 3 \cdot \Delta x = 1 + 3\left(\frac{3}{5}\right) = 2.8$$
  $$x_5 = 1 + 4 \cdot \Delta x = 1 + 4\left(\frac{3}{5}\right) = 3.4$$

  $$L_5 = \sum_{i=1}^{5} f(x_i)\Delta x = \sum_{i=1}^{5} \log_2(x_i) \cdot \frac{3}{5} = \frac{3}{5}\sum_{i=1}^{5} \log_2(x_i)$$
  $$= \frac{3}{5}[\log_2(1) + \log_2(1.6) + \log_2(2.2) + \log_2(2.8)$$
  $$+ \log_2(3.4)]$$
  (See Graphing Calculator Hints, page 286)
  $$\approx \frac{3}{5}(0 + .67807 + 1.13750 + 1.48543 + 1.76553)$$
  $$= \frac{3}{5}(5.06654) = 3.03992$$

(b) Upper Rieman sum U_5
Right end points are

$$x_1 = 1 + 1 \cdot \Delta x = 1 + 1\left(\frac{3}{5}\right) = 1$$

$$x_2 = 1 + 2 \cdot \Delta x = 1 + 2\left(\frac{3}{5}\right) = 2.2$$

$$x_3 = 1 + 3 \cdot \Delta x = 1 + 3\left(\frac{3}{5}\right) = 2.8$$

$$x_4 = 1 + 4 \cdot \Delta x = 1 + 4\left(\frac{3}{5}\right) = 3.4$$

$$x_5 = 1 + 5 \cdot \Delta x = 1 + 5\left(\frac{3}{5}\right) = 4$$

$$U_5 = \sum_{i=1}^{5} f(x_i)\Delta x = \sum_{i=1}^{5} \log_2(x_i) \cdot \frac{3}{5} = \frac{3}{5} \sum_{i=1}^{5} \log_2(x_i)$$

$$= \frac{3}{5}[\log_2(1.6) + \log_2(2.2) + \log_2(2.8) + \log_2(3.4) + \log_2(4)]$$

$$\approx 4.23992$$

(c) Average $= \dfrac{L_5 + U_5}{2} \approx \dfrac{3.03992 + 4.23992}{2} = 3.63992$

(d) Largest possible error $= \dfrac{\text{upper sum} - \text{lower sum}}{2}$

$$= \dfrac{U_5 - L_5}{2} \approx \dfrac{4.23992 - 3.03992}{2} = 0.6$$

Graphing Calculator Hints

- Your calculator has features that will quickly evaluate each term of a Riemann sum. You can try this using EXAMPLE 12, page 384. The Riemann sum is

$$\sum_{i=1}^{n} f(x_i)\Delta x = \sum_{i=1}^{5} \frac{1}{2^{x_i}}\left(\frac{3}{5}\right)$$

Thus, the interior expression is

$$\frac{1}{2^{x_i}}\left(\frac{3}{5}\right)$$

which can be entered as

'(1/(2^X))*.6'   or   (1/(2^x))*.6

depending on your calculator.

HP-48G

Enter the expression as the calculator's EQ variable. (Using the PLOT feature is one way.) Get into the SOLVR menu (which is not listed on the keyboard), by pressing ⁊, SOLVE, [ROOT], [SOLVR]. Enter a value for x_i as the X variable and obtain the result by pressing [EXPR=]. See SOLVR environment in your *User's Guide* index. *Note*: the **HP-48S** is similar and actually easier. (Further exploration: Get into PLOT and graph the function)

TI-85

Enter the expression under GRAPH, [y(x) =]. Press 2nd, SOLVER, choose the appropriate function name at the bottom of the view screen. Press ENTER for the next screen. Move the cursor to the x= and enter the value for x_i. Move the cursor to exp= and press [SOLVE] (the F5 key) to obtain the result. See SOLVER in your *Guidebook* index. (Further exploration: Press the GRAPH (F1) key at the bottom of the view screen.)

TI-82

Enter the expression under Y= and plot using ,GRAPH. Bring up the CALC menu screen with 2nd, CALC. Choose value, press ENTER. Enter the desired $x - value$, press ENTER, and read the results from the listed coordinates. See Variables, CALC menu in your *Guidebook* index.

- This will help you with logarithmic functions, like $f(x) = \log_2(x)$ from Problem 8.

  The only logarithmic functions available on your calculator are

  LN for the natural logarithm (base e), and
  LOG for the Common logarithm (base 10).

To enter the function $f(x) = \log_2(x)$, it must be converted to one of the above forms using some of the general properties of logarithms (pages 62-67). For this demonstration, we will convert to the natural logarithm.

Let $y = \log_2(x)$, then
$$2^y = 2^{\log_2(x)} = x \implies \ln 2^y = \ln x \implies y \ln 2 = \ln x$$

$$\implies y = \frac{\ln x}{\ln 2}$$

But $y = f(x) = \log_2(x)$, therefore
$$f(x) = \log_2(x) = \frac{\ln x}{\ln 2}$$

which can be entered into your calculator as

HP-48x

'LN(X)/LN(2)'

TI-8x

ln(x)/ln(2)

Study Guide for Section 6.2
Computing Definite Integrals – Part 2

Key Terms & Notation

Riemann sum	Lower Riemann sum	Upper Riemann sum
Regular partition	$\sum_{i=1}^{n} f(x_i)\Delta x$	$\int_a^b f(x)\,dx$
Definite integral	Riemann integrable	Integrand
Limits of integration		Variable of integration
Additivity property		Linearity property
Constant multiple property		Additive function property
Comparison property		

Questions for Understanding

- Do Definition 1 and EXAMPLE 11 help you better understand why limits are so important in calculus? (pages 383-384)

- Does it seem safe to say that, graphically, a definite integral is basically the signed (+ or –) area between a function's graph and the x-axis? (pages 386)

- Considering the *Linearity Property*, would you say these equalities are correct?

$$\int_{-4}^{7} \pi(x-4)^2\sqrt{2}\,dx + \sqrt{2}\int_{-4}^{7} x^3\pi\,dx - \pi\int_{-4}^{7} \sqrt{2}(x-3)\,dx$$

$$= \pi\sqrt{2}\int_{-4}^{7} \left[(x-4)^2 + x^3 - (x-3)\right]\,dx$$

$$= \pi\sqrt{2}\int_{-4}^{7} (x^3 - x^2 - 8x + 19)\,dx$$

Which of the three forms do you think would be easiest to evaluate? (pages 388-389)

- Let $f(x) = \begin{cases} \sqrt{4-(x-1)^2}, & a \leq x \leq b \\ (x-3), & b \leq x \leq c \end{cases}$. Using Figure 26, how would you say the following are related?

$$\int_a^c f(x)\, dx \quad \text{and} \quad \frac{1}{4}\pi r^2 + \frac{1}{2}(c-b)d$$

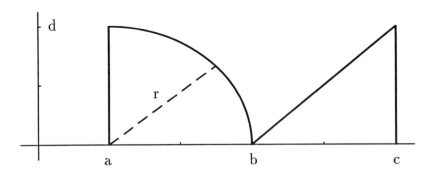

Figure 26: Graph of $y = f(x)$

Suggested Exercises

§6.2 (pages 391-395): 1, 4, 5, 12, 14, 16, 24 (Is this the same as 23?), 29, 34

Notes on Exercises

- Study Guide for Section 6.2 – Part 1, will help.

- Problem 29
  Once you formulate the Riemann sum, it helps to know the relationship
  $$\sum_{i=1}^n i^2 = \frac{n(n+1)(2n+1)}{6}$$

Graphing Calculator Hints

- Same as Study Guide for Section 6.2 – Part 1.

Study Guide for Section 6.3
What is an Antiderivative?

Key Terms & Notation

Antidifferentiation Initial condition Slope/direction field
Antiderivative

Questions for Understanding

- Suppose $F' = f$ and $G(x) = F(x) - 11$. What is G'? What is $(G(x) + x)'$? (pages 221, 396)

- What does each of the small line segments in a *slope field* represent? What does each of them approximate? (pages 397-398)

- In EXAMPLE 16, what does $f(0) = 2$ mean in terms of $F(0)$, if $F' = f$? (pages 397-398)

- What is the importance of an *initial condition* in terms of antiderivatives? (pages 396, 398)

- What is the physical representation of an initial condition in the subsection "Speed-ometers, odometers, and clocks?" (pages 398-400)

Suggested Exercises

§6.3 (pages 401-404): 6, 7, 13, 20, 26, 27

Notes on Exercises

- See Graphing Calculator Hints for ways to use your calculator to generate slope fields.

- When generating a slope field on your calculator, use the default viewing rectangle first. If necessary, try to reset the viewing range to include all the genreral structure of the field. Finally, rescale to the interval asked for in the particular problem being worked. Problems 5 and 6 are good examples of why.

Graphing Calculator Hints

- The **HP-48G** has a built-in slope field generator. To use it

    Get to the PLOT menu. Highlight TYPE: and press [CHOOS]. Arrow down to **Slopefield** (the ninth one down), and press [OK]. Press [OPTS], NXT, [RESET], [OK], arrow to **Reset plot**, press [OK], [OK] to obtain default settings and return to the PLOT menu. Enter the desired expression in the EQ: area, then plot. See "SLOPEFIELD" in your *User's Guide* index.

- The **HP48S** requires two programs to generate a slope field. They are named **SLPF** and **TSEG**. These may be available from your instructor for infrared transfer. A listing of each program follows this lesson. Once the programs are loaded in your VAR menu, they are both ran from the SLPF program as follows:

    Under the PLOT, [PLOTR] menu, make sure the **Plot type:** is FUNCTION, and press NXT, [RESET] to obtain default settings. Bring up the VAR menu, enter the desired expression on the stack and press **SLPF** to plot the slope field.

- HP-48x calculators can transfer programs via infrared beam at close range. At the top of the calculator, there is a small arrow molded into the plastic, located a little left of center. The two calculators involved in the transfer must be placed no more than 2" apart, with these arrows lined up. It is easiest to put them right next to each other, with no gap. Both calculators are then given their respective commands to receive or send. Generally, the receiving calculator must be in receiving mode before the final command to send is given to the sending calculator. (Note: Data can also be transferred by cable, so you must make sure your calculator is set to receive by infrared beam. See "transmit modes" in your *User's Guide* index. The HP-48G and HP-48S can also cross-transfer, but each has its own way to begin the process.)

HP-48G

RECEIVE	TRANSMIT
I/O, arrow to Get from HP 48, [OK]	I/O, arrow to Send to HP 48 ..., [OK], enter SLPF, press [SEND]

Note: I/O is GREEN ARROW SHIFT, 1

HP-48S

RECEIVE	TRANSMIT
I/O, [RECV]	Enter SLPF on stack, I/O, [SEND]

Note: I/O is ORANGE ARROW SHIFT, PRG

- The **TI-85** requires its own version of the SLPF program. It may be available from your instructor for transfer by link cable. A listing of the program follows this lesson. Once the program is loaded in your calculator, it is run as follows:

    Under MODE, make sure you are in Func mode. Under GRAPH, [ZOOM], reset the plot parameters to [ZSTD]. Enter the desired expression as y1=, then get back to the HOME screen by pressing EXIT. Press PRGM, [NAMES], and select [SLPF]. Press ENTER to execute the program.

- The **TI-82** requires its own version of the SLPF program. It may be available from your instructor for transfer by link cable. A listing of the program follows this lesson. Once the program is loaded in your calculator, it is run as follows:

    Under MODE, make sure you are in Func mode. Under ZOOM, reset the plot parameters to ZStandard. Enter the desired expression as Y1, press PRGM, select SLPF, and press ENTER to execute the program.

- The TI-85 and TI-82 are *not* compatible when it comes to program transfers. See "Communication Link" in your *Guidebook* table of contents. The transfer procedures for each calculator are:

TI-85

RECEIVE   TRANSMIT

LINK, [RECV]   LINK, [SEND], [PRGM], arrow to SLPF, [SELECT], [XMIT]

Note: LINK is 2nd, x-VAR

TI-82

RECEIVE   TRANSMIT

LINK, arrow to RECEIVE, ENTER   LINK, arrow to 2:SelectAll-..., ENTER, arrow to SLPF, ENTER arrow to TRANSMIT, ENTER

Note: LINK is 2nd, X,T,θ

A Slope Field Program for the HP-48S

There are two programs necessary: **TSEG** and **SLPF**. **TSEG** is a subprogram called by **SLPF**. Both programs can be transferred to the HP-48S via infrared beam. The HP-48S and HP-48G can transfer programs to each other.

TSEG

```
<< pt m xdel ydel
  <<
IF 'ABS(m)>ydel / xdel'
THEN ydel m 2 * / ydel 2 / R→C
ELSE xdel 2 / xdel m * 2 / R→C
END DUP pt SWAP - SWAP pt + LINE
  >>
```

'TSEG', STO stores the program under your VAR menu.

SLPF

```
<< ERASE DRAX 'PPAR(1)'
EVAL C→R 'PPAR(2)' EVAL C→R
0 0 0 0 0 → lft bot rht top der
xpxl ypxl x0 y0
  <<
  rht lft - 130 / 'xpxl' STO
top bot - 63 / 'ypxl' STO
'der(X,Y)' SWAP =
DEFINE { # 0d # 0d } PVIEW 1 8
   FOR I 1 12
     FOR J 'lft+J*11*xpsl-6*xpsl' EVAL
'x0' STO 'bot+ I*8*ypxl-4*ypxl' EVAL
'y0' STO x0 y0 R→C 'der(x0,y0)' EVAL
xpsl 10 * ypxl 7 * TSEG
     NEXT
   NEXT GRAPH
  <<
<<
```

'SLPF', STO stores the program under your VAR menu.

A Slope Field Program for the TI-85 and TI-82

- Name the program SLPF
- Each line begins with a colon (:)
- You get a new line by pressing ENTER

TI-85	**TI-82**
:ClDrw	:ClrDraw
:FnOff	:FnOff
:(yMax-yMin)/8→V	:(YMax-YMin)/8→V
:(xMax-xMin)/12→H	:(XMax-XMin)/12→H
:1→I	:1→I
:Lbl A	:Lbl 1
:1→J	:1→J
:Labl B	:Labl 2
:xMin+(J-1*H+H/2→x	:XMin+(J-1)*H+H/2→X
yMin+(I-1)*V+V/2→y	YMin+(I-1)*V+V/2→Y
:y1→M	:y1→M
:y-M*H/2→S	:Y-M*H/2→S
:y+M*H/2→Z	:Y+M*H/2→Z
:x-H/2→P	:X-H/2→P
:x+H/2→Q	:X+H/2→Q
:If abs(Z-S)>V	:If abs(Z-S)>V
:Goto D	:Goto 3
:Lbl E	:Lbl 4
:Line(P,S,Q,Z)	:Line(P,S,Q,Z)
:IS>(J,12)	:IS>(J,12)
:Goto B	:Goto 2
:IS>(I,8)	:IS>(I,8)
:Goto A	:Goto 1
:Stop	:Stop
:Lbl D	:Lbl 3
:y+V/2→Z	:Y+V/2→Z
:y-V/2→S	:Y-V/2→S
:(Z-Y)/M+x→Q	:(Z-Y)/M+X→Q
:(S-Y)/M+x→P	:(S-Y)/M+X→P
:Goto E	:Goto 4

Study Guide for Section 6.4
Computing Indefinite Integrals

Key Terms & Notation

Arbitrary constant $\quad \int f(x)\,dx \quad\quad$ Pattern matching
Indefinite integral $\quad$ Linearity rule $\quad\quad$ Area function

Questions for Understanding

- What role does C play in indefinite integrals? $\hfill$ (page 405)

- The y-axis represents the real number line $\mathbb{R}$. How many numbers does it contain? How many antiderivatives might be associated with $\hfill$ (page 405)

$$\int x^2\,dx$$

- How are an *initial condition* and C related? $\hfill$ pages (405-406)

- How would you write the most general antiderivative of $\hfill$ (pages 407-408)

$$\int \frac{1}{x^3}\,dx$$

- How does the *Linearity rule for antiderivatives* compare with those for derivatives and limits? $\hfill$ (pages 152, 221, 409)

- Once you have found an antiderivative, you can *always* tell if it is correct. How? $\hfill$ (page 410)

- How many antiderivative formulas are listed in this section? $\hfill$ (pages 410-412)

- Traditionally, most of integral calculus was learning techniques and the formulas at the end of theis section. Would it be worthwhile to make a list of them and practice? $\hfill$ (pages 410-412)

- Why can't we use the formula $\dfrac{x^{r+1}}{r+1} + C$ for $\int \dfrac{1}{x}\,dx$? (page 411)

- What does "patern matching" have to do with antidifferentiation? (page 413)

- Can we use what we know about definite integrals, area, and indefinite integrals to investigate an *area function*? (page 414)

Suggested Exercises

§6.4 (pages 413-416): 2, 10, 12, 13, 14, 18, 20

Notes on Exercises

- Problems 1-20

$$\int \left(10x^2 + 5e^x - \frac{20}{x^{2/3}} - 15\sin x\right) dx$$

$$= 10\int x^2\,dx + 5\int e^x\,dx - 20\int x^{2/3}\,dx - 15\int \sin x\,dx$$

$$= 10\frac{x^3}{3} + 5e^x - 20\frac{x^{1/3}}{1/3} - 15(-\cos x) + C$$

$$= \frac{10}{3}x^3 + 5e^x - 60x^{1/3} + 15\cos x + C$$

Graphing Calculator Hints

- Each calculator has its own numerical integration capabilities. Suppose we want to find the numerical value of the definite integral

$$\int_a^b f(x)\, dx \;=\; \int_1^4 \sqrt{1 + [\cosh(x)]^2}\, dx$$

HP-48G

Under [r], SYMBOLIC, Integrate, enter the integrand '√(1+COSH(X)^2)' for EXPR:. Enter X, 1, 4 for VAR:, LO:, HI:, respectively. Change RESULT to NUMERIC and press OK to evaluate.

(Note: COSH is found under MTH, [HYP])

HP-48S

Use the EQUATION writer to enter the complete integral or enter the expression '∫(1,4,√(1+COSH(X)^2),X)' on the stack. Press [r], → NUM to evaluate.

(Note: COSH is found under MTH, [HYP])

TI-85

Under 2nd, CALC, [fnInt], enter √(1+(cosh x)²),x,1,4), and press ENTER to evaluate.

(Note: cosh is found under MATH, [HYP])

TI-82

Under MATH, arrow to 9:fnInt, then press ENTER to select. Enter
√(1+(cosh X)²),X,1,4), then press ENTER to evaluate.

(Note: cosh is found under MATH, arrow to HYP, 2:cosh)

Study Guide for Section 6.5
The Fundamental Theorems of Calculus

Key Terms & Notation

Fundamental theorem of arithmetic

First fundamental theorem of calculus $\quad A(x) = \int_a^x f(t)dt$

Second fundamental theorem of calculus $\quad \int_a^b f(x)\,dx = F(b) - F(a)$

$$F(x)\Big]_{x=a}^{x=b}$$

Questions for Understanding

- A *theorem* is a type of general statement about a concept and/or a relationship. The word *fundamental* generally indicates something of deep significance to existence or success. What does this infer about the two *fundamental theorems of calculus*?

- How does an *area function* relate to indefinite integrals?
  (pages 417-418)

- If $A(x) = \int_0^x f(t)dt$, how are A' and f related? pages (418-421)

- How are $A(x) = \int_0^x f(t)dt$ and $F(x) = \int_{-2}^x f(t)dt$ related?
  (page 421)

- If F is an antiderivative of the continuous function f, how are $f(x), F(b),$ and $F(a)$ related? (page 422)

- Suppose $A(x), F(x)$ are two antiderivatives of a continuous function f. Does there exist some C such that (page 424)

$$A(x) + C = F(x)\ ?$$

Study Guide for Section 6.5 299

- Considering the problem solving power of differential and integral calculus, would you consider it coincidental that the beginning of the First Industrial Revolution is dated around 1750? (page 424-425)

Suggested Exercises

§6.5 (pages 425-428):   4, 7, 8, 14, 18, 26

Notes on Exercises

- **Problem 3**

   The first fundamental theorem of calculus tells us that for any F that is an antiderivative of f, we know

   $$F(x) = \int_a^x \cos^2(t)\, dt$$

   Using the initial condition $F(-\pi) = 0$, we have

   $$0 = F(-\pi) = \int_a^{-\pi} \cos^2(t)\, dt$$

   The second fundamental theorem of calculus tells us that

   $$\int_a^{-\pi} \cos^2(t)\, dt = F(-\pi) - F(a)$$

   $$\implies\quad 0 = F(-\pi) = F(-\pi) - F(a) \implies F(a) = 0$$

   is the condition we must satisfy. The correct choice is $a = -\pi \implies$

   $$F(x) = \int_{-\pi}^x \cos^2(t)\, dt$$

   is *an* antiderivative satisfying the initial condition.

- Problem 9

$$y = x^2 G(x) \implies \frac{dy}{dx} = 2xG(x) + x^2 G'(x) \quad \text{(product rule)}$$

$$G(x) = \int_2^x \frac{\sin(2t)}{1+t^2}\, dt, \quad \text{so if we let} \quad f(t) = \frac{\sin(2t)}{1+t^2}, \quad \text{then}$$

$$G'(x) = f(x) = \frac{\sin(2x)}{1+x^2}$$

$$\text{Thus,} \quad \frac{dy}{dx} = 2x \int_2^x \frac{\sin(2t)}{1+t^2}\, dt + x^2 \left(\frac{\sin(2x)}{1+x^2} \right)$$

Graphing Calculator Hints

- There are just some things you can't get a machine to do.

Study Guide for Section 6.6
Numerical Integration Techniques

Key Terms & Notation

Left endpoint rule   Midpoint rule   Simpson's rule
Right endpoint rule   Trapezoidal rule   Tangent rule

Questions for Understanding

- Suppose $f(x) = \ln x$, $a = 1$, $b = 2$, $n = 10$. What is the Riemann sum associated with these conditions? What is the numerical value of the fourth term of this Riemann sum using   (pages 429-433)

    1. the left endpoint rule

    2. the midpoint rule

    3. the right endpoint rule

- For which of the following functions will the *right rectangle rule* estimate produce a better approximation for $\int_0^\pi f(x)\,dx$ than the *left rectangle rule* estimate?   (pages 429-433)

    $f(x) = x^2$ $f(x) = \sin x$ $f(x) = \cos x$

- Your text lists a formula
    $$\text{TRAP} = (0.5)(\text{LEFT} + \text{RIGHT})$$
    for using the *trapezoidal rule*. Write a similar one for *Simpson's rule*. Can you make the necessary adaptations to store both of these in your calculator for use with the RSUM program at the end of this lesson?   (pages 433-436)

- From Figure 6.40, can you determine how the *trapezoidal rule* was named?

                                                                   (page 434)

- What do each of the following symbols mean? Use both the English words definition and the mathematical notation definition.

$$L_n \qquad R_n \qquad M_n \qquad T_n \qquad S_{2n}$$

  What do the *odd* subscripts for x and y correspond to? The *even* subscripts? Why does S have the subscript $2n$? (pages 436-437)

Suggested Exercises

§6.6 (pages 438-444): 1, 5, 7, 11, 14, 16, 27

Notes on Exercises

- Here is a synopsis of the steps to setting up a Riemann sum, from page 429.

> **Step 1.** Choose a number n of subintervals in the regular partition of $[a, b]$.
>
> **Step 2.** Calculate $\Delta x = \dfrac{b-a}{n}$.
>
> **Step 3.** Locate the n inputs $x_1, x_2, \ldots, x_n$.
>
> **Step 4.** Evaluate f at each input x_i and find the Riemann sum
>
> $$\sum_{i=1}^{n} f(x_i) \Delta x$$

- Numerically estimate the definite integral $\int_0^1 e^{-x^2}$ using the RSUM program for your calculator. Which number of partitions $n = 2, 6, 10, 14$ guarantees the estimate to be correct out to two decimal places?

  The Riemann sum to use is $\sum_{i=1}^{n} e^{-x_i^2} \Delta x$.

$n = 2 \implies \Delta x = 1 \implies 1 \cdot \sum_{i=1}^{2} e^{-x_i^2}$

$\left. \begin{array}{l} \text{Left endpoint} \implies .88940 \\ \text{Right endpoint} \implies .57334 \end{array} \right\} \implies$

$$\text{greatest error} \approx \frac{.88940 - .57334}{2} = .15803$$

$n = 6 \implies \Delta x = \frac{1}{6} \implies \frac{1}{6} \sum_{i=1}^{6} e^{-x_i^2}$

$\left. \begin{array}{l} \text{Left endpoint} \implies .79780 \\ \text{Right endpoint} \implies .69244 \end{array} \right\} \implies$

$$\text{greatest error} \approx \frac{.79780 - .69244}{2} = .05268$$

$n = 10 \implies \Delta x = \frac{1}{10} \implies \frac{1}{10} \sum_{i=1}^{10} e^{-x_i^2}$

$\left. \begin{array}{l} \text{Left endpoint} \implies .77782 \\ \text{Right endpoint} \implies .71460 \end{array} \right\} \implies$

$$\text{greatest error} \approx \frac{.77782 - .71460}{2} = .03161$$

$n = 14 \implies \Delta x = \frac{1}{14} \implies \frac{1}{14} \sum_{i=1}^{14} e^{-x_i^2}$

$\left. \begin{array}{l} \text{Left endpoint} \implies .76909 \\ \text{Right endpoint} \implies .72394 \end{array} \right\} \implies$

$$\text{greatest error} \approx \frac{.76909 - .72394}{2} = .02257$$

Hence, any n must actually be *greater* than 14 to guarantee accuracy out to two decimal places. (In fact, n must be ≥ 32)

Graphing Calculator Hints

- There is a RSUM program for each of the calculators. Both types of the HP-48's use the same one. The programs for the TI-85 and TI-82 are slightly different from each other. These programs may be available from your instructor for direct transfer from calculator to calculator. If they are not, there are listings for them on the pages following this §6.6 Study Guide.

Transferring Programs for the HP-48x, TI-8x

HP-48G

RECEIVE	TRANSMIT
I/O, arrow to Get from HP 48, [OK]	I/O, arrow to Send to HP 48 ..., [OK], enter RSUM, press [SEND]

Note: I/O is GREEN ARROW SHIFT, 1

HP-48S

RECEIVE	TRANSMIT
I/O, [RECV]	Enter RSUM on stack, I/O, [SEND]

Note: I/O is ORANGE ARROW SHIFT, PRG

TI-85

RECEIVE	TRANSMIT
LINK, [RECV]	LINK, [SEND], [PRGM], arrow to RSUM, [SELECT], [XMIT]

Note: LINK is 2nd, x-VAR

TI-82

RECEIVE
LINK, arrow to RECEIVE,
ENTER

TRANSMIT
LINK, arrow to 2:SelectAll-...,
ENTER, arrow to RSUM,
ENTER arrow to TRANSMIT,
ENTER

Note: LINK is 2nd, X,T,θ

Using the RSUM program on your graphing calculator:

RSUM will calculate the Riemann SUM approximating

$$\int_a^b f(x)\,dx$$

for a regular partition of the interval $[a, b]$ into n subintervals and a choice of left endpoint, right endpoint, or midpoint from each interval.

In the RSUM program:

    A represents a B represents b N represents n

The variable R is used to represent the choice of point:

    0 for left endpoint,   0.5 for midpoint,   1 for right endpoint

The variable D represents the subinterval length: $D = \dfrac{B-A}{N}$.

To run the program on the TI machines:

1. Enter the function $f(x)$ as **y1** (or **Y1**)
2. Press **PRGM** and select **RSUM**
3. Press **ENTER** to execute
4. Enter the appropriate numbers at the prompts

To run the program on the HP machines:

1. Store the function $f(x)$ in **EQ**
2. Press **VAR** and press **[RSUM]**
3. Enter the appropriate numbers at the prompts

Example: To find the Riemann sum approximating $\int_1^2 x^2$ with the midpoints chosen from 20 equal subintervals: Store x^2 (as **y1** (or **Y1**) on the TI's; in **EQ** on the HP's), run **RSUM** and enter

    1 for A   2 for B   20 for N   0.5 for midpoint

You should get **2.33125** for the result.

A Riemann Sum Program for the HP-48x

(NOTE: RSUM can be transferred via infrared beam)

Listing	Comments
<< "ENTER A" ""	
INPUT OBJ→	Prompt for A (left endpoint)
"ENTER B" "" INPUT	Prompt for B (right endpoint)
OBJ→ "ENTER N" ""	Prompt for N (# of subintervals)
INPUT OBJ→	
"ENTER 0 FOR LEFT	Prompt for choice of point
0.5 FOR MIDPOINT	
1 FOR RIGHT"	
"" INPUT OBJ→	
0→ A B N R D	Set up local variables
<< B A - N / 'D'	Compute subinterval length D
STO 0 1 N	Initialize sum and set up loop
FOR I 'A+(I +R-1)*D' EVAL	
'X' STO EQ EVAL D * +	Calculate Riemann sum
NEXT 'X' PURGE	Purge global variable X
>>	
>>	

'RSUM', STO stores the program under your VAR menu.

A Riemann Sum Program for the TI-8x
(NOTE: RSUM can be transferred via link cable)

Name the program RSUM. Each new line begins with a colon (:), and you get a new line by pressing ENTER. Here's the listing for RSUM on the TI-82, with hints where to find certain commands. Changes necessary for the TI-85 are in **bold**.

Listing	Comments
:Disp "ENTER A"	Find Disp under PRGM, I/O menu
:Input A	Find Input under PRGM, I/O menu
:Disp "ENTER B"	
:Input B	
:Disp "ENTER N"	
:Input N	
:Disp "ENTER POINT"	
:Disp "0 FOR LEFT ENE"	Be sure to use 0 (zero), not O (oh)
:Disp "0.5 FOR MIDDLE"	
:Disp "1 FOR RIGHT END"	
:Input R	
:(B-A)/N→D	Use STO> to get →
:0→S	
:1→I	
:Lbl 1	Find Lbl under PRGM, CTL menu
	(**use E instead of 1 on TI-85**)
:A+(I+R-1)*D→X	(**use x on TI-85**)
:Y₁*D+S→S	Find Y₁ under Y-Vars
	(**use y1 on TI-85**)
:IS>(I,N)	Find IS> under PRGM, CTL menu
:Goto 1	Find Goto under PRGM, CTL menu
	(**use E instead of 1 on TI-85**)
:Disp S	

Study Guide for Section 6.7
Methods of Substitution

Key Terms & Notation

Method of substitution $\quad \int f'(x) \dfrac{du}{dx}\, dx \quad$ Differential

Questions for Understanding

- How do the integral formulas of this section compare with those of §6.4?

  (pages 410-411, 444-445)

- Why is the **CAUTION** of page 445 necessary? (pages 445-446)

- From the discussion following EXAMPLE 37, do you think that the symbol $\dfrac{du}{dx}$ is the same as any fraction of real numbers $\left(\text{like } \dfrac{3}{4}\right)$?

  (pages 447-448)

- On page 448, your text states, "Good choices of substitutions to try are ..." Does this suggest that you may need to try more than one to solve an integral?

- What should you hope to see from **Step 5**? (page 448)

- Consider **Step 1** and **Step 2**. Does this suggest learning to look for both $u = g(x)\, dx$ and $du = g'(x)\, dx$ at the same time? (page 448)

- How can we apply the methods of this section to *definite* integrals?

  (pages 450-451)

Suggested Exercises

§6.7 (pages 453-456):   2, 12, 14, 16, 26, 32, 34, 42, 52, 56

Notes on Exercises

- **Problem 19**
  Two things should come to mind when you look at this problem, they are

  $$\int e^u \, du = e^u + C \qquad \text{and} \qquad \frac{d\, ax^3}{dx} = 3ax^2$$

  These suggest a u-substitution of $\quad u = 4x^3 \implies du = 12x^2 \, dx$.

  Method 1
  $$du = 12x^2 \, dx \implies dx = \frac{du}{12x^2} \implies$$

  $$\int x^2 e^{4x^3} \, dx = \int x^2 e^u \left(\frac{du}{12x^2}\right) = \int \frac{x^2}{12x^2} e^u \, du = \frac{1}{12} \int e^u \, du$$

  Method 2
  need to create $du = 12x^2 \, dx$
  $$\int x^2 e^{4x^3} \, dx = \int x^2 e^u \, dx = \int e^u \left(x^2 \, dx\right)$$
  $$= \int e^u \left(\frac{12}{12} x^2 \, dx\right) = \int e^u \frac{1}{12} \left(12x^2 \, dx\right)$$
  $$= \int e^u \frac{1}{12} du = \frac{1}{12} \int e^u \, du$$

  which leads to
  $$\frac{1}{12} \int e^u \, du = \frac{1}{12} e^u + C = \frac{1}{12} e^{4x^3} + C$$

- **Problem 25**
  $$\int \frac{e^x}{1 + e^{2x}} \, dx = \int \frac{e^x}{1 + (e^x)^2} \, dx = \int \frac{1}{1 + (e^x)^2} \left(e^x \, dx\right)$$

  We know that $\int \frac{1}{1+u^2} \, du = \arctan(u) + C$ which suggests a u-substitution of $u = e^x \implies du = e^x \, dx$.

Study Guide for Section 6.7

Method 1
$$du = e^x \, dx \implies dx = \frac{du}{e^x} \implies$$

$$\int \frac{1}{1+(e^x)^2} (e^x \, dx) = \int \frac{1}{1+u^2} e^x (\, dx)$$

$$= \int \frac{1}{1+u^2} e^x \left(\frac{du}{e^x}\right) = \int \frac{1}{1+u^2} du$$

Method 2
need to create $du = e^x \, dx$

$$\int \frac{e^x}{1+e^{2x}} \, dx = \int \frac{1}{1+(e^x)^2} (e^x \, dx) = \int \frac{1}{1+u^2} du$$

which leads to

$$\int \frac{1}{1+u^2} du = \arctan(u) + C = \arctan(e^x) + C$$

- **Problem 37**

  Looking for patterns helps to recall that

  $$\sec(u) + C = \int \sec(u) \tan(u) du \quad \text{and} \quad \frac{d \ln(x)}{dx} = \frac{1}{x}$$

  These suggest a u-substitution of $u = \ln(x) \implies du = \frac{1}{x} dx$.

  Method 1
  $$du = \frac{1}{x} dx \implies dx = x \, du \implies$$

  $$\int \frac{\sec(\ln(x)) \tan(\ln(x))}{x} dx = \int \frac{\sec(u) \tan(u)}{x} (x \, du)$$
  $$= \int \sec(u) \tan(u) du$$

Method 2

need to create $du = \frac{1}{x}$

$$\int \frac{\sec(\ln(x))\tan(\ln(x))}{x} dx = \int \sec(u)\tan(u)\left(\frac{1}{x}dx\right)$$

$$= \int \sec(u)\tan(u)du$$

which leads to

$$\int \sec(u)\tan(u)du = \sec(u) + C = \sec(\ln(x)) + C$$

Graphing Calculator Hints

- Once an antiderivative $F(x) = \int f(x)\,dx$ is found, using the first fundamental theorem of calculus $F'(x) = f(x)$ is the surest way to check your answer. However, the numerical integration feature of your calculator can also be used, along with the second fundamental theorem of calculus

$$\int_a^b f(x)\,dx = F(b) - F(a)$$

Evaluate $\int_a^b f(x)\,dx$ on your calculator, then use the antiderivative $F(x)$ to find $F(b) - F(a)$.

Study Guide for Section 7.1
Using Definite Integrals to Measure Area

Key Terms & Notation

Cavalieri's principle $\qquad \int_a^b |f(x) - g(x)|\, dx$

Questions for Understanding

- How are *Cavalieri's principle* and Riemann sums related? (page 458)

- Your calculator has a numerical integration routine. Suppose you used it to calculate $\int_1^3 x^2\, dx$ and obtained the value 4. How could you check if this is reasonable? (pages 459-460)

- Suppose you calculated $\int_0^1 x^3\, dx$ using your calculator and the result was 0.5. Plotting $y = x^3$ may not help you determine if this is reasonable. How can you use the techniques of Chapter 6 to check your calculator?

- Evaluate $\int_{-2}^2 x^3\, dx$ directly, then find the area enclosed by $y = x^3$, the x-axis, $x = -2$, and $x = 2$. How can you reconcile the two? (pages 460-461)

- What does
$$\int_a^b |f(x) - g(x)|\, dx = \begin{cases} \int_a^c (f(x) - g(x))\, dx, & a \leq x \leq c \\ \int_c^b (g(x) - f(x))\, dx, & c \leq x \leq b \end{cases}$$
mean graphically? (pages 461-462)

- The area enclosed by $y = x^2$, the x-axis, $x = 1$, and $x = 2$, is $\int_1^2 x^2\, dx$. If this were written in the form $\int_1^2 |f(x) - g(x)|\, dx$, what would the logical choices for $f(x)$ and $g(x)$ be?

- When finding area through integration on an interval $[a, b]$, we normally evaluate $\int_a^b f(x)\, dx$, placing the smaller value a as the lower limit. If we integrate over an interval of y-values $[c, d]$, what values would you place as the limits of integration α, β in $\int_\alpha^\beta h(y)\, dy$? (page 462)

- Using the *symmetry* of the function $y = x^2$, find the value of a that makes the following equation true

$$\int_{-2}^{2} x^2\, dx = 2\int_{a}^{2} x^2\, dx$$

Suggested Exercises

§7.1 (pages 462-466):   11, 14, 16, 20, 23, 24

Notes on Exercises

- Problem 9
  The points of intersection satisfy the equation

  $$8 - y^2 = -y + 2 \implies y^2 - y - 6 = 0 \implies (y-3)(y+2) = 0$$

  $$y = 3 \implies x = -1, \quad y = -2 \implies x = 4$$

  $\implies$ the two curves intersect at the points $(-1, 3)$ and $(4, -2)$.
  If $y = 0$, then $8 - y^2 = 8 > 2 = -y + 2 \implies$

  $$\int_{-2}^{3} \left[(8 - y^2) - (-y + 2)\right] dy = \int_{-2}^{3} (y - y^2 + 6)\, dy$$

  $$= \left(\frac{y^2}{2} - \frac{y^3}{3} + 6y\right)\Bigg|_{-2}^{3} = \left[\frac{9}{2} - \frac{27}{3} + 18\right] - \left[\frac{4}{2} + \frac{8}{3} - 12\right]$$

  $$= \frac{5}{2} - \frac{35}{3} + 30 = \frac{125}{6} \approx 20.83333$$

- Problem 15
  The points of intersection satisfy the equation

  $$\cos(x) = \frac{x+1}{3} \implies \cos(x) - \left(\frac{x+1}{3}\right) = 0$$

  Use one of the root finding features of your calculator to find that the x-coordinates of the points of intersection are:

  $$x \approx -3.63796, \ -1.86236, \ 0.88947.$$

  $\cos(-\pi) = -1 > \dfrac{-\pi+1}{3} \implies \cos(x)$ is the upper curve on the interval $[-3.63796, -1.86236]$

  $\cos(0) = 0 < \dfrac{0+1}{3} \implies \dfrac{x+1}{3}$ is the upper curve on the interval $[-1.86236, 0.88947]$.

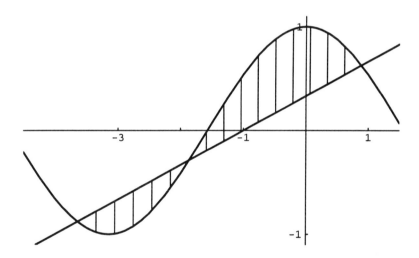

Figure 27: Problem 15

Therefore,

$$\text{Area} = \int_{-3.63796}^{0.88947} \left|\cos(x) - \left(\frac{x+1}{3}\right)\right| dx$$

$$= \int_{-3.63796}^{-1.86236} \left[\cos(x) - \left(\frac{x+1}{3}\right)\right] dx$$

$$+ \int_{-1.86236}^{0.88947} \left[\left(\frac{x+1}{3}\right) - \cos(x)\right] dx$$

$$\approx 1.662$$

Question: Would problem 15 be easier if you found that your calculator can integrate the *absolute value* of the difference between two functions?

- Problem 17

  We want to find c such that the line $y = c$ divides the *area* into two equal portions. So, we need c such that

$$\int_0^c \sqrt{y}\, dy = \int_c^4 \sqrt{y}\, dy \implies$$

$$\left.\frac{y^{3/2}}{3/2}\right]_0^c = \left.\frac{y^{3/2}}{3/2}\right]_c^4 \implies \left.\frac{2}{3}y^{3/2}\right]_0^c = \left.\frac{2}{3}y^{3/2}\right]_c^4$$

$$\implies c^{3/2} - 0 = (4)^{3/2} - c^{3/2} \implies 2c^{3/2} = 8$$

$$\implies c^{3/2} = 4 \implies c = (4)^{3/2} = \sqrt[3]{16}$$

(See Figure 28, next page.)

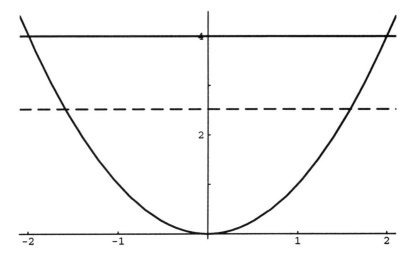

Figure 28: Problem 17

Graphing Calculator Hints

- Finding the x-coordinates where two functions $f(x), g(x)$ intersect is the same as finding the roots of the expression $f(x)-g(x)$ (or $g(x)-f(x)$), which is the same as solving the equation $f(x) - g(x) = 0$. One way to do this is to plot $f(x) - g(x)$ on your calculator and use the root finding feature of the plotter.

HP-48x
Plot the expression for $f(x) - g(x)$. For each root, use the arrow keys to move the plot cursor near the root. Press [FCN], [ROOT].

TI-85
Plot the expression for $f(x) - g(x)$. For each root, use the arrow keys to move the plot cursor near the root. Press [MATH], [ROOT].

TI-82
Plot the expression for $f(x) - g(x)$. For each root, use the arrow keys to move the plot cursor near the root. Press 2nd, TRACE to bring up the CALC menu. Arrow down to 2:root, press ENTER.

- Your calculator can plot two expressions at the same time. To illustrate this, suppose we want to plot $y = x^2$ and $y = 2x$.

    HP-48x
    Enter the two expressions in the form   {'X^2' '2*X'}

    TI-8x
    Enter x^2 for **y1 =** (or **Y1 =**) and $2x$ for **y2 =** (or **Y2 =**). Make sure both are selected and plot.

Study Guide for Section 7.2
Using Definite Integrals to Measure Volume – Part 1

Key Terms & Notation

Cross-sectional area   "Disk"   Cylindrical shell
Solid of revolution   "Washer"

Questions for Understanding

- In §7.1, we estimated the area between two curves by adding up the *areas* of thin rectangles. Suppose we extended to three dimensions, then each rectangle also has a depth. What would we be adding up now?   (page 466)

- Given a function $f(x)$ and a particular x_0, $|f(x_0)|$ can be thought of as the height (or length) of a perpendicular line from the x-axis to the point $(x_0, f(x_0))$. In a sense, §7.1 showed us that area is the sum of the lengths of all such lines contained in an interval $[a, b]$. If we let $\ell(x)$ represent the length $|f(x)|$ for each x, then area becomes $\int_a^b \ell(x)\,dx$. How is this idea extended to three dimensions in this section?   (page 466)

- How closely does the "deck of square cards" analogy fit in with the above two items?   (pages 466-467)

- What would Figure 7.7 look like if it represented EXAMPLE 5?   (pages 466-467)

- What type of measurment does $A(x)$ represent in $\int_a^b A(x)\,dx$? Does $A(x)$ refer to something in the xy-plane?   (pages 466-467)

- In EXAMPLE 6, what is the physical description of $A(x)$? What is the mathematical formula used to determine the value of $A(x)$? How is that formula obtained?   (pages 467-468)

- In a *solid of revolution*, what kind of geometric shape are we guaranteed to have describing $A(x)$? (page 469)

- In a *solid of revolution*, what are the only two types of cross sections that are possible? (pages 469-470)

Suggested Exercises

§7.2 (pages 474-477):   2, 3, 4, 7, 10

Notes on Exercises

- Problem 6

A = area of an isosceles right triangle

$= \frac{1}{2}bh$ with $b = h = e^{3x} - 1$

$\implies A(x) = \frac{1}{2}(e^{3x} - 1)^2$

$= \frac{1}{2}(e^{6x} - 2e^{3x} + 1) \implies$

$V = \int_0^1 A(x)\,dx$

$= \int_0^1 \frac{1}{2}(e^{6x} - 2e^{3x} + 1)\,dx$

$= \frac{1}{2}\int_0^1 (e^{6x} - 2e^{3x} + 1)\,dx$

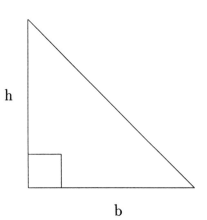

Figure 29: Problem 6

$= \frac{1}{2}\int_0^1 e^{6x}\,dx - \frac{1}{2}\int_0^1 2e^{3x}\,dx + \frac{1}{2}\int_0^1 dx$

$= \frac{1}{2}\left[\frac{1}{6}e^{6x} - \frac{2}{3}e^{3x} + x\right]_0^1 = \frac{1}{2}\left[\left(\frac{1}{6}e^6 - \frac{2}{3}e^3 + 1\right) - \left(\frac{1}{6} - \frac{2}{3} + 0\right)\right]$

$= \frac{1}{2}\left(\frac{1}{6}e^6 - \frac{2}{3}e^3 + \frac{3}{2}\right) \approx 55.34777$

Study Guide for Section 7.2 – Part 1

- Problem 9, rotated about the x-axis

(a) Figure 30 shows the graph of $y = \dfrac{3}{1+x^2}$ and $y = e^{x^2}$ on the same plot. It is used to determine the upper and lower curves. Figure 31 shows the graph of $y = e^{x^2} - \dfrac{3}{1+x^2}$. It is used with the root finding feature of your calculator's plotter to find the two points of intersection.

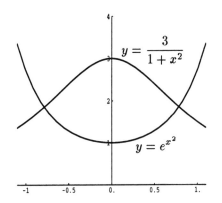

Figure 30: Problem 9   Figure 31: Problem 9

(b) $y = \dfrac{3}{1+x^2}$ is the upper curve on the bounded region. The solid of revolution is a washer with outside radius $R(x) = \dfrac{3}{1+x^2}$ and inside radius $r(x) = e^{x^2}$

$$\Longrightarrow\quad A(x) = \pi(R(x))^2 - \pi(r(x))^2 = \pi\left(\dfrac{3}{1+x^2}\right)^2 - \pi(e^{x^2})^2$$

$$= \pi\left(\left(\dfrac{3}{1+x^2}\right)^2 - e^{2x^2}\right)$$

The x-coordinates of the two intersection points are approximately $x = -.78590, .78590$

$$\Longrightarrow\quad V \approx \int_{-.78590}^{.78590} \pi\left(\left(\dfrac{3}{1+x^2}\right)^2 - e^{2x^2}\right)\,dx$$

$$= \pi \int_{-.78590}^{.78590} \left(\left(\frac{3}{1+x^2} \right)^2 - e^{2x^2} \right) dx$$

Since the antiderivative of e^{x^2} cannot be expressed *as a nice functional expression*, it must be numerically estimated. Hence the definite integral must be numerically estimated. Using a numerical integrator gives $V \approx 24.55715$.

Graphing Calculator Hints

- Here is a quick reminder on numerical integration of $\int_a^b f(x)\,dx$.

  HP-48G
  SYMBOLIC, Integrate, enter '$f(x)$'. Enter X, a, b for VAR:, LO:, HI:. Change RESULT to NUMERIC, press [OK].

  HP-48S
  Use the EQUATION writer to enter the complete integral or enter the expression '∫(a,b,$f(x)$,X)' on the stack. → NUM to evaluate.

  TI-85
  2nd, CALC, [fnInt]. Enter $f(x)$,x,a,b). Press ENTER.

  TI-82
  MATH, 9:fnInt, ENTER. Enter $f(x)$,X,a,b). Press ENTER.

Study Guide for Section 7.2
Using Definite Integrals to Measure Volume – Part 2

Key Terms & Notation

Cross-sectional area   "Disk"   Cylindrical shell
Solid of revolution   "Washer"

Questions for Understanding

- In Figure 7.14, what is R in terms of x? What is r in terms of x?
  (pages 469-470)

- In EXAMPLE 9, the cross section looks like a washer. The area of the washer is the area of the outer circle minus the area of the inner circle. What are the expressions for the outer circle area and the inner circle area? Can you substitute these expressions into the following "definite integral" to obtain the correct volume integral?
  (pages 470-471)

$$V = \int_1^2 [\text{outer area} - \text{inner area}]\, dx$$

- In EXAMPLE 10, why are two definite integrals necessary? Is there an analogy to finding the *area* between two curves, which is what we studied in §7.1? (pages 471-472)

- How is the expression $2\pi r h$ related to the rectangle pictured in Figure 7.17? If we extend to three dimensions, what would Δx represent? (page 474)

- How does the picture of nested shells in Figure 7.17 relate to the expression (page 474)

$$\sum_{i=1}^{n} (2\pi r(x_i) h(x_i)) \Delta x$$

- Conceptually, how are the volume calculating techniques of *solids of revolution* and *cylindrical shells* the same? How are they different?

Suggested Exercises

§7.2 (pages 474-477):   12, 15, 16, 18, 22

Notes on Exercises

- Problem 9, rotated about $y = -1$

  (a) From Study Guide for Section 7.2 – Part 1, we know $y = \dfrac{3}{1+x^2}$ is the upper curve, $y = e^{x^2}$ is the lower curve, and the x-coordinates of the two intersection points are $x \approx -.78590, .78590$.

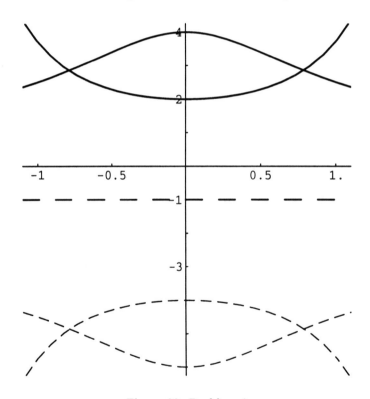

Figure 32: Problem 9

(b) By examining the graph of the two functions together, we see that
$$R(x) = \frac{3}{1+x^2} - (-1) = \frac{3}{1+x^2} + 1$$
and
$$r(x) = e^{x^2} - (-1) = e^{x^2} + 1 \implies$$
$$A(x) = \pi(R(x))^2 - \pi(r(x))^2 = \pi \left(\frac{3}{1+x^2} + 1\right)^2 - \pi(e^{x^2}+1)^2$$
$$= \pi \left(\left(\frac{3}{1+x^2} + 1\right)^2 - (e^{x^2}+1)^2\right) \implies$$
$$V \approx \int_{-.78590}^{.78590} \pi \left(\left(\frac{3}{1+x^2} + 1\right)^2 - (e^{x^2}+1)^2\right) dx \approx 37.31923$$

(How does this compare with the case of rotation about $y = 0$, from Study Guide for Section 7.2 – Part 1?)

- Problem 21

    The region to be rotated is a rectangle $\implies$ the surface of revolution is a solid right cylinder $\implies$

    $$V = \int_0^5 A(x)\, dx = \pi r^2 h = \quad \text{Volume formula for a right cylinder with } r = 3, h = 5$$

    $$\implies V = \pi(3^2)5 = 45\pi$$

- (See next page for Problem 19)

- Problem 19

    (a) First, consider the circle with radius r, centered at the origin. It is composed of two functions: one for the upper hemisphere $f(x)$, and one for the lower hemisphere $g(x)$. This comes from the circle equation

    $$x^2 + y^2 = r^2 \implies y = \begin{cases} f(x) = +\sqrt{r^2 - x^2} \\ g(x) = -\sqrt{r^2 - x^2} \end{cases}$$

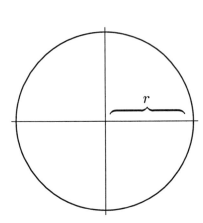

Figure 33: Problem 19

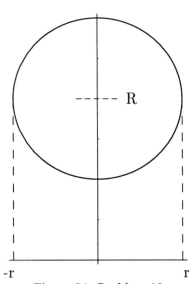

Figure 34: Problem 19

    (b) If we raise the center of the circle above the x-axis by R units, then the corresponding y-values become

    $$y = \begin{cases} R + \sqrt{r^2 - x^2}, & \text{above the circle center} \\ R - \sqrt{r^2 - x^2}, & \text{below the circle center} \end{cases}$$

    If we integrate with respect to x-values, then the cross section of the surface of revolution (the torus) is a washer with outside and inside radii given by the above split function. Further, x-values will range from $-r$ to r $\implies$

Study Guide for Section 7.2 – Part 2

$$V = \int_{-r}^{r} \pi \left((R+\sqrt{r^2-x^2})^2 - (R-\sqrt{r^2-x^2})^2\right) dx$$

$$= \pi \int_{-r}^{r} \left[R^2 + 2R\sqrt{r^2-x^2} + r^2 + x^2\right.$$

$$\left. -(R^2 - 2R\sqrt{r^2-x^2} + r^2 + x^2)\right] dx$$

$$= \pi \int_{-r}^{r} 4R\sqrt{r^2-x^2}\, dx$$

$$= 4\pi R \left[\frac{x}{2}\sqrt{r^2-x^2} + \frac{r^2}{2}\arcsin\left(\frac{x}{r}\right)\right]_{-r}^{r}$$

$$V = 2\pi R\left[\left(r\sqrt{r^2-x^2} + r^2\arcsin\left(\frac{r}{r}\right)\right)\right.$$

$$\left. - \left(r\sqrt{r^2-x^2} + r^2\arcsin\left(\frac{-r}{r}\right)\right)\right]$$

$$= 2\pi R\left[0 + r^2\left(\frac{\pi}{2}\right) - \left(0 + r^2\left(\frac{3\pi}{2}\right)\right)\right]$$

$$= 2\pi R(\pi r^2)) = 2\pi^2 R r^2$$

Graphing Calculator Hints

- See Study Guide for Section 7.2 – Part 1.

Study Guide for Section 7.3
Using Definite Integrals to Measure Averages

Key Terms & Notation

Average value
Mean value

Monte Carlo technique
Moving average

Questions for Understanding

- In the Study Guide for Section 7.2 – Part 1, the idea of an integral as a sum of line lengths ($\ell(x)$) was mentioned. How would the *average length* of these line lengths relate to the discussion preceding Definition 1? (pages 477-478)

- Suppose F is an antiderivative of f. Is F a continuous function? How does the Mean value theorem *for functions* relate to the Mean value theorem *for integrals* in terms of F and f?
(pages 258, 283, 478-479)

- Consider the area of a quarter circle contained in the unit square $[0, 1] \times [0, 1]$. Let A_C = area of the quarter circle, A_S = area of the unit square, and (x_0, y_0) be any chosen point in the square. How is the ratio $\dfrac{A_C}{A_S}$ related to the probability that (x_0, y_0) lies within the quarter circle? (pages 480-481)

- Let h represent the average value of $f(x)$. Why does knowing the following two facts allow us to estimate $\displaystyle\int_a^b f(x)\,dx$? (pages 481-482)

    1. $h = \dfrac{\int_a^b f(x)\,dx}{b-a}$
    2. $h \approx c = \dfrac{1}{n}\sum_{i=1}^n f(x_i)$ for large n

Study Guide for Section 7.3

- In the stock market example for a *moving average*, what would A represent? If $A = 30$ and x represented dates in the month of July, what would the moving averages for $x = 5, 6, 7, 8$ describe?
(pages 482-483)

Suggested Exercises

§7.3 (pages 483-486):   8, 10, 12, 14, 18 (Explain why), 21, 24

Notes on Exercises

- Problem 18
Remember that a border is a circumference, which can be thought of as a line segment with its ends joined.

- Problem 21
$$\text{Surface area of a sphere} = 4\pi r^2$$
$$\text{Circumference of the earth} = 25{,}000 \text{ miles}$$

Graphing Calculator Hints

- The *Monte Carlo* technique can be tedious and prone to errors when done by hand. Fortunately, your calculator can make things much easier. The following pages illustrate how to do this for each type of machine.

Monte Carlo Method for the HP-48G, HP-48S
to approximate any definite integral.

$$\int_a^b f(x)\,dx \approx (b-a)\,\frac{f(x_1)+f(x_2)+\ldots+f(x_n)}{n}$$

The left side is a general approach. The right side is an example. First, create the RAB program that generates a random number from the interval $[a,b]$ by entering '<< RAND B A - * A + >>' then storing with 'RAB', STO. You enter RAND by using MTH, [PROB], [RAND].

General $\int_a^b f(x)\,dx$ **Example** $\int_2^5 x^2\,dx$

- Store a as A, b as B
- Define the function by entering 'F(X)=$f(x)$', then pressing LEFT SHIFT, DEF.
- Move to the VAR menu by pressing VAR.
- Press the two keys RAB, F in succession a total of n times to generate the values $f(x_1), f(x_2), \ldots, f(x_n)$
- Press + a total of $n-1$ times to obtain the sum

$$f(x_1)+f(x_2)+\ldots+f(x_n)$$

- Press n, $\div$, $(b-a)$, × to get the approximation

$$(b-a)\,\frac{f(x_1)+f(x_2)+\ldots+f(x_n)}{n}$$

- Store 2 as A, 5 as B
- Define the function by entering 'F(X)=X^2', then pressing LEFT SHIFT, DEF.
- Move to the VAR menu by pressing VAR.
- Press the two keys RAB, F in succession a total of 10 times to generate the values $x_1^2, x_2^2, \ldots, x_{10}^2$
- Press + a total of 9 times to obtain the sum

$$x_1^2 + x_2^2 + \ldots + x_{10}^2$$

- Press 10, $\div$,3, × to get the approximation

$$3 \cdot \frac{x_1^2 + x_2^2 + \ldots + x_{10}^2}{10}$$

(The actual value is $\int_2^5 x^2\,dx = \frac{117}{3} = 39$.)

Study Guide for Section 7.3 331

Monte Carlo Method for the TI-85
to approximate any definite integral.

$$\int_a^b f(x)\, dx \approx (b-a)\, \frac{f(x_1) + f(x_2) + \ldots + f(x_n)}{n}$$

The left side is the general approach. The right side is an example. You enter RAND by using 2nd, MTH, [PROB], [RAND]. You enter y1 by using 2nd, ALPHA, Y, 1.

General $\displaystyle\int_a^b f(x)\, dx$ **Example** $\displaystyle\int_2^5 x^2\, dx$

- Under GRAPH, [y1=], enter
  y1=a+(b − a)*rand
  y2=f(y1)
- Press 2nd, QUIT to return to the HOME screen.
- Press 0, ENTER to initialize.
- Press +, 2nd, ALPHA, Y, 2 to get
  ANS + y2
  on the screen.
- Press ENTER a total of n times to obtain the cumulative sum
  $$f(x_1) + f(x_2) + \ldots + f(x_n)$$
- Press $(b-a)$, ×, 2nd, ANS, /, n to get
  $(b-a)$*ANS/n
  on the screen.
- Press ENTER to get the approximation
  $$(b-a)\, \frac{f(x_1) + f(x_2) + \ldots + f(x_n)}{n}$$

- Under GRAPH, [y1=], enter
  y1=2+3*rand
  y2=y1^2
- Press 2nd, QUIT to return to the HOME screen.
- Press 0, ENTER to initialize.
- Press +, 2nd, ALPHA, Y, 2 to get
  ANS + y2
  on the screen.
- Press ENTER a total of 10 times to obtain the cumulative sum
  $$x_1^2 + x_2^2 + \ldots + x_{10}^2$$
- Press 3, ×, 2nd, ANS, /, 10 to get
  3*ANS/10
  on the screen.
- Press ENTER to get the approximation
  $$3 \cdot \frac{x_1^2 + x_2^2 + \ldots + x_{10}^2}{10}$$

(The actual value is $\displaystyle\int_2^5 x^2\, dx = \frac{117}{3} = 39$.)

Monte Carlo Method for the TI-82
to approximate any definite integral.

$$\int_a^b f(x)\, dx \approx (b-a)\frac{f(x_1) + f(x_2) + \ldots + f(x_n)}{n}$$

The left side is the general approach. The right side is an example. You enter RAND by using MATH, arrow to PRB, select 1:rand, and press ENTER. You enter Y1 with ALPHA, Y, 1.

General $\int_a^b f(x)\, dx$ **Example** $\int_2^5 x^2\, dx$

- Under Y1=, enter
  $$Y1 = f(x)$$
- Press 2nd, QUIT to return to the HOME screen.
- Press 0, ENTER to initialize.
- Enter
  $$\text{ANS} + Y1(a + (b-a)*\text{rand})$$
  on the screen.
- Press ENTER a total of n times to obtain the cumulative sum
  $$f(x_1) + f(x_2) + \ldots + f(x_n)$$
- Enter
  $$(b-a)*\text{ANS}/n$$
  on the screen and press ENTER to get the approximation
  $$(b-a)\frac{f(x_1) + f(x_2) + \ldots + f(x_n)}{n}$$

- Under Y1=, enter
  $$Y1 = X\wedge 2$$
- Press 2nd, QUIT to return to the HOME screen.
- Press 0, ENTER to initialize.
- Enter
  $$\text{ANS} + Y1(2 + 3*\text{rand})$$
  on the screen.
- Press ENTER a total of 10 times to obtain the cumulative sum
  $$x_1^2 + x_2^2 + \ldots + x_{10}^2$$
- Enter
  $$3*\text{ANS}/10$$
  on the screen and press ENTER to get the approximation
  $$3 \cdot \frac{x_1^2 + x_2^2 + \ldots + x_{10}^2}{10}$$

(The actual value is $\int_2^5 x^2\, dx = \frac{117}{3} = 39$.)

Study Guide for Section 7.4
Applications to Particle Motion

Key Terms & Notation

Net distance $\quad \int_a^b v(t)\,dt$

Total distance $\quad \int_a^b |v(t)|\,dt$

Questions for Understanding

- Let $s(t)$, $v(t)$, $a(t)$ represent position, velocity and acceleration, respectively. (pages 486-487)
    1. What are $v(t)$ and $a(t)$ in terms of $s(t)$?
    2. What are $s(t)$ and $a(t)$ in terms of $v(t)$?
    3. What are $s(t)$ and $v(t)$ in terms of $a(t)$?

- In EXAMPLE 15, why are two constants, C_1 and C_2, used?
  (page 487)

- What is *net distance* in English and mathematical symbols? What is *total distance* in English and mathematical symbols? (page 487)

- What is the physical interpretation of the sign $(+/-)$ of $v(t)$ that is represented in Figure 7.23? (page 488)

- Is the following equation *always* true? Sometimes true? Never true?
  (page 489)

$$\left| \int_a^b f(x)\,dx \right| = \int_a^b |f(x)\,dx|$$

Suggested Exercises

§7.4 (pages 489-490): 2, 4, 10, 12

Notes on Exercises

- **Problem**

  Suppose an automobile is clocked at 20 kph (kilometers per hour) when it passes a police officer. If the car is constantly *accelerating* at 10 kph, when will it reach the legal speed of 90 kph? How far will it be from the officer at that time?

  Want:

  (a) Time t to reach 90 kph given initial velocity = 20 kph

  (b) Position from officer when velocity is 90 kph

  Need:

  (a) Velocity equation $v(t)$ and solve for t when $v(t) = 90$

  (b) Position equation $x(t)$ evaluated with the t that gives $v(t) = 90$

  Know:

  Since all measurements are requested in relation to the officer, $t = 0$, $v(0) = 20$ kph, $s(t) = 0$, and $v(t_f) = 90$ (where t_f represents the final time desired). Acceleration $a(t)$ is constant at 10 kph $\Longrightarrow$

  $$v(t) = \int_a^b a(t)\, dt = \int_0^{t_f} 10\, dt$$

  $$s(t) = \int_a^b v(t)\, dt = \int_0^{t_f} v(t)\, dt$$

  $\Longrightarrow$ ... (The details are left to you) ... $\Longrightarrow$

  $$t_f = 4.5 \text{ hours} \qquad s(t_f) = 191.25 \text{ miles}$$

  Do these answers make sense within the context of the problem? (Consider that automobiles can generally accelerate from 0 to 90 kph in 10 to 20 *seconds*.)

Graphing Calculator Hints

- Once the proper form of a *definite* integral is found, you may be able to evaluate it on your calculator. Otherwise, your machine can only help you with arithmetic.

Study Guide for Section 7.5
Using Definite Integrals to Measure Arc Length

Key Terms & Notation

Arclength $\qquad \int_C ds \qquad$ Surface of revolution

Questions for Understanding

- Have you noticed a pattern, yet?
    1. Recognize a quantity we want to add up.
    2. Establish a summing expression.
    3. Turn that into an _____

- What does it mean to say a curve is "polygonal?" How can this concept be used to estimate the length of a curve that is smooth?
(pages 490-492)

- What does "a differentiable function is approximately locally linear" mean?
(pages 174, 185, 491)

- Using the Pythagorean theorem, how can we find the length of a very small line segment Δs? (page 491)

- What does Δs approximate in terms of $\frac{\Delta y}{\Delta x}$? What happens as $n \to \infty$?
(pages 491-492)

- Suppose the values of x and y are dependent on time t. How is the chain rule for derivatives used to express arc length in terms of t?
(pages 492-493)

- In §7.2, we used $\int_a^b A(x)\,dx$ to find the *volume* of a surface of revolution. In that case, the integrand $A(x)$ represented the area of a circular cross section. What if we replaced the area expression with one for the *circumference* of the circular cross section? (pages 494-495)

Suggested Exercises

§7.5 (pages 496-497): 2, 8, 12, 18, 20, 26

Notes on Exercises

- Problem 11

$$\Delta s \approx \sqrt{(\Delta x)^2 + (\Delta y)^2} = \sqrt{(\Delta x)^2 \left(1 + \frac{(\Delta y)^2}{(\Delta x)^2}\right)}$$

$$= \sqrt{1 + \left(\frac{\Delta y^2}{\Delta x^2}\right)}\,\Delta x$$

$$\implies \int_1^4 \sqrt{1 + ([\sinh(x)]')^2}\,dx$$

$$[\sinh(x)]' = \left[\frac{e^x - e^{-x}}{2}\right]' = \frac{e^x + e^{-x}}{2} = \cosh(x)$$

$$\implies \int_1^4 \sqrt{1 + ([\sinh(x)]')^2}\,dx = \int_1^4 \sqrt{1 + [\cosh(x)]^2}\,dx$$

$$= \quad \ldots \text{ see Graphing Calculator Hints } \ldots \quad \approx 26.43715$$

- Problem

    A home owner wants to outline a flower bed next to her house with bendable wooden edging (See Figures 35 and 36). The curved portion will follow the curve $y = \cos(x)$, and the two end pieces will each be 4 feet long. The net length of the flower bed is 19 feet. Find the linear amount of edging needed to enclose the area, including the two end pieces, when none is used against the house. (Ignore the thickness and height of the edging.)

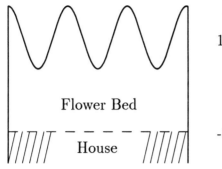
Figure 35: Yard plan

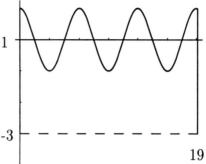
Figure 36: Graph

Want:
Length of left and right sides plus length of curve across the 19 foot span.

Need:
Expressions for lengths of both sides and the curve length

Know:
Length of both sides $= 4 + 4 = 8$ ft

$$\text{Arc length} = \int_0^{19} \sqrt{1 + ([\cos(x)]')^2} \, dx$$

$$= \int_0^{19} \sqrt{1 + (-\sin(x))^2} \, dx$$

$$= \int_0^{19} \sqrt{1 + \sin^2(x)} \, dx \approx 23.07219$$

$\implies$ Total length $\approx 8 + 23.07219 = 31.07219$

Suppose you made a mistake and integrated

$$\int_0^{19} \sqrt{1 - (\sin(x))^2} \, dx = \int_0^{19} \sqrt{\cos^2(x)} \, dx = \int_0^{19} \cos(x) \, dx$$

$$\approx .14988$$

How would you know a mistake had been made by looking at the numerical value of your answer?

Graphing Calculator Hints

- Here is a quick reminder on numerical integration of $\int_a^b f(x)\,dx$.

   HP-48G

   SYMBOLIC, Integrate, enter '$f(x)$'. Enter X, a, b for VAR:, LO:, HI:. Change RESULT to NUMERIC, press [OK].

   HP-48S

   Use the EQUATION writer to enter the complete integral or enter the expression '∫(a,b,$f(x)$,X)' on the stack. → NUM to evaluate.

   TI-85

   2nd, CALC, [fnInt]. Enter $f(x)$,x,a,b). Press ENTER.

   TI-82

   MATH, 9:fnInt, ENTER. Enter $f(x)$,X,a,b). Press ENTER.

Study Guide for Section 7.6
Applications to Physics – Part 1

Key Terms & Notation

Work
$F \Delta x$
$\int_a^b F(x)\, dx$

Hooke's law
$F(x) = kx$

Density of a fluid
Weight of a fluid

Questions for Understanding

- What do we have to do to apply definite integration as a measurement tool? Steps 1–4 apply to which part of this? (page 498)

- What is the general definition of *work* in English and mathematical symbols? If *force* varies relative to the variation in x, how do we write the work done on an interval $[x, x + \Delta x]$? (pages 499-500)

- As $n \to \infty$, what does Δx approach? How is total work *defined*? What does the "outside" part $\underbrace{\int_a^b dx}$ of the integral represent in a physical sense? (page 500)

- If the force required to move a particle along the x-axis is described by $G(x) = e^{-x}$, what is the work expression for moving the particle from $x = 3$ to $x = 7$? Is the work increasing or decreasing over the entire interval? Is the work done on the interval $[3, 5]$ larger or smaller than the work done on the interval $[5, 7]$? (page 500)

- How is force represented when a spring is displaced? How is work represented? (pages 500-501)

- What are the major differences, if any, between work done on a spring and work done lifting a weight? (pages 500-503)

Suggested Exercises

§7.6 (pages 505-507):   4, 8, 12, 18, 21, 22, 24, 26

Notes on Exercises

- Here is a synopsis of the steps to identifying the appropriate function over an interval.

> **Step 1.** Pick some arbitrary value x in the interval $[a, b]$ and focus your attention on a tiny subinterval $[x, x + \Delta x]$.
>
> **Step 2.** Calculate the desired quantity just for this subinterval to get an expression of the form
>
> $$f(x)\Delta x$$
>
> for some function f.
>
> **Step 3.** Calculate $\displaystyle\int_a^b f(x)\,dx$.
>
> **Step 4.** Check the reasonableness of the result.

- Problem 20, modified

Suppose the cone is filled to only half its height, similar to problem 13. First represent the problem graphically.

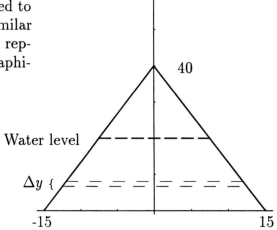

Figure 37: Problem 20, modified

Study Guide for Section 7.6 – Part 1

Step 1. Since we are lifting, we will integrate with respect to y. Therefore, subintervals will be of the form $[y, y + \Delta y]$. The total interval under consideration will be the cone height *plus* 20 $\Longrightarrow 40 + 20 = 60$.

Step 2. (a) Tank is filled to half-height $\Longrightarrow$ from $0 \leq y \leq \dfrac{40}{2}$ we have the force for a subinterval = weight of a Δy-thick layer = (density)(volume of layer) = ρV_y. V_y = (area of layer)(height of layer) = $A_y \Delta y$. A_y = area of a circle = πr^2 with r determined by the circle's position in the cone. From the graph, we see that $r = x$, as well. By using *similar triangles*, we observe the two ratios

$$\frac{r}{40-y} = \frac{30}{40} \Longrightarrow r = \frac{3}{4}(40-y) \Longrightarrow$$

$$A_y = \pi \left[\frac{3}{4}(40-y)\right]^2 = \frac{9}{16}\pi(40-y)^2 \Longrightarrow$$

$$V_y = \frac{9}{16}\pi(40-y)^2 \Delta y \quad \text{for } 0 \leq y \leq 20$$

$\Longrightarrow$ (since the weight of a Δy-thick layer = ρV_y)

work to lift a Δy-thick layer through y inches = $\rho V_y \, y$

$$= \left[\rho \frac{9}{16}\pi(40-y)^2 \Delta y\right] y = \left[\rho \frac{9}{16}\pi(40-y)^2 y\right] \Delta y$$

$$\Longrightarrow f_1(y) = \rho \frac{9}{16}\pi(40-y)^2 y \quad \text{for } 0 \leq y \leq 20$$

(b) For $20 \leq y \leq 60$, the volume V_y never changes $\Longrightarrow$ weight never changes $\Longrightarrow$ work to lift a Δy-thick layer when $20 \leq y \leq 60$ = (weight of layer)(distance)

$$= \rho V_y(60-20) = \rho V_y 40$$

$$\Longrightarrow f_2(y) = \rho \frac{9}{16}\pi(40-y)^2 40 \quad \text{for } 20 \leq y \leq 60$$

To obtain $f(y)\Delta y$ for $0 \leq y \leq 60$, we *add* the results from (a) and (b) to obtain

$$f(y)\Delta y = [f_1(y) + f_2(y)]\Delta y = \ldots = \left[\rho \frac{9}{16}\pi(40-y)^2(y+40)\right]\Delta y$$

Step 3.

$$\int_a^b f(y)\,dy = \int_0^{60} \left[\rho \frac{9}{16}\pi(40-y)^2(y+40)\right] dy$$

$$= \rho \frac{9}{16}\pi \int_0^{60} (40-y)^2(y+40)\,dy$$

$$= \ldots \text{after algebra and the power rule for antiderivatives} \ldots =$$

$$= \rho \frac{9}{16}\pi \left(64000\,y - 800\,y^2 - \frac{80}{3}y^3 + y^4\right)\Big]_0^{60}$$

$$= \frac{9}{16}\rho\pi\left(13,920,000 - \frac{80}{3}(216,000)\right) = \rho\pi(7,506,000)$$

Recall that $\rho = $ density of water in this problem

$$= 6235 \text{ lbs/ft}^3 = \frac{62.5 \text{ lbs}}{1 \text{ ft}^3} \cdot \frac{1 \text{ ft}^3}{(12 \text{ in})^3} \approx 0.0362 \text{ lbs/in}^3$$

$$\implies \text{Work} \approx 853,624.76 \text{ in lbs}$$

Step 4. Note that the y^4 term has in^4 units and density has lbs/in^3 units $\implies$ final units of in lbs which is a unit of work. Does the magnitude of this answer seem reasonable? Could you get a feel for this by looking at the results from other problems?

Graphing Calculator Hints

- As the Notes on Exercises indicate, your calculator can play only a minor role in these types of problems. It can be useful for arithmetic and, perhaps, numerical integration.

Study Guide for Section 7.6
Applications to Physics – Part 2

Key Terms & Notation

Work
$F\Delta x$
$\int_a^b F(x)\,dx$

Hooke's law
$F(x) = kx$

Density of a fluid
Weight of a fluid

Questions for Understanding

- What does it mean for a cable to be "uniform?" (page 502)

- Integration is used for what part of the work done in EXAMPLE 24? Why isn't it necessary for the other part(s)? (pages 502-503)

- What does $\dfrac{60 \text{ lbs}}{40 \text{ ft}}$ represent in EXAMPLE 24? (page 502)

- If you use the 4 steps in this section, what would the $f(x)$ be for **Step 2** in EXAMPLE 24? What about EXAMPLES 25 and 26? (pages 502-505)

- Would a Δx-long segment from EXAMPLE 24 correspond to a Δy-thick layer in EXAMPLE 25? What about EXAMPLE 26? (pages 502-505)

- **NOTE:** In EXAMPLE 26, near the top of page 505, the sentence

    The work required to lift it x feet ...

  should read

    The work required to lift it y feet ...

- In EXAMPLE 26, the expression $50\pi y(16-y^2)\Delta y$ is derived. What does the 50 represent? What does $\pi(16-y^2)$ represent? What does the y term represent? (pages 504-505)

Suggested Exercises

§7.6 (pages 505-507):   4, 8, 12, 18, 21, 22, 24, 26

Notes on Exercises

- Problem 23
  Can you use the 4 steps on page 498 and EXAMPLE 24 to do this one?

- Problem 27
  For a subinterval $[R, R - \Delta R]$

  $$W_{\Delta R} = (\text{Force})(\text{change in distance}) = F \Delta R$$

  $$\implies \text{total work} = W = \lim_{n \to \infty} \sum_{i=1}^{n} F(R_i) \Delta R = \int_{10}^{10^{-6}} F(R)\, dR$$

  $$= \int_{10}^{10^{-6}} \frac{kQ_1Q_2}{R^2}\, dR = kQ_1Q_2 \int_{10}^{10^{-6}} R^{-2}\, dR = kQ_1Q_2 \left(\frac{R^{-1}}{-1} \right) \Bigg]_{10}^{10^{-6}}$$

  $$= \frac{-kQ_1Q_2}{R} \Bigg]_{10}^{10^{-6}} = -\frac{kQ_1Q_2}{10^{-6}} - \left[-\frac{kQ_1Q_2}{10} \right]$$

  $$= kQ_1Q_2 \left(\frac{1}{10} - 10^6 \right)$$

  Notice that the answer in your text is slightly different. Can you tell why by looking at the limits of integration? What would this mean in a physical sense?

Graphing Calculator Hints

- See Study Guide for Section 7.6 – Part 1.

Study Guide for Section 7.7
Improper Integrals

Key Terms & Notation

Improper integral   Converges/diverges $\int_a^\infty f(x)\,dx$

Horizontal type improper integral $\lim\limits_{b\to\infty} \int_a^b f(x)\,dx$

Vertical type improper integral $\lim\limits_{a\to c^+} \int_a^b f(x)\,dx$

Questions for Understanding

- Why are certain integrals called *improper*? (page 508)

- What is the general defining characteristic of *horizontal type* improper integrals? (page 509)

- In EXAMPLES 27-31, which part of the fundamental theorems of calculus is used? (pages 509-512)

- How would you handle $\int_{-\infty}^{\infty} \dfrac{1}{x^4}\,dx$? (page 511)

- What was "exploited" to evaluate the integral of EXAMPLE 31? (pages 511-512)

- What is the general defining characteristic of *vertical type* improper integrals? (pages 512-513)

- Which of these two improper integrals is the horizontal type and which is the vertical type?

$$\int_{-1}^{0} \frac{1}{x}\,dx \qquad \int_{-\infty}^{-1} \frac{1}{x}\,dx$$

- Why is $\int_0^2 \frac{1}{x-3}\,dx$ a proper integral, but $\int_0^2 \frac{1}{x-1}\,dx$ is an improper one? (pages 512-513)

- How do we determine if an improper integral *converges* or *diverges*? (pages 511, 514)

Suggested Exercises

§7.7 (pages 515-518):   2, 10, 11, 14, 16, 19, 21, 28, 36, 37

Notes on Exercises

- Problem 25

$$\int_{-\infty}^{\infty} x\,dx = \lim_{a \to -\infty} \int_a^0 x\,dx + \lim_{b \to \infty} \int_0^b x\,dx$$

$$= \lim_{a \to -\infty} \left[\frac{x^2}{2}\right]_a^0 + \lim_{b \to \infty} \left[\frac{x^2}{2}\right]_0^b$$

$$= \lim_{a \to -\infty} \left[\frac{0}{2} - \frac{a^2}{2}\right] + \lim_{b \to \infty} \left[\frac{b^2}{2} - \frac{0}{2}\right]$$

$$= \lim_{a \to -\infty} \left(\frac{-a^2}{2}\right) + \lim_{b \to \infty} \left(\frac{b^2}{2}\right)$$

Both integrals in the sum diverge $\implies \int_{-\infty}^{\infty} x\,dx$ diverges.

(Note: This would still be true if only *one* of them diverged.)

- Problem 45

$$\int_{-a}^{a} x\,dx = \left.\frac{x^2}{2}\right]_{-a}^{a} s = \frac{a^2}{2} - \frac{(-a)^2}{2} = \frac{a^2}{2} - \frac{a^2}{2} = 0$$

$$\implies \lim_{a\to\infty}\int_{-a}^{a} x\,dx = 0 \quad \text{for any positive integer } a$$

Compare this with problem 25 and see pages 517-518.

Graphing Calculator Hints

- Use your numerical integrator to evaluate $\int_a^b f(x)\,dx$ with several very large values for b. Does $\int_a^b f(x)\,dx$ get progressively larger as b does (diverges), or does it seem to approach a particular value (converges)?

Study Guide for Section 8.1
Integration by Parts

Key Terms & Notation

Integration by parts $\qquad \int u\,dv \qquad uv - \int v\,du$

Questions for Understanding

- In the equation $C = b - F(a)$, what do each of the symbols C, b, $F(a)$, a stand for? (page 520)

- How can we use the fundamental theorems of calculus to solve a differential equation even if an antiderivative cannot be expressed in algebraic form? (pages 521-522)

- How is the *Integration by parts formula* related to the product rule for derivatives? (page 522)

- When you use the Integration by parts techniques, is it likely that you may need to make more than one try on a problem? What will it take to become skilled at this technique? (page 523)

- In EXAMPLE 3, suppose we let $u = e^x$ and $dv = x\,dx$. How does $\int v\,du$ compare with the original $\int xe^x\,dx$? Would you say it is a simpler expression? Does this suggest something to watch for? (page 523)

- Would you say there is a definite advantage to having the various integration rules memorized?

Suggested Exercises

§8.1 (pages 525-526):   1, 4, 12, 14, 16, 18, 28, 36, 48

Study Guide for Section 8.1

Notes on Exercises

- Here is a synopsis of the steps outlining the method of integration by parts.

> **Step 1.** Let $u = f(x)$ and $dv = g(x)\,dx$, where $f(x)g(x)$ is the original integrand.
> **Step 2.** Compute $du = f'(x)\,dx$ and $v = \int g(x)\,dx$.
> **Step 3.** Substitute $u, v, du,$ and dv into the formula
> $$\int u\,dv = uv - \int v\,du.$$
> **Step 4.** Calculate $uv - \int v\,du$. (If $\int v\,du$ is impossible to integrate, go back to Step 1 and consider other choices for u and dv.)
> **Step 5.** Check your solution by differentiating and comparing to the original integrand.

- Problem 2
  We have y as the particular antiderivative with $y(0) = 1$.
  $y' = \ln(x^2 + 1)$ is continuous for any $x \implies$
  $$y(x) = b + \int_a^x f(x)\,dx = 1 + \int_0^x \ln(x^2 + 1)\,dx$$
  $$\implies y(3) = 1 + \int_0^3 \ln(x^2 + 1)\,dx \approx 4.406$$

- Problem 17 (This may help with problem 14)
  $$\left.\begin{array}{r} u = x^2 \implies du = 2x\,dx \\ dv = \sin(x)\,dx \implies v = -\cos(x) \end{array}\right\} \implies$$
  $$uv - \int v\,du = -x^2\cos(x) - \int(-2x\cos(x))\,dx$$
  $$= -x^2\cos(x) + \underbrace{\int 2x\cos(x)\,dx}_{\text{apply integration by parts again}}$$

$$\left.\begin{array}{l} u = 2x \implies du = 2\ dx \\ dv = \cos(x)\ dx \implies v = \sin(x) \end{array}\right\} \implies$$

$$\int 2x \cos(x)\ dx = 2x \sin(x) - \int 2\sin(x)\ dx$$
$$= 2x\sin(x) + 2\cos(x) + C$$

Therefore,

$$\int x^2 \sin(x)\ dx = -x^2 \cos(x) + (2x\sin(x) + 2\cos(x)) + C$$

Graphing Calculator Hints

- As the EXAMPLES indicate, your calculator can play only a limited role in these types of problems. It can be useful for arithmetic, graphing, and numerical integration for an exact value.

Study Guide for Section 8.2
The Exponential Model

Key Terms & Notation

Linear model
$y' = ky$

Exponential model
$y(0) = b$ $\qquad y = be^{kt}$

Questions for Understanding

- What numerical evidence indicates the solution to a differential equation conforms to the *linear model*? How is this observed graphically? (page 527)

- What type of solution function does the term *exponential model* suggest to you?

- What is the numerical evidence indicating an exponential model applies? (pages 527-529)

- What function y is equal to its own derivative and also satisfies the equation
$y(0) = 1$? (page 528)

- How are the following equations related? (page 529)

$$y' = ky \quad \text{and} \quad y' = kbe^{kx}$$

- What are the *parameters* associated with the exponential model? What does each one represent? (page 530)

- What does the *proportionality constant* tell us about? What does the y-intercept represent? (page 530)

Suggested Exercises

§8.2 (pages 530-531): 4, 6, 12, 14, 16, 20

Notes on Exercises

- Problem 10

$$y' = \frac{dy}{dx} = \frac{y}{x} \implies \frac{1}{y} dy = \frac{1}{x} dx \implies \int \frac{1}{y} dy = \int \frac{1}{x} dx$$

$$\implies \ln|y| = \ln|x| + C \implies e^{\ln|y|} = e^{\ln|x|+C} = e^{\ln|x|} e^C$$

$$\implies |y| = |x| e^C$$

$$y(-1) = 0.75 \implies |y(-1)| = |0.75| = |-1| e^C$$

$$\implies e^C = 0.75 \implies |y| = 0.75|x|$$

But the equation $|y| = 0.75|x|$ is not a *function*, as evidenced by

$x = -1 \implies .075|x| = 0.75$. If $|y| = 0.75$, then $y = 0.75$ or $y = -0.75$.

Therefore, there are two y-values associated with the single x-value, $x = -1$. The initial condition $y(-1) = 0.75$ and $y' = \frac{y}{x} = \frac{0.75}{-1} \implies y = -0.75x$ is a solution for all $x < 0$. $y' = \frac{y}{x} \implies x \neq 0$. Since we need a continuous function as a solution on an interval containing $x = -1$, we have $y = -0.75x$ with a domain $D = (-\infty, 0)$ as the solution we seek.

- Problem 19

$$[\log_2(x) + 5]' = [\log_2(x)]' + [5]' = [\log_2(x)]' + 0 = [\log_2(x)]'$$

To compute $[\log_2(x)]'$, recall that

$$y = \log_2(x) \implies 2^y = 2^{\log_2(x)} = x$$

$$\implies \ln 2^y = \ln x \implies y \ln 2 = \ln x$$

$$\implies y = \frac{1}{\ln 2} \ln x$$

Study Guide for Section 8.2

$$\implies [\log_2(x)]' = y' = \frac{1}{\ln 2}[\ln x]'$$

$$= \frac{1}{\ln 2}\frac{1}{x} = \frac{1}{x \ln 2}$$

A common convention is to find $y(0)$ or $y(1)$ for the initial condition. Since $y(0)$ is undefined, we use $y(1)$.

$$y(1) = \log_2(1) = 0 \quad \text{for} \quad y = \log_2 x$$

$$y(1) = \log_2(1) + 5 = 5 \quad \text{for} \quad y = \log_2 x + 5$$

Graphing Calculator Hints

- Graphing the slopefield of a differential equation can give clues to its type of model. See Study Guide for Section 6.3 to review how to use your calculator for this.

Study Guide for Section 8.3
Applications of the Exponential Model

Key Terms & Notation

$$y(t) = bc^t \qquad \text{half-life} \qquad C(t) = A \cdot \left(\frac{1}{2}\right)^{t/k}$$

$$A(t) = P\left(1 + \frac{r}{n}\right)^{nt} \qquad y(t) = A \cdot \left(\frac{1}{2}\right)^{t/N} \qquad i(t) = Ac^t + T$$

$$A(t) = Pe^{rt}$$

Questions for Understanding

- Suppose $y(t) = 7(2^t)$. Can you also express $y(t)$ in terms of e? What would that version look like? \hfill (page 532)

- Why does an exponential model work well for EXAMPLE 10? \hfill (page 532)

- What physical quantity does A represent in EXAMPLE 10? \hfill (page 532)

- What physical quantity does A represent in the discussion of compound interest? \hfill (page 534)

- How are the following expressions related? \hfill (page 534)

$$P\left(1 + \frac{r}{n}\right)^{t/N} \qquad \text{and} \qquad Pe^{rt}$$

- In the equation $y(t) = A(\frac{1}{2})^{t/N}$, on page 535, does the notation $A(\frac{1}{2})$ mean A represents a function in the "variable" $\frac{1}{2}$?

- What physical quantity does A represent in the *half-life* equation for radioactive decay? \hfill (page 535)

Study Guide for Section 8.3 355

- In the context of radioactive decay, if

$$\frac{1}{2}P_0 = P_0 e^{kt} = A \cdot \left(\frac{1}{2}\right)^{t/N}$$

What would P_0 represent? What would $\frac{1}{2}P_0$ represent? What does N represent? How are e^k and $\left(\frac{1}{2}\right)^{1/N}$ related? (page 535)

- In EXAMPLE 14, what physical quantity does A represent?
(page 537)

- Why must the parameter T be included in the *heat transfer* equation?
(pages 536-537)

Suggested Exercises

§8.3 (pages 537-539):   2, 4, 6, 8, 10, 16

Notes on Exercises

- Problem 3
  Want:
  Temperature of object at $t = 1$ minute and $t = 2$ minutes, given water is maintained at $40°F$.
  Need:
  Expression for object temperature $i(t)$ and values for applicable parameters.
  Know:
  $y(t)$ = difference between object and water temperatures
  $= i(t) - T = Ac^t \implies$
  $i(t) = Ac^t + T = Ac^t + 40$

From problem 2,
$$i(0) = Ac^0 + 40 \approx 100.44 \implies A \approx 60.44$$

$c = e^k$ is the constant of proportionality for heat transfer in water. From problem 1, we can derive $c \approx 0.804$. Therefore,
$$i(t) \approx (60.44)(0.804)^t + 40 \implies$$
$$i(1) \approx (60.44)(0.804)^1 + 40 = 88.59$$
$$i(2) \approx (60.44)(0.804)^2 + 40 = 79.07$$

- Problem 5

  Want:

  half-life of radioactive substance

  Need:

  Expression for amount of substance left after t years, and values for applicable parameters.

  Know:
  $$y(t) = \text{amount present} = A \cdot \left(\frac{1}{2}\right)^{t/N}$$
  $$A = \text{initial amount of substance}$$
  $$N = \text{half-life of substance}$$
  $$y(10) = 30\% \text{ of initial amount} = 0.3A$$
  $$= A \cdot \left(\frac{1}{2}\right)^{10/N}$$
  $$\implies 0.3A = A \cdot \left(\frac{1}{2}\right)^{10/N} \implies \frac{3}{10} = \left(\frac{1}{2^{10}}\right)^{1/N}$$
  $$\implies \left(\frac{3}{10}\right)^N = \frac{1}{2^{10}} \implies 2^{10} = \left(\frac{10}{3}\right)^N$$
  $$\implies 10 \ln 2 = N \ln\left(\frac{10}{3}\right) = N [\ln 10 - \ln 3]$$
  $$\implies N = \frac{10 \ln 2}{\ln 10 - \ln 3} \approx 5.75717 \text{ years}$$

Graphing Calculator Hints

• If you fully understand the generic meanings of each of the parameters in
$$y(t) = Ac^t \quad \text{and} \quad y(t) = be^{kt}$$
Then you can store either (or both) of the equations in your calculator and use the solver feature.

HP-48G

Enter the expression as the calculator's EQ variable. (Using the PLOT feature is one way.) Get into the SOLVR menu (which is not listed on the keyboard), by pressing [↑], SOLVE, [ROOT], [SOLVR]. Enter a value for x_i as the X variable and obtain the result by pressing [EXPR=]. See SOLVR environment in your *User's Guide* index. *Note*: the **HP-48S** is similar and actually easier. (Further exploration: Get into PLOT and graph the function)

TI-85

Enter the expression under GRAPH, $y(x) =$. Press 2nd, SOLVER, choose the appropriate function name at the bottom of the view screen. Press ENTER for the next screen. Move the cursor to the x= and enter the value for x_i. Move the cursor to exp= and press [SOLVE] (the F5 key) to obtain the result. See SOLVER in your *Guidebook* index. (Further exploration: Press the GRAPH (F1) key at the bottom of the view screen.)

TI-82

Enter the expression under Y= and plot using ,GRAPH. Bring up the CALC menu screen with 2nd, CALC. Choose **value**, press ENTER. Enter the desired $x - value$, press ENTER, and read the results from the listed coordinates. See Variables, CALC menu in your *Guidebook* index.

Study Guide for Section 8.4
Differential Equations

Key Terms & Notation

Separation of variables   "equation of differentials"

Questions for Understanding

- What does the notation $G(x, y)$ represent (aside from dy/dx)?

  (page 540)

- In EXAMPLE 15, the sentence beginning

  Choosing points with integer coordinates ...

  should read

  Choosing points with integer coordinates and computing the slope at each point (for example, $y' = -(-2/3) = 2/3$ at the point $(-2, 3)$) gives the slope field pictured in Figure 8.5.

  (page 541)

- In EXAMPLE 15, neither initial condition provides the whole circle. Why is this? (pages 541-542)

- In EXAMPLE 16, the differential equation under consideration is
  $$\frac{dv}{dt} = (15 - 0.2v) \text{ m/s}^2$$
  What physical quantity does the number 15 represent? What does $-0.2v$ represent? (page 542)

- In the solution to EXAMPLE 16, your text states

  Note that in this case, the slope dv/dt depends on v but not on t ...

Study Guide for Section 8.4

How is this reflected in the equation for dv/dt? What is the corresponding characteristic of the slope field? How does this compare with EXAMPLE 17? (pages 542-543)

- Does the term *separation of variables* seem to fit its definition? (page 544)

- In EXAMPLE 18, your text *multiplies* by dx. Does this mean that $\dfrac{dy}{dx}$ is a simple fraction like $\dfrac{7}{5}$? (page 544)

- How are the following related? (pages 544, 546)

$$\text{Antidifferentiating both sides of } y\,dy = -x\,dx$$

$$\int y\,dy = \int -x\,dx$$

$$\text{Integrating both sides of } y\,dy = -x\,dx$$

- In EXAMPLES 18 and 19, how are C, C_1, and C_2 related? Are they necessarily the same from one example to another? (pages 544-546)

Suggested Exercises

§8.4 (pages 546-548):   4, 5, 10*, 14, 17, 28, 31

* Do problem 10 using a slopefield *and* by solving the differential equation

Notes on Exercises

- Problem 29

$$\frac{dy}{dx} = \frac{1+x}{\sqrt{y}} \implies \sqrt{y}\,dy = (1+x)\,dx$$

$$\implies \int \sqrt{y}\,dy = \int (1+x)\,dx$$

(a) $\displaystyle \int \sqrt{y}\,dy = \int y^{1/2}\,dy = \frac{y^{3/2}}{3/2} + C_1 = \frac{2}{3}y^{3/2} + C_1$

(b) $\int (1+x)\,dx = x + \dfrac{x^2}{2} + C_2$

Therefore,

$$\dfrac{2}{3}y^{3/2} + C_1 = x + \dfrac{x^2}{2} + C_2 \implies \dfrac{2}{3}y^{3/2} = x + \dfrac{x^2}{2} + C$$

(where $C = C_2 - C_1$)

$y(2) = 9 \implies y = 9$ when $x = 2 \implies$

$\dfrac{2}{3}(9)^{3/2} = 2 + \dfrac{2^2}{2} + C \implies \dfrac{2}{3}(27) = 4 + C \implies C = 18 - 4 = 14$

$\implies \dfrac{2}{3}y^{3/2} = x + \dfrac{x^2}{2} + 14 \implies \ldots$ (after some algebra) $\ldots \implies$

$$y = \left(\dfrac{3x^2 + 6x + 84}{4}\right)^{2/3}$$

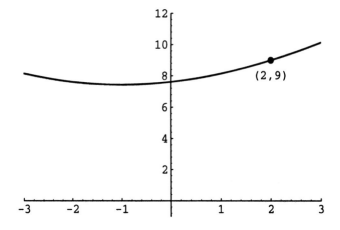

Figure 38: Problem 29

Graphing Calculator Hints

- Remember that the Study Guide for Section 6.3 introduced how to use your calculator to generate slope fields. Here is a reprint of the basics.

HP-48G

Get to the PLOT menu. Highlight TYPE: and press [CHOOS]. Arrow down to Slopefield (the ninth one down), and press [OK]. Press [OPTS], NXT, [RESET], [OK], arrow to Reset plot, press [OK], [OK] to obtain default settings and return to the PLOT menu. Enter the desired expression in the EQ: area, then plot. See "SLOPEFIELD" in your *User's Guide* index.

HP-48S

Under the PLOT, PLOTR menu, make sure the Plot type: is FUNCTION, and press NXT, [RESET] to obtain default settings. Bring up the VAR menu, enter the desired expression on the stack and press SLPF to plot the slope field.

TI-85

Under MODE, make sure you are in Func mode. Under ZOOM, reset the plot parameters to ZSTD. Enter the desired expression as y1=, then get back to the HOME screen by pressing EXIT. Press PRGM, [NAMES], and select [SLPF]. Press ENTER to execute the program.

TI-82

Under MODE, make sure you are in Func mode. Under ZOOM, reset the plot parameters to ZStandard. Enter the desired expression as Y1, press PRGM, select SLPF, and press ENTER to execute the program.

Study Guide for Section 9.5
L'Hôpital's Rule

Key Terms & Notation

Indeterminate quotients L'Hôpital's rule for 0/0
Indeterminate difference L'Hôpital's rule for ∞/∞
Squeezing principle Indeterminate exponentials

Questions for Understanding

- If $\lim_{x \to 3} f(x) = e$, $\lim_{x \to 3} g(x) = \pi$, what is $\lim_{x \to 3} \frac{g(x)}{f(x)}$?
  (pages 596-597)

- Why is it "that $\lim_{x \to a} \frac{f(x)}{g(x)}$ is undefined in the first four lines of the third column of the table?" (page 597)

- Is the expression $\frac{0}{0}$ a real number? (page 598)

- Let $f(x) = m_1(x - a)$, $g(x) = m_2(x - a)$. What is $\frac{f(x)}{g(x)}$? What is $\frac{f'(x)}{g'(x)}$? What does this *suggest* about (pages 599-600)
$$\lim_{x \to a} \frac{f(x)}{g(x)} \quad \text{and} \quad \lim_{x \to a} \frac{f'(x)}{g'(x)}$$

- To see why Hypothesis 1 is necessary to L'Hôpital's rule, suppose $f(a) = 0 = g(a)$. We know one of the general definitions of the derivative is
$$f'(a) = \lim_{x \to a} \frac{f(x) - f(a)}{x - a}$$
Under these conditions, how does $\frac{f'(a)}{g'(a)}$ compare to $\lim_{x \to a} \frac{f(x)}{g(x)}$? How does this relate to $\lim_{x \to a} \frac{f'(x)}{g'(x)}$? Does all of this seem to hold true if

Study Guide for Section 9.5

$f(a) \neq 0$ or $g(a) \neq 0$? (pages 598, 600-601)

- How are $\lim\limits_{x \to \infty} \dfrac{f(x)}{g(x)}$ and $\lim\limits_{h \to 0+} \dfrac{f(1/h)}{g(1/h)}$ related? (pages 601-602)

- What are the various *indeterminate forms* under which L'Hôpital's rule can be applied?

- How can L'Hôpital's rule be used in cases of an *indeterminate difference*? (page 603)

- What is the general form of a limit involving *indeterminate exponentials*? What are the steps for handling them? If f is continuous, then why is (pages 604-605)

$$\exp\left(\lim_{x \to a} \ln(f(x))\right) = \lim_{x \to a} f(x) \;?$$

- Would it be worthwhile to review the linearity rules for limits? (page 152)

Suggested Exercises

§9.5 (pages 605-607): 2, 5, 6, 11, 18, 22, 24

Notes on Exercises

- Problem 14

    $\lim\limits_{x \to \infty} \dfrac{x^{1/x}}{x}$ may seem to approach $\dfrac{1}{\infty}$, because $x^{1/\infty} \to x^0 = 1$.

    Therefore, the limit is not in one of the applicable indeterminate forms. However, we know that

    $$\lim_{x \to \infty} \frac{x^{1/x}}{x} = \lim_{x \to \infty} x^{1/x} \cdot \frac{1}{x} = \left(\lim_{x \to \infty} x^{1/x}\right)\left(\lim_{x \to \infty} \frac{1}{x}\right)$$

    if both limits exist

(a) $\lim_{x \to \infty} x^{1/x}$ has the indeterminate form ∞^0

Step 1. $\ln\left(\lim_{x \to \infty} x^{1/x}\right) = \lim_{x \to \infty}\left(\ln x^{1/x}\right) = \lim_{x \to \infty} \frac{1}{x} \ln x$

$$= \lim_{x \to \infty} \frac{\ln x}{x}$$

which has the indeterminate form $\frac{\infty}{\infty}$

$\implies$ L'Hôpital's rule applies.

Step 2. $\lim_{x \to \infty} \frac{\ln x}{x} = \lim_{x \to \infty} \frac{1/x}{1} = \lim_{x \to \infty} \frac{1}{x} = 0$

Step 3. $\lim_{x \to \infty} x^{1/x} = \exp\left(\ln(\lim_{x \to \infty} x^{1/x})\right) = \exp\left(\lim_{x \to \infty} \ln x^{1/x}\right)$

$$= \exp(0) = e^0 = 1$$

(b) It is easy to see that $\lim_{x \to \infty} \frac{1}{x} = 0$

Therefore, (a) and (b) show that both limits exist and we have

$$\lim_{x \to \infty} \frac{x^{1/x}}{x} = \left(\lim_{x \to \infty} x^{1/x}\right)\left(\lim_{x \to \infty} \frac{1}{x}\right) = 1 \cdot 0 = 0$$

Graphing Calculator Hints

- Graphing the function under consideration often gives clues about what to expect. For example, a quick glance at problem 7 and EXAMPLE 26, might suggest that $\lim_{x \to \infty} x^{-1/x} = -1$. But a graph of the function does not agree.

- If you apply L'Hôpital's rule, what do you get for $\lim_{x \to 0} \frac{\cos(x)}{1 + 2x}$? What does the graph indicate? How can you explain the discrepancy? (*Hint*: what are the initial hypotheses for L'Hôpital's rule?)

Differentiation Quiz
(SAMPLE)

For exercises 1-10: Find the indicated derivative. It is not necessary to simplify your answer. **No calculators, notes, or books are permitted for this quiz.** (For trigonometric functions, assume that angles are measured in radians.)

1. Find $\dfrac{dy}{dx}$ where $y = (-12x^5 - 4x^4 + 3)^{3/2}$.

2. Find $s'(t)$ where $s(t) = \tan(7t^3 - 5t)$.

3. Find y' where $y = \sqrt[5]{2 - x^2}$.

4. Find $\dfrac{dy}{dx}$ where $y = \sqrt[3]{\dfrac{3 - x^2}{2 + x^3}}$.

5. Find $h'(x)$ where $h(x) = x \cos(x) \cot(2x)$.

6. Find $\dfrac{dQ}{dt}$ where $Q(t) = (x^{12} + 4x^3 - 9)^{2/3}$.

7. Find $\dfrac{d}{dx}\left[\cos^3(x^2 + \sqrt{x})\right]$.

8. Find $f'(x)$ where $f(x) = (3 - 7x + 4x^3)^4 (x + 3)(x^2 - 7)^{1/2}$.

9. Find $\dfrac{df(x)}{dx}$ where $f(x) = \sqrt{\dfrac{13}{x} + \sqrt{\dfrac{x}{3}}}$.

10. Find $A'(x)$ where $A(x) = \dfrac{\sqrt{x^2 - 3x}}{\cos x^2}$.

Integration Quiz
(SAMPLE)

For exercises 1-10: Find the indicated antiderivative. It is not necessary to simplify your answer. **No calculators, notes, or books are permitted for this quiz.** (For trigonometric functions, assume that angles are measured in radians.)

1. Find $\displaystyle\int \frac{2x^2 - x}{4x^3 - 3x^2}\, dx$

2. Find $\displaystyle\int \sin^4(3x)\, dx$

3. Find $\displaystyle\int \sin^4 x \cos^3 x\, dx$

4. Find $\displaystyle\int \frac{2x + 3}{x^2 + 4}\, dx$

5. Find $\displaystyle\int \frac{x + 1}{x^2 + 4x - 5}\, dx$

6. Find $\displaystyle\int x^2 e^{-4x}\, dx$

7. Find $\displaystyle\int \frac{dx}{8 + 2x^2}$

8. Find $\displaystyle\int (2x + 3)e^{x^2 + 3x + 1}\, dx$

9. Find $\displaystyle\int \frac{x^3 [\ln(x^4 - 3)]^2}{x^4 - 3}\, dx$

10. Find $\displaystyle\int \frac{dx}{x^2 \sqrt{x^2 + 9}}$

Differentiation Quiz
(Solutions to SAMPLE)

Here are solutions to the sample Differentiation Quiz. Since many of the problems may have more than one approach, these solutions may not be in the same form as yours. However, if your work is correct, algebraic manipulation will show them to be equivalent.

1. $\dfrac{3}{2}\left(-12x^5 - x^4 + 3\right)^{1/2}\left(-60x^4 - 4x^3\right)$

2. $\left(21t^2 - 5\right)\sec^2\left(7t^3 - 5t\right)$

3. $-\dfrac{2}{5}x\left(2 - x^2\right)^{-4/5}$

4. $\dfrac{1}{3}\left(\dfrac{3 - x^2}{2 + x^3}\right)^{-2/3}\left(\dfrac{-2x(2 + x^3) - 3x^2(3 - x^2)}{(2 + x^3)^2}\right)$

5. $-2x\cos(x)\csc^2(2x) + \cot(2x)(\cos x - x\sin x)$

6. $\dfrac{2}{3}\left(x^{12} + 4x^3 - 9\right)^{-1/3}\left(12x^{11} + 12x^2\right)$

7. $-3\left(2x + \dfrac{1}{2\sqrt{x}}\right)\cos^2\left(x^2 + \sqrt{x}\right)\sin\left(x^2 + \sqrt{x}\right)$

8. $\left(3 - 7x + 4x^3\right)^4\left(\dfrac{x(x + 3)}{(x^2 - 7)^{1/2}} + x(x^2 - 7)^{1/2}\right)$
$\qquad\qquad\qquad\qquad\qquad\qquad + (12x^2)(x + 3)(x^2 - 7)^{1/2}$

9. $\dfrac{1}{2}\left(\dfrac{13}{x} + \sqrt{\dfrac{x}{3}}\right)^{-1/2}\left(\dfrac{-13}{x^2} + \dfrac{1}{6}\left(\dfrac{x}{3}\right)^{-1/2}\right)$

10. $\dfrac{\dfrac{2x - 3}{2\sqrt{x^2 - 3x}}\cos x^2 + 2x\sin x^2\sqrt{x^2 - 3x}}{\cos^2 x^2}$

Integration Quiz
(Solutions to SAMPLE)

Here are solutions to the sample Integration Quiz. Since many of the problems may have more than one approach, these solutions may not be in the same form as yours. However, if your work is correct, algebraic manipulation will show them to be equivalent.

1. $\dfrac{1}{6}\ln|4x^3 - 3x^2| + C$

2. $\displaystyle\int \dfrac{1}{4}\left(1 - 2\cos 6x + \cos^2 6x\right)\,dx$

$$= \dfrac{x}{4} - \dfrac{1}{2}\left(\dfrac{\sin 6x}{6}\right) + \dfrac{1}{4}\left(\dfrac{x}{2} + \dfrac{\sin 12x}{24}\right) + C$$

3. $\dfrac{\sin^5 x}{5} - \dfrac{\sin^7 x}{7} + C$

4. $\displaystyle\int \dfrac{2x}{x^2+4}\,dx + \int \dfrac{1}{x^2+4}\,dx = \ln|x^2+4| + \dfrac{3}{2}\arctan\left(\dfrac{x}{2}\right) + C$

5. $\displaystyle\int \left(\dfrac{2/3}{x+5} + \dfrac{1/3}{x-1}\right)dx = \dfrac{2}{3}\ln|x+5| + \dfrac{1}{3}\ln|x-1| + C$

6. $-\dfrac{1}{4}x^2 \exp^{-4x} - \dfrac{1}{8}x\exp^{-4x} - \dfrac{1}{32}\exp^{-4x} + C$ (parts)

7. $\dfrac{1}{4}\arctan\left(\dfrac{x}{2}\right) + C$

8. $\exp^{x^2+3x+1} + C$

9. $\dfrac{1}{12}\left[\ln(x^4 - 3)\right]^3 + C$

10. $\dfrac{-\sqrt{x^2+9}}{9x} + C$